PROBLEM SUPPLEMENT

to accompany

PHYSICS

for

SCIENTISTS & ENGINEERS

DOUGLAS C. GIANCOLI

JOHN ESSICK
Reed College

Upper Saddle River, NJ 07458

Associate Editor: Christian Botting
Senior Editor: Erik Fahlgren
Editor-in-Chief, Science: Dan Kaveney
Editorial Assistant: Jessica Berta
Executive Managing Editor: Kathleen Schiaparelli
Assistant Managing Editor: Karen Bosch
Production Editors: Gina Cheselka and Diane Hernandez
Supplement Cover Manager: Paul Gourhan
Supplement Cover Designer: Christopher Kossa
Manufacturing Buyer: Ilene Kahn
Manufacturing Manager: Alexis Heydt-Long
Senior Managing Editor, Art Production and Management: Patricia Burns
Manager, Production Technologies: Matthew Haas
Managing Editor, Art Management: Abigail Bass
Art Production Editor: Eric Day
Illustrations: GEX

© 2006 Pearson Education, Inc.
Pearson Prentice Hall
Pearson Education, Inc.
Upper Saddle River, NJ 07458

All rights reserved. No part of this book may be reproduced in any form or by any means, without permission in writing from the publisher.

Pearson Prentice Hall™ is a trademark of Pearson Education, Inc.

The author and publisher of this book have used their best efforts in preparing this book. These efforts include the development, research, and testing of the theories and programs to determine their effectiveness. The author and publisher make no warranty of any kind, expressed or implied, with regard to these programs or the documentation contained in this book. The author and publisher shall not be liable in any event for incidental or consequential damages in connection with, or arising out of, the furnishing, performance, or use of these programs.

> **This work is protected by United States copyright laws and is provided solely for teaching courses and assessing student learning. Dissemination or sale of any part of this work (including on the World Wide Web) will destroy the integrity of the work and is not permitted. The work and materials from it should never be made available except by instructors using the accompanying text in their classes. All recipients of this work are expected to abide by these restrictions and to honor the intended pedagogical purposes and the needs of other instructors who rely on these materials.**

Printed in the United States of America

10 9 8 7 6 5 4 3 2 1

ISBN 0-13-153048-8

Pearson Education Ltd., *London*
Pearson Education Australia Pty. Ltd., *Sydney*
Pearson Education Singapore, Pte. Ltd.
Pearson Education North Asia Ltd., *Hong Kong*
Pearson Education Canada, Inc., *Toronto*
Pearson Educación de Mexico, S.A. de C.V.
Pearson Education—Japan, *Tokyo*
Pearson Education Malaysia, Pte. Ltd.

Contents

Preface VII

1 Introduction, Measurement, Estimating — 1
- Section 1–3 1
- Sections 1–4 and 1–5 1
- Section 1–6 1
- Section 1–7 2
- General Problems 2

2 Describing Motion: Kinematics in One Dimension — 4
- Sections 2–1 to 2–3 4
- Sections 2–5 and 2–6 5
- Section 2–7 6
- Section 2–8 7
- General Problems 8

3 Kinematics in Two Dimensions; Vectors — 10
- Sections 3–1 to 3–5 10
- Section 3–6 10
- Sections 3–7 and 3–8 10
- Section 3–9 12
- Section 3–10 12
- General Problems 13

4 Dynamics: Newton's Laws of Motion — 15
- Sections 4–4 to 4–6 15
- Section 4–7 16
- General Problems 18

5 Further Applications of Newton's Laws — 20
- Section 5–1 20
- Sections 5–2 and 5–3 22
- Section 5–4 24
- General Problems 24

6 Gravitation and Newton's Synthesis — 26
- Sections 6–1 to 6–3 26
- Section 6–4 28
- Section 6–5 29
- General Problems 30

7 Work and Energy — 34
- Section 7–1 34
- Section 7–2 34
- Section 7–3 34
- Section 7–4 35
- Section 7–5 36
- General Problems 36

8 Conservation of Energy — 38
- Sections 8–1 and 8–2 38
- Sections 8–3 and 8–4 38
- Sections 8–5 and 8–6 39
- Section 8–7 40
- Section 8–8 41
- Section 8–9 41
- General Problems 42

9 Linear Momentum and Collisions — 44
- Section 9–1 44
- Section 9–2 44
- Section 9–3 45
- Sections 9–4 and 9–5 45
- Section 9–6 45
- Section 9–7 46
- Section 9–8 47
- Section 9–9 47
- General Problems 48

10 Rotational Motion about a Fixed Axis — 50
- Sections 10–1 to 10–3 50
- Sections 10–6 and 10–7 51
- Section 10–8 52
- Section 10–9 53
- Section 10–10 54
- Section 10–11 55
- General Problems 55

11 General Rotation — 57
- Section 11–1 57
- Section 11–3 57
- Sections 11–4 and 11–5 57
- Section 11–6 59
- Section 11–7 59
- Section 11–8 61
- Section 11–9 62
- General Problems 62

12 STATIC EQUILIBRIUM; ELASTICITY AND FRACTURE — 64

- Sections 12–1 to 12–3 — 64
- Section 12–5 — 69
- Section 12–6 — 69
- General Problems — 70

13 FLUIDS — 71

- Section 13–1 — 71
- Sections 13–2 to 13–5 — 71
- Section 13–6 — 73
- Sections 13–7 to 13–9 — 74
- Section 13–11 — 75
- General Problems — 75

14 OSCILLATIONS — 77

- Sections 14–1 and 14–2 — 77
- Section 14–3 — 80
- Section 14–5 — 81
- Section 14–6 — 81
- Section 14–7 — 82
- Section 14–8 — 82
- General Problems — 82

15 WAVE MOTION — 85

- Sections 15–1 and 15–2 — 85
- Section 15–3 — 85
- Section 15–4 — 85
- Section 15–5 — 86
- Section 15–7 — 86
- Section 15–9 — 87
- Section 15–10 — 88
- General Problems — 88

16 SOUND — 90

- Section 16–1 — 90
- Section 16–2 — 90
- Section 16–3 — 90
- Section 16–4 — 91
- Section 16–5 — 92
- Section 16–6 — 92
- Section 16–7 — 93
- General Problems — 93

17 TEMPERATURE, THERMAL EXPANSION, AND THE IDEAL GAS LAW — 94

- Section 17–1 — 94
- Section 17–2 — 94
- Section 17–4 — 94
- Section 17–5 — 95
- Sections 17–7 and 17–8 — 95
- Section 17–9 — 96
- General Problems — 96

18 KINETIC THEORY OF GASES — 99

- Section 18–1 — 99
- Section 18–2 — 99
- Section 18–4 — 100
- Section 18–5 — 101
- Section 18–6 — 101
- General Problems — 102

19 HEAT AND THE FIRST LAW OF THERMODYNAMICS — 105

- Section 19–1 — 105
- Sections 19–3 and 19–4 — 105
- Section 19–5 — 105
- Sections 19–6 and 19–7 — 106
- Section 19–9 — 106
- Section 19–10 — 107
- General Problems — 108

20 SECOND LAW OF THERMODYNAMICS — 111

- Section 20–3 — 111
- Section 20–4 — 111
- Sections 20–5 and 20–6 — 112
- Section 20–9 — 114
- General Problems — 114

21 ELECTRIC CHARGE AND ELECTRIC FIELD — 116

- Section 21–5 — 116
- Sections 21–6 to 21–8 — 116
- Section 21–10 — 118
- Section 21–11 — 118
- General Problems — 119

22 GAUSS'S LAW — 122

- Section 22–2 — 122
- Section 22–3 — 122
- General Problems — 125

23 ELECTRIC POTENTIAL — 129

- Section 23–2 — 129
- Section 23–3 — 130

Section 23–4	130	
Section 23–7	131	
Section 23–8	131	
General Problems	132	

24 Capacitance, Dielectrics, Electric Energy Storage — 134

- Section 24–1 — 134
- Section 24–2 — 134
- Section 24–4 — 135
- Section 24–5 — 137
- General Problems — 137

25 Electric Currents and Resistance — 141

- Sections 25–2 and 25–3 — 141
- Section 25–4 — 141
- Sections 25–5 and 25–6 — 143
- Section 25–7 — 143
- Section 25–8 — 143
- General Problems — 143

26 DC Circuits — 146

- Section 26–2 — 146
- Section 26–3 — 147
- Section 26–4 — 148
- Section 26–5 — 149
- General Problems — 150

27 Magnetism — 153

- Section 27–3 — 153
- Section 27–4 — 153
- Section 27–5 — 154
- Section 27–8 — 155
- Section 27–9 — 156
- General Problems — 156

28 Sources of Magnetic Field — 160

- Sections 28–1 and 28–2 — 160
- Sections 28–4 and 28–5 — 160
- Section 28–6 — 161
- Section 28–9 — 163
- General Problems — 163

29 Electromagnetic Induction and Faraday's Law — 166

- Sections 29–1 and 29–2 — 166
- Section 29–3 — 167
- Section 29–4 — 168
- Section 29–6 — 168
- Section 29–7 — 168
- General Problems — 168

30 Inductance; and Electromagnetic Oscillations — 172

- Section 30–2 — 172
- Section 30–3 — 172
- Section 30–4 — 172
- Section 30–5 — 173
- Section 30–6 — 174
- General Problems — 174

31 AC Circuits — 177

- Sections 31–1 to 31–4 — 177
- Section 31–5 — 177
- Section 31–6 — 179
- General Problems — 179

32 Maxwell's Equations and Electromagnetic Waves — 182

- Section 32–1 — 182
- Section 32–3 — 182
- Section 32–5 — 182
- Section 32–6 — 183
- Section 32–7 — 183
- Section 32–8 — 183
- Section 32–9 — 183
- General Problems — 184

33 Light: Reflection and Refraction — 186

- Section 33–2 — 186
- Section 33–3 — 186
- Section 33–4 — 186
- Section 33–5 — 187
- Section 33–6 — 187
- Section 33–7 — 188
- Section 33–8 — 189
- General Problems — 189

34 Lenses and Optical Instruments — 191

- Sections 34–1 and 34–2 — 191
- Section 34–3 — 192
- Section 34–4 — 192
- Section 34–5 — 192

Section 34–6	193
Section 34–7	193
Section 34–8	193
Section 34–9	193
Section 34–10	193
General Problems	194

35 WAVE NATURE OF LIGHT; INTERFERENCE — 195

Section 35–3	195
Section 35–5	196
Section 35–6	197
Section 35–7	197
General Problems	198

36 DIFFRACTION AND POLARIZATION — 201

Section 36–1	201
Section 36–2	201
Section 36–3	202
Sections 36–4 and 36–5	202
Section 36–7	202
Section 36–9	203
Section 36–10	203
Section 36–11	203
General Problems	204

37 SPECIAL THEORY OF RELATIVITY — 205

Sections 37–4 to 37–6	205
Section 37–8	205
Section 37–9	206
Section 37–11	206
Section 37–12	206
General Problems	206

38 EARLY QUANTUM THEORY AND MODELS OF THE ATOM — 208

Section 38–1	208
Section 38–2	208
Section 38–3	209
Section 38–6	209
Section 38–9 and 38–10	209
General Problems	209

ANSWERS TO ODD-NUMBERED PROBLEMS — A–211

PREFACE

This Problem Supplement is a companion to the text *Physics for Scientists and Engineers, Third Edition* by Douglas C. Giancoli. Its purpose is to supply a broad range of problems that complement and enrich the collection in the text. The notation and conventions (e.g., terminology, significant figures, and unit abbreviations) used here are consistent with *Physics for Scientists and Engineers, Third Edition*.

I believe that with a working knowledge of basic physics, one is equipped to understand a wide variety of physical phenomena in the world around us. My goal in writing this Problem Supplement is to provide support for this belief. In the following pages, situations encountered in everyday life, contemporary scientific research, and modern technology are analyzed using concepts appropriate for the introductory course. Much effort has been expended in distilling the essence of these (sometimes quite complex) situations into tractable problems and in using accurate numerical values for relevant quantities. My hope is that, in working these problems, the meaning and power of introductory physics will become clearer as we explore the inner workings of interesting and useful applications.

Presenting real-world problems in a manner appropriate for an introductory physics course provided me with a great challenge. One must accurately but succinctly describe the relevant situation, convey how it is to be idealized in a theoretical model, and alert the reader about what influences to neglect. To meet this challenge, many of the problems in this supplement are written in the following style: First, ample text is devoted to describing situations that, for many students, will be unfamiliar. These descriptions also define how each situation is to be idealized in a model, including explicit instructions about seemingly relevant quantities that can be safely ignored. My intent is to provide clear and well-defined problem statements, so that students waste no time misunderstanding the situation or assumptions to be made. Second, figures are generously used throughout this book to illustrate the circumstances of particular problems. Third, many of the difficult calculations are presented as multipart problems with intermediate answers given in each step. These intermediate answers are meant to be "rest stops" along the journey to the final answer, providing a reassuring sense of accomplishment and keeping the calculation on the proper track. Finally, the order-of-magnitude of physical quantities and the sense of how one influence can dominate over another in a given situation are important components of these problems. Many problems conclude the story they tell by instructing students to evaluate derived quantities using realistic numerical values for input parameters. The results obtained are usually intriguing and sometimes surprisingly unintuitive. Other problems take advantage of the access and skill most students now have with the graphing capabilities of calculators and computers (notably, spreadsheet programs). Students generate plots of derived formulas to visualize how a particular quantity varies over a range of input values, and to understand more fully the meaning of a derived relation.

Also included in this supplement are problems that explore further the physical theories presented in Giancoli's text. Some of these problems involve the derivation of a useful relation, or an instructive example from the historical development of a physical theory. Many others develop skills that are in the "tool chest" of all practicing physicists, such as determining the validity and limits of applicability of an approximation, designing an experimental test of a given theory, verifying a theoretical model through graphical analysis of experimental data, and extracting numerical values from a graph for a theoretical model's parameters.

Finally, scattered throughout these pages you will find some whimsical problems in which fanciful situations are used to illustrate physical principles. Students in my classes over the years have found these problems entertaining and educational. I hope you do too.

At Prentice Hall, I would like to thank Erik Fahlgren for envisioning this Problem Supplement and entrusting me with his vision, Christian Botting for ably assisting me in carrying out this project and coordinating the myriad of components involved, and Gina Cheselka for her skilled work in production. I'd like also to thank the reviewers and problem solvers whose hard work and helpful feedback refined many problems in this text. At home, I'd like to thank my family—Katie, Eden, Wade, and Devon—for their support and understanding through my many late-night and early-morning work sessions.

I hope you find this collection of problems instructive, interesting, and even fun. If you have comments, suggestions, or corrections, please contact me at jessick@reed.edu.

John Essick
Portland, Oregon

CHAPTER 1

Introduction, Measurement, Estimating

Section 1–3

1. (II) To the correct number of significant figures, determine the ratio of the mass of the proton compared with the mass of the electron using the
   (a) "Approximate Value" for each of these quantities listed on the front cover of the text.
   (b) "Current Best Value" for each of these quantities listed on the front cover of the text (ignore numbers in parentheses, which indicate experimental uncertainties of masses).

2. (II) To the correct number of significant figures, use the information on the front cover of the text to determine the ratio of
   (a) the surface area of Earth compared with the surface area of the Moon.
   (b) the volume of Earth compared with the volume of the Moon.

3. (III) For small angles θ, the numerical value of $\sin(\theta)$ is approximately the same as the numerical value of $\tan(\theta)$. From $\theta = 0°$ to $\theta = 25°$ in 1° increments, evaluate sine and tangent to two significant figures and determine the threshold angle above which sine and tangent values always disagree. [You will observe occasional disagreement below the threshold angle due to "rounding error."]

Sections 1–4 and 1–5

4. (I) The American Lung Association gives the following formula for an average person's expected lung capacity:

   $$V = 0.041H - 0.018A - 2.69$$

   where V, H, and A are the person's lung capacity (in liters), height (in centimeters), and age (in years), respectively. In this formula, what are the units of the numbers 0.041, 0.018, and 2.69?

5. (II) The meter was defined originally as one ten-millionth of the distance from Earth's equator to the North Pole. Given this definition, determine the circumference C and radius R of Earth to three significant figures. Determine the percent difference between the value that you calculate for R and the current value given on the front cover of the text.

6. (II) The universe is expanding. Astronomers have found that the length of space in each direction increases by roughly $2 \times 10^{-16}\%$ every second. At that rate of expansion, by how much will the length of one meter of space expand over one decade? According to Table 1–1 in the text, this change in length is comparable in size with what type of object in nature?

7. (II) In the British system of units, an acre of land is defined to have an area of 160 square rods, where 1 rod is 16.5 feet. An American football field is 91.4 m long and 48.8 m wide.
   (a) Show that the area of an American football field is approximately equal to an acre.
   (b) How many percent larger is the area of an American football field than an acre?

8. (II) A computer tycoon in the state of Washington, who makes five gigadollars (US) per year by thinking about computer problems during all waking hours, spies a one-hundred dollar bill on the ground.
   (a) Should he spend three seconds of his time to pick up the bill or would he be better off, financially, to work on computer problems during those three seconds?
   (b) By what factor is that decision better? Assume the tycoon sleeps eight hours every night.

9. (II) The latest findings in astrophysics indicate that the observable universe can be modeled as a sphere of radius $R = 14 \times 10^9$ light-years containing mass with an average density of 1×10^{-26} kg/m^3, where only 4% of the universe's total mass is due to "ordinary" matter (such as protons, neutrons, and electrons). Use this information to estimate the total mass of ordinary matter in the observable universe (1 light-year = 9.5×10^{15} m).

10. (II) A certain audio compact disc (CD) contains 783.216 megabytes of digital audio information. Each byte consists of exactly 8 bits. When played, a CD player reads out the CD's digital information at a constant rate of 1.4 megabits per second. How many minutes does it take for the entire CD to be read?

Section 1–6

11. (II) After inhaling as much air as possible, you deeply exhale into an empty balloon, inflating it to a spherical shape. You measure the inflated balloon's circumference to be 63 cm. Use this information to estimate how many liters of air your lungs hold when fully inflated, given that 1 liter (L) = 1000 cm^3.

12. (II) A New World Order decrees that Earth's dry land surface be equally divided among all of the world's (approximately) 6.5 billion living people. Estimate the number of square meters each person receives as a result of this order.

13. (II) Approximately how many helium balloons (each of the size you might buy at an amusement park) would you have to tie together in order to lift a 10 year-old child off of the ground?

14. (II) I have agreed to hire you for one month and I let you decide between two possible methods of payment: either I pay you $1,000 dollars per day each of the 30 days or I pay you one penny on the first day, two pennies on the second day, and continue to double your daily pay on each subsequent day up to day 30. Using quick estimation to make your decision, how do you wish to be paid?

15. (II) A gallon of paint covers about 30 m^2 of wall area. Estimate the thickness (in mm) of the paint on the wall.

16. (II) Over an entire lifetime, will most people alive today walk a distance equivalent to at least one time around Earth?

17. (III) At a west-facing beach on the equator, a physicist enjoys a beautiful sunset. As she watches the sun set over the ocean, the physicist carries out the following peculiar sequence of events (see Fig. 1–1): While lying on her stomach with her eyes very close to the level of the sand, she waits until the top of the sun just disappears below the horizon, at which point, she quickly stands up, while simultaneously glancing at the second hand of her wristwatch. With her new vantage point a distance $h = 1.8 \text{ m}$ above the sand, a bit of the sun is visible again; she waits until the top of the sun again dips below the horizon, at which point she finds from her watch that 10 seconds have elapsed since she stood up. Knowing that one complete revolution of Earth takes 24 hours, the physicist then quickly deduces the radius R of Earth. What value does she obtain for R?
[*Hint*: Use the fact that $h^2 + 2hR \approx 2hR$, after explaining why it is a very good approximation.]

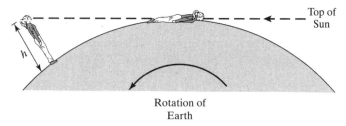

FIGURE 1–1

Section 1–7

18. (II) Three fundamental constants of nature—the Gravitational constant G, Planck's constant h, and the speed of light c—have the dimensions of $\left[\dfrac{L^3}{MT^2}\right]$, $\left[\dfrac{ML^2}{T}\right]$, and $\left[\dfrac{L}{T}\right]$, respectively. The Planck time t_P is defined as follows:

$$t_P = \sqrt{\dfrac{Gh}{c^5}}$$

(a) Demonstrate through dimensional analysis that this mathematical way of combining these constants does indeed yield a quantity with dimensions of time.
(b) Using the values given for fundamental constants on the front cover of the text, calculate the Planck time. The Planck time is thought to be the earliest time, after the creation of the universe, at which the currently known laws of physics can be applied.

19. (III) As an object moves along a path within Earth's atmosphere, it has to push aside air, an action that causes an air resistance force (called the "drag force" F_D) to act on the object. One might conjecture that the drag force depends on only three quantities: air's density ρ (kg/m^3), the object's cross-sectional area A (m^2), and the object's speed v (m/s), which have dimensions of $\left[\dfrac{M}{L^3}\right]$, $[L^2]$, and $\left[\dfrac{L}{T}\right]$, respectively. In Chapter 4 of the text, we will see that force $F = ma$, where m is mass and a is acceleration; so force has dimensions of $\left[\dfrac{ML}{T^2}\right]$. Start by writing the conjectured drag force as the product $F_D = k\rho^\alpha A^\beta v^\gamma$, where α, β and γ, and k are numbers without dimension and then obtain a formula for F_D based solely on dimensional analysis.

20. (III) In Problem 18, a mathematical combination of the fundamental constants G, h, and c was given that had the dimensions of time. Find the mathematical combination of these three fundamental constants that has the dimensions of length. This combination is called the "Planck length" λ_P. Determine the numerical value of λ_P.

21. (III) Except for the tiniest ripples, wave motion on the surface of the ocean does not depend on the properties of water such as density and surface tension. The primary "return force" for water piled up in the wave crests is due to the gravitational attraction of Earth. Thus, the speed v (m/s) of these surface water waves depends on the acceleration due to gravity g (m/s^2). It is reasonable to expect that v might also depend on water depth h (m) and the wave's wavelength λ (m). Wavelength is the distance between adjacent wave crests (see Fig. 1–2). Assume wave speed is given by the functional form $v = kg^\alpha h^\beta \lambda^\gamma$, where α, β and γ, and k are numbers without dimension.
(a) In deep water, well-submerged water does not affect the motion of waves at the surface. Thus, v becomes independent of depth h (i.e., $\beta = 0$). Using only dimensional analysis, determine the formula for the speed of surface water waves in deep water.
(b) In shallow water, the speed of surface water waves is found experimentally to be independent of the wavelength (i.e., $\gamma = 0$). Using only dimensional analysis, determine the formula for the speed of surface water waves in shallow water.

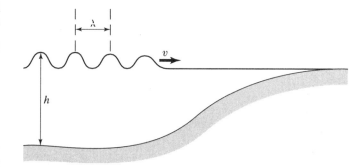

FIGURE 1–2

General Problems

22. (II) The density of an object is defined as its mass divided by its volume. Suppose the mass and the volume of a rock are measured to be 8 g and 2.8325 cm^3, respectively. To the correct number of significant figures, determine the rock's density.

23. (II) To the correct number of significant figures, use the information at the front of the text to determine the ratio of
    (a) the mass of Earth compared with the mass of the Moon.
    (b) the mean Earth-Sun distance compared with the mean Earth-Moon distance.
24. (II) One mole of jellybeans consists of 6.02×10^{23} individual jellybeans. If you wanted to count all of these jellybeans and wanted to complete this counting project within a time interval equal to the age of the universe, how many jellybeans would you have to count every second? The age of the universe is 13.7 billion years.
25. (II) One mole of atoms consists of 6.02×10^{23} individual atoms. If a mole of atoms were spread uniformly over the surface of Earth, how many atoms would there be per square meter?
26. (II) Computer programs that use "UNIX time," including those written in the popular C language, are going to "run out of time" at some date in the future, potentially causing catastrophes like those (unrealized) predictions for the Y2K problem. In UNIX time, a 32-bit binary number is used to count the number of seconds elapsed since 12:00 a.m. January 1, 1970 (an arbitrary date, roughly equal to the birthday of UNIX). In this 32-bit number, the first bit is used to indicate a plus or minus sign, and the remaining 31 bits represent its size. Thus, the maximum number of decimal seconds that can be counted is 2,147,483,647 $(= 2^{31} - 1)$. One second later, UNIX time will reset itself (somewhat like an automobile odometer rollover from 999999 to 000000). Show why this computer glitch is nicknamed the "Y2038 Problem."
    [*Hint*: There is no need to account for leap years.]
27. (II) If you only used a keyboard to enter data for storage, how many years would it take to fill up the hard drive within your computer? Assume that a "normal" work day is 8.0 hours, that the hard drive can store 120 gigabytes of data, that one byte is required to store one keyboard character, and that you can type 200 characters per minute.
28. (II) How many sheets of paper can be produced from a single 0.5 m diameter tree that is 10 m tall?
29. (II) A nearly complete skeleton of a Brachiosaurus, one of the largest known dinosaurs, is on display in the Humboldt University Natural History Museum in East Berlin. This skeleton suggests that the midsection of a Brachiosaurus's body was about 4 m in diameter and 6 m long, while its neck and tail were both about 1 m in diameter with lengths of 10 m and 7 m, respectively.
    (a) Model the body of a Brachiosaurus as consisting of three cylinders with the given dimensions to estimate its volume. Then, assuming as is true for modern-day animals, that the density of a dinosaur was about equal to the density of water (1000 kg/m^3), estimate the mass of a Brachiosaurus.
    (b) Compare (as a ratio) the estimated mass of a Brachiosaurus with that of a modern-day male African Elephant (5000 kg).
30. (II) Over a broad range in sizes of four-legged mammals, an animal's mass m in kg has been found to be related to the combined circumference C in mm of its front (humerus) and rear (femur) leg bones by the following relation:

$$m = \left(0.000084 \frac{\text{kg}}{\text{mm}^{2.73}}\right) C^{2.73}$$

    (a) Rewrite this expression so that when C is in cm, m will be given in metric tons. One metric ton equals 1000 kg.
    (b) Assume the relation between m and C for modern-day mammals also applies to ancient four-legged animals. What mass (in metric tons) is predicted for a Brachiosaurus, one of the largest known dinosaurs? Measurements on fossilized skeletons of this animal have determined that $C = 140$ cm.

CHAPTER 2

Describing Motion: Kinematics in One Dimension

Sections 2–1 to 2–3

1. (II) A bicyclist climbs a hill at 12 km/h and then descends at a constant speed, returning to her starting point. Her average speed for the trip is 16 km/h. What is her downhill speed?

2. (II) You have a postcard (15 cm wide by 10 cm high) with a picture of our Milky Way Galaxy completely filling its width (Fig. 2–1). This postcard is held on your bulletin board by a pushpin that makes a pinhole of 1 mm diameter on the Galaxy's image. Given that the diameter of the Milky Way Galaxy is 100,000 light-years $\left(1 \text{ light-year} = 9.45 \times 10^{15} \text{ m}\right)$, determine the distance D within our Galaxy that the pinhole represents; then calculate (to one significant figure) how long it would take to traverse this "pinhole" distance D in a commercial jetliner. Commercial jetliners travel at an average speed of 900 km/h.

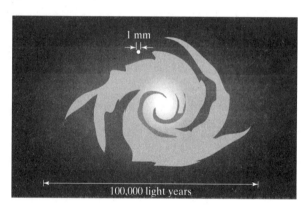

FIGURE 2–1

3. (II) A digital video (DV) image is a collection of small dots called *pixels*, each of varying brightness and color. The pixels are arranged on a rectangular-shaped grid, which point-by-point paints a picture. The location of a particular pixel is given by its Cartesian coordinates (x, y), where the origin is located in the upper left-hand corner of the image and x and y are integers in the range 1 to 720 and 1 to 480, respectively (see Fig. 2–2). One DV still image is called a *frame*. In a DV movie, you view a succession of 30 frames per second and your eye and brain connect the images together to create the illusion of continuous motion. Assume you are given a video clip consisting of a sequence of three successive frames that documents the horizontal motion of a person. In Frame 1, you find that the tip of the person's nose appears at pixel location (50, 200), while in Frame 2 the tip is at (110, 200) and in Frame 3 the tip is at (170, 200). Also in the clip, a horizontally oriented meter stick (length 100 cm) appears in each frame and you find that the zero-centimeter and the 100-centimeter tips of the meter stick are at pixel locations (100, 300) and (400, 300), respectively. Using the fact that neighboring video frames are captured at times 1/30 of a second apart, what is the person's constant speed in m/s?

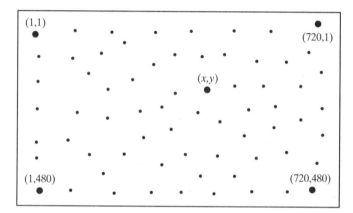

FIGURE 2–2

4. (II) On a 12.0-cm diameter audio compact disc (CD), digital bits of information are encoded sequentially along an outward spiraling path.
(a) The spiral starts at radius $R_1 = 2.5 \text{ cm}$ and winds its way out to radius $R_2 = 5.8 \text{ cm}$. As the path spirals outward, the distance between the centers of neighboring spiral-windings is always $1.6 \, \mu\text{m} \left(= 1.6 \times 10^{-6} \text{ m}\right)$. Determine the total length of the spiraling path. [*Hint:* Imagine "unwinding" the spiral into a straight path of width $1.6 \, \mu\text{m}$, and note that the original spiral and the straight path both occupy the same area.]
(b) To read information, a CD player adjusts the rotation of the CD so that the player's readout laser moves along the spiral path at a constant speed of 1.20 m/s. What is the total playing time of the CD, i.e., how long does it take the laser to scan along the entire path length?

5. (II) At many schools, a motion sensor is used to measure distance in instructional physics lab experiments. This device can accurately measure the distance from itself to an object of interest via the "sonar" technique used in older autofocus cameras. A short pulse of high-frequency sound is emitted from the motion sensor and, as the sound-wave pulse travels away, it reflects from any objects it encounters (Fig. 2–3). These reflected waves travel back to the motion sensor and, upon arrival, are detected as "echo" pulses. The motion sensor is programmed to measure the time interval T between

the emission of the original sound pulse and the arrival back to the motion sensor of the first echo.

(a) Given that a sound wave travels at a constant speed of $v = 343$ m/s in room-temperature air, show that the distance d from the motion sensor to the nearest (sound-reflecting) object is

$$d = \frac{1}{2}vT.$$

(b) The time interval T can be measured with high precision only if it is at least 1.0 millisecond ($= 1.0 \times 10^{-3}$ s) long. What is the smallest distance that can be measured with the motion sensor?

(c) By default, the motion sensor makes 20 distance measurements every second (i.e., it emits 20 sound pulses per second at evenly spaced time intervals). Thus, the measurement of T must be completed within the time interval between the emissions of successive pulses. What is the largest distance that can be measured with the motion sensor?

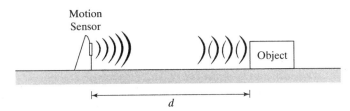

FIGURE 2–3

6. (III) No experimentally determined quantity x can be measured with perfect accuracy; there is always some uncertainty Δx (called the *absolute uncertainty*) in any measurement. A motion sensor (see Fig. 2–3) determines the distance d to an object by measuring the round trip time T it takes a sound wave to travel (at constant speed $v = 343$ m/s) from the sensor to the object, then back again. According to its manufacturer, the motion sensor can measure T with an uncertainty of $\Delta T = 1.0$ microsecond ($= 0.0000010$ s). This uncertainty in T means that there will be an uncertainty Δd in the motion sensor's distance determination. Let T_1 and T_2 be two values of T, separated by the accuracy level of the motion sensor, i.e., $\Delta T = T_2 - T_1$. Using your knowledge of how the motion sensor works, find the two distances d_1 and d_2 associated with T_1 and T_2, respectively. Then, defining $\Delta d = d_2 - d_1$, show that the motion sensor is able to measure the distance from itself to a target object to an uncertainty of $\Delta d = 0.17$ mm. [Note that Δd is independent of how far the target object is away from the motion sensor.]

7. (III) A lifeguard, standing at the side of a swimming pool, spots a child in distress, as shown in Fig. 2–4. The guard runs with average speed v_R along the pool's edge for a distance x, then jumps into the pool and swims with average speed v_S on a straight path to the child.

(a) Show that the total time t it takes the lifeguard to get to the child is given by

$$t = \frac{x}{v_R} + \frac{\sqrt{D^2 + (d-x)^2}}{v_S}.$$

(b) Assume $v_R = 4.0$ m/s and $v_S = 1.5$ m/s. Use a graphing calculator or a computer to plot the expression in part (a) and, from this plot, determine the optimal distance x the lifeguard should run before jumping into the pool (i.e., find the value of x that minimizes the time t to get to the child).

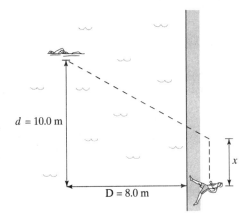

FIGURE 2–4

Sections 2–5 and 2–6

8. (II) Two trains, one traveling at 30 m/s and the other at 40 m/s, are headed toward one another along a straight level track. When they are 1000 m apart, each engineer sees the other's train and applies the brakes. The brakes decelerate each train at a rate of 1.0 m/s².
(a) How long after the brakes are applied do the two trains collide?
(b) What is the speed of each train just before the collision?

9. (II) Mary and Sally are running in a foot race as shown in Fig. 2–5. When she is 20 m from the finish line, Mary has a speed of 4.0 m/s and is 5.0 m behind Sally, who has a speed of 5.0 m/s. Sally thinks she has an easy win and so, during the remaining portion of the race, decelerates at a constant rate of 0.50 m/s² to the finish line. What constant acceleration a should Mary undergo during the remaining portion of the race if she wishes to cross the finish line side-by-side with Sally?

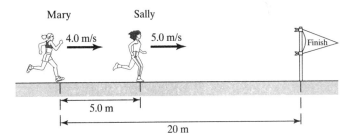

FIGURE 2–5

10. (II) If an object moves in one dimension with constant acceleration a, its velocity $v(t)$ as a function of time t will be given by

$$v = v_o + at$$

where v_o is the object's velocity at $t = 0$. Assume you measure an object's velocity v at several times t as it moves in one dimension. If you then analyze these data points by plotting them on a velocity v (y-axis) versus time t (x-axis) graph, explain why you expect the points to lie on a straight line if the observed object was indeed moving with constant acceleration. What information of physical interest can be determined from the straight line's slope and y-intercept?

11. (II) For the situation in Problem 3, assume instead that the tip of the person's nose is at the following three pixel locations in a sequence of three successive video frames: (50, 200), (92, 200), and (138, 200). Assume each image is taken very quickly (e.g., in 1/500 of a second) at the beginning of the 1/30 second time interval represented by each frame. Again, the zero-centimeter and the 100-centimeter tips of a meter stick are at pixel locations (100, 300) and (400, 300), respectively. Taking the first frame as time $t = 0$, find the person's initial velocity and acceleration over the time interval ($= 2/30$ of a second) documented by these three frames.

12. (II) As shown in Fig. 2–6, an automatic sliding door in a supermarket slides open along a straight line after it detects the presence of an approaching person. Assume the door is set to detect a person 1.5 m away and that the person walks toward the door at the moderate (constant) speed of 1.5 m/s. The door must travel from its fully closed to its fully open position (a distance of 1.0 m) during the time interval from when the approaching person is first detected until he or she passes through the plane of the doorway.
(a) Assume the door is accelerated from rest at constant acceleration $+a$ over the first half of its travel distance and then decelerated to rest at constant acceleration $-a$ over the second half of its travel. This scenario will require the least powerful driving motor. Show that the required value for a is 4.0 m/s^2.
(b) What maximum speed does the door attain during its opening motion?
(c) If a person walks toward the door at a speed $v > 1.5$ m/s, the door will not be fully open when he or she arrives at the plane of the doorway. If the person needs an opening of at least 0.60 m so as not to collide with the moving door, what is the maximum value for v so that the person can pass unimpeded through the doorway?

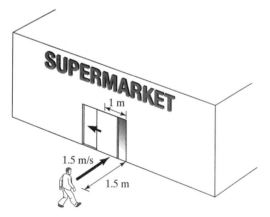

FIGURE 2–6

13. (III) In Problem 12, the magnitude of the door's acceleration was 4.0 m/s^2 at all times during its opening motion. Consider the following alternate way in which the door could be programmed to open. When the approaching person is detected, the door is accelerated from rest at 6.0 m/s^2 (a rate 50% higher than in Problem 12) to a speed of u. The door then travels at the constant speed u, until it reaches the proper location where it is then decelerated at -6.0 m/s^2 to rest. The total travel distance of the door is still 1.0 m. In comparison with the simpler method of Problem 12, what percentage change is there in the door's opening time T by this more complicated pattern of door accelerations if (a) $u = 1.0$ m/s; (b) $u = 2.0$ m/s?
(c) Which of these two possible values of u should one choose if the goal is to reduce T?

14. (III) An automatic sliding door in a commercial building slides open along a straight line after it detects the presence of an approaching person (Fig. 2–6). Consider the following way in which the door could be programmed to open: When the door detects an approaching person, it is accelerated from rest at 6.0 m/s^2 to a speed of u. The door then travels at the constant speed u, until it reaches the proper location where it is then decelerated at -6.0 m/s^2 to rest. The total travel distance of the door is 1.0 m. Assume that the door always moves in the same direction while opening.
(a) In this scenario, there is a maximum possible choice for u. Find the value of u_{max}.
(b) Show that the general relation between the door's opening time T (s) and the choice of u (m/s) is given by $T = \dfrac{u}{6} + \dfrac{1}{u}$. Graph, possibly with the help of a graphing calculator or a computer, T (y-axis) vs. u (x-axis).
(c) Use your graph to answer these questions: If one's goal in using this door-opening scenario is to open the sliding door in a time quicker than 1.0 second, what range of u values will accomplish this feat? Which u value yields the minimum T attainable by the scenario and what is the value of T_{min}?

15. (III) In Problem 14, we found that the sliding door's opening time T (s) as a function of speed u (m/s) is given by $T = \dfrac{u}{6} + \dfrac{1}{u}$. By differentiating this expression with respect to u and setting $\dfrac{dT}{du} = 0$, show that the choice $u = \sqrt{6}$ m/s yields the minimum value for the door opening time T. [Hint: If f and g are two functions of u, then $\dfrac{d}{du}[f + g] = \dfrac{df}{du} + \dfrac{dg}{du}$. Also $\dfrac{d}{du}(Au^n) = Anu^{n-1}$ where A and n are constants.]

Section 2–7

16. (II) Roger, the staff resident in a college dorm, sees that water balloons are falling past his window (see Fig. 2–7). He is unable to lean out far enough to see the culprit, but he notices that each balloon strikes the sidewalk 0.83 s after passing his window. Roger's room is on the third floor, 15 m above the sidewalk.
(a) How fast are the balloons traveling when they pass Roger's window?

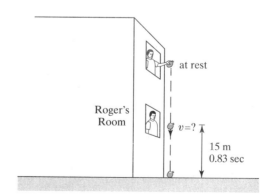

FIGURE 2–7

(b) Assuming the balloons are being released from rest, from what floor are they being released? Each floor of the dorm is 5.0 m high.

17. (II) If you jump off the top of a wall, gravity will accelerate you at 9.8 m/s^2 during your brief free fall to the ground. Imagine that you could maintain this constant acceleration a for a long period of time. With $a = 9.8 \text{ m/s}^2$, how long (in days) would it take to speed up from rest to a final speed of $3.0 \times 10^6 \text{ m/s}$ ($= 1.0\%$ of the speed of light)? [For larger final speeds, relativity effects would have to be taken into account.]

18. (II) The choice of positive direction, although a necessary first step in solving a problem, is completely arbitrary and does not affect the ultimate conclusion of a calculation. To illustrate this fact, consider the following problem: A person throws a ball vertically upward from the ground with a speed of 20 m/s. After the ball has traveled to its peak height and is on its downward flight, find the time t when it is 10 m above the ground, and then use $v = v_o + at$ to predict the velocity of the ball at this time.
    (a) Solve this problem by arbitrarily defining the positive direction to be upward, with the origin at the ground.
    (b) Solve this problem by arbitrarily defining the positive direction to be downward, with the origin at the ground.
    (c) Do parts (a) and (b) yield the same prediction for the time and velocity at the point of interest on the ball's trajectory?

19. (II) You—the video game creator famous for making your games consistent with the laws of physics—are animating your latest release, working in the 720×480 DV format. In the scene you are working on presently, you draw the ground as a horizontal line so that it includes all of the pixels with y-coordinates equal to 400 (see Fig. 2–2 for the coordinate system used in this problem). One of your game characters, a little man named Mario, holds a magic coin momentarily motionless at pixel location $(x, y) = (250, 344)$ and then accidentally drops it in a frame labeled "Frame A." If you want the coin to fall from rest to the ground as if accelerated by gravity at Earth's surface, what should its pixel locations be in the subsequent frames (i.e., Frames B, C, D, etc., until the frame where the coin hits the ground)? Each DV frame is separated from its preceding frame by 1/30 second and a frame shows a snapshot of the scene that is assumed to have been taken very quickly at the beginning of the frame's 1/30 of a second time period. Also, in your story, you have conceived of Mario as 50 cm tall. In this scene, when he stands up straight, there are 200 vertical pixels from the bottom of his feet to the top of his head.

20. (III) Continuing with the video game scene in Problem 19, Mario picks up the coin and lifts it to pixel location $(x, y) = (300, 260)$, where he tosses the coin straight upward in a frame labeled "Frame 0." If we label the subsequent frames as Frame 1, Frame 2, Frame 3, etc., then the time t of the Nth frame is $t_N = N \Delta t$, where $\Delta t = 1/30$ s. You want the flight of the coin to appear as if accelerated by gravity at Earth's surface.
    (a) If, in Frame 4, you want the coin to be at pixel location (300, 140) and still be moving upward, what is the initial velocity of the coin in Frame 0, in units of m/s?
    (b) In which frame does the coin finally hit the ground?

21. (III) Pretend that you only know that the value for the acceleration due to gravity g is in the range of 8 to 12 m/s^2 and that you are designing an experiment to more accurately determine this physical quantity. No experimentally determined quantity x can be measured with perfect accuracy; there is always some uncertainty Δx (called the *absolute uncertainty*) in any measurement. You know that an object falling freely from rest obeys the equation $y = \frac{1}{2}gt^2$, where t is the time it takes for the object to fall a distance y; so, you decide to design an experimental apparatus that exploits this equation to measure g to an accuracy of 5%, i.e., $\frac{\Delta g}{g} = 0.05$. Assume that you have a very accurate distance-measuring instrument, so the uncertainty in measuring the object's "fall distance" y can be ignored. However, you are only able to measure the object's "fall time" t with a precision of $\Delta t = 0.01$ s.
    (a) Show that the change of g with respect to change in t is given by
$$\frac{dg}{dt} = -\frac{4y}{t^3}.$$
    The minus sign here means that the change in g (its uncertainty) becomes smaller when t is increased (the object is allowed to fall for a longer time).
    [Hint: The derivative with respect to x of the quantity Ax^n, where A and n are constants, is $\frac{d(Ax^n)}{dx} = Anx^{n-1}$.]
    (b) The uncertainty Δg in the acceleration due to gravity g will be related to the uncertainty Δt in the fall time t by
$$\Delta g \approx \left|\frac{dg}{dt}\right| \Delta t.$$
    We take the absolute value of the derivative because we only want to know the magnitude of Δg, not whether it increases or decreases when t is increased. Starting with this relation, show that
$$\frac{\Delta g}{g} \approx 2\frac{\Delta t}{t}.$$
    (c) Given your goal of determining g to an accuracy of $\frac{\Delta g}{g} = 0.05$, how long does the object's "fall time" have to be in your experiment? Since you start off knowing that the ballpark value for g is 10 m/s^2, over what distance should you plan for the object to fall in order to yield a value for g of the desired accuracy?

Section 2–8

22. (III) For an object moving in one dimension, let $x(t)$ be the function that describes the object's position x as time t increases.
    (a) First, assume $x(t)$ is given by the following second-order polynomial
$$x = A + Bt + Ct^2$$
where A, B, and C are constants. Show that the object's velocity is then
$$v = B + 2Ct$$
and its acceleration is
$$a = 2C;$$
then show that the kinematic equations for motion with constant acceleration obey this mathematical pattern. What physical quantities are represented by A, B, and C?

(b) Second, assume $x(t)$ is given by the following third-order polynomial

$$x = A + Bt + Ct^2 + Dt^3$$

where A, B, C, and D are constants. Determine the object's velocity and its (nonconstant) acceleration.

General Problems

23. (II) On an audio compact disc (CD), digital bits of information are encoded sequentially along a spiraling path. Through a clever coding scheme, each bit occupies a length on the spiral equivalent to $0.278\,\mu\text{m}$. To read its information, a CD player rotates the CD so that the player's readout laser scans along the spiral's sequence of bits at a constant speed of $1.20\,\text{m/s}$.
    (a) Determine the number N of digital bits that a CD player reads every second.
    (b) To faithfully reproduce stereo sound, signal processing theory dictates that one needs to update the audio information sent to the two speakers 88,200 times per second. In a digital audio system, each one of these updates requires 16 bits and so, for a stereo system, one would (at first glance) think the required bit rate for a CD player is

$$N_o = 88{,}200\,\frac{\text{updates}}{\text{second}} \times 16\,\frac{\text{bits}}{\text{update}} = 1.41 \times 10^6\,\frac{\text{bits}}{\text{second}}.$$

The number N of bits read per second by a CD player is greater than N_o. The excess number $(= N - N_o)$ is required by the peculiar encoding and error-correction schemes used on audio CDs. What percentage of the total bits on a CD is dedicated to these encoding and error-correction schemes?

24. (II) The world's fastest top-fuel dragsters can, starting from rest, accelerate to the end of a 1/4 mile ($= 402\,\text{m}$) course in 4.6 s. During such a race, the driver tries to keep the dragster at its maximum level of acceleration; let's analyze this motion assuming the acceleration remains constant.
    (a) Determine the dragster's (assumed) constant acceleration (m/s^2).
    (b) Under the assumption of constant acceleration, what is the dragster's final velocity (km/h) at the finish line?
    (c) In an actual race, the "finish-line" velocity of the fastest dragster is measured to be about 525 km/h. How close does your predicted value in part (b) match a dragster's actual final velocity? Comment on what might be causing any discrepancy between these two values.

25. (II) You borrow your parent's minivan for a trip to the video store, and while stopped at a red light, a Porsche pulls alongside you. One thing leads to another and a few moments later, you are in a street race with the Porsche to the next intersection. Assuming that you and the Porsche start side-by-side at rest, that the distance to the next intersection is 200 m away, and that you both undergo constant acceleration at your maximum possible rates during the entire race, by how many car lengths do you lose the race? A Porsche and a minivan can accelerate from 0 km/h to 97 km/h ($= 60\,\text{mph}$) in 3.7 and 9.5 s, respectively. A typical car length is 4.5 m.

26. (II) In an atomic clock, the frequency of a microwave source is varied until it matches the natural resonance frequency of the cesium atom. The microwave frequency is then used to define the time unit of one second. In this frequency-tuning process a given sample of cesium atoms must interact with the microwaves at two times, t_1 and t_2, which are separated by the time interval $\Delta t = t_2 - t_1$. To increase the precision of the clock, Δt must be large. In an atomic fountain clock (accurate to about one second over 20 million years), a burst of laser light gently tosses a small spherical cloud of very cold cesium atoms upward in a vertical vacuum chamber. On its upward journey, the cloud passes through a "microwave cavity," with a launch speed v_o, and moves under the influence of gravity to a maximum height of 30 cm above the cavity. The cloud then falls under the influence of gravity, returning to the microwave cavity (see Fig. 2–8).
    (a) Find the launch speed v_o of the cloud of atoms.
    (b) Find Δt for the atomic fountain clock.
    (c) By virtue of their very cold temperature, the cesium atoms have "thermal motion," which causes them to move in all directions within the cloud with an average (constant) speed of 1.0 cm/s. Due to this thermal motion, how much does the cloud's radius expand during its journey within the vacuum chamber, which occurs over the time interval Δt?

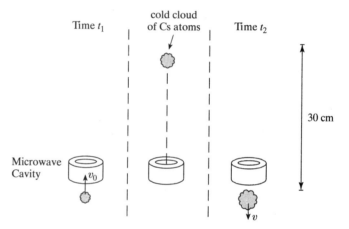

FIGURE 2–8

27. (II) The human body has all sorts of limitations; let's investigate one of them. Estimate the maximum height that a person, standing on the ground, can be expected to throw an object into the air. To make this estimate, you will have to make an educated guess about the maximum possible speed that a person is capable of tossing an object upward.

28. (II) If fired straight up, a bullet will travel to heights where it is hazardous to aircraft. Once it reaches its maximum height, it then falls back to the ground, arriving with a dangerous velocity. Depending on the gun, bullets are fired with an initial speed in the range of 150 to 1400 m/s. Because these speeds are large, air resistance significantly affects the motion of bullets, but this effect is somewhat counteracted by their aerodynamic design. Let's estimate the flight of a bullet by ignoring the effect of air resistance, an approximation that will be most valid for bullets with the slowest initial speed. Assuming that only gravity acts on a bullet and it is fired straight up with an initial speed of 150 m/s,
    (a) what maximum height (km) does the bullet attain?
    (b) how long is the bullet in the air before it returns to the ground?
    (c) with what speed does it return to the ground? (A speed of 60 m/s can be lethal.)
    [A sophisticated calculation that takes into account air resistance yields maximum-height, flight-time, and return-speed

values that differ by factors of 0.77, 0.42, and 0.68, respectively, from the values that you obtained.]

29. (III) When is it acceptable to neglect the effect of air resistance on a falling object? In the idealized case of zero air resistance, gravity accelerates all objects with the same acceleration $g = 9.8 \text{ m/s}^2$. In the presence of air, a falling object's acceleration a (m/s^2) can be described by the following relation:

$$a = g - \alpha v^2$$

where v (m/s) is the object's instantaneous speed and α is a constant. The velocity-squared term describes the effect of air resistance and it becomes more and more significant as the object attains greater and greater speeds during its fall. The constant α depends on the size and shape of the object and is largest for lightweight objects with large surface areas (such as a leaf). For a 1.0-kg steel ball (radius of about 3.0 cm), $\alpha_{ball} = 0.00092 \text{ m}^{-1}$, while a raindrop with radius 2.0 mm has $\alpha_{raindrop} = 0.11 \text{ m}^{-1}$. As an arbitrary (but reasonable) criterion, let's say that when a falling object reaches the speed at which its acceleration has been diminished to 9.7 m/s^2, then air resistance can no longer be neglected. Use this criterion to estimate (to one significant figure) the maximum distance over which an object can fall and one can still safely assume that the object has been accelerated at (close to) the constant value of g. Do this (a) when the object is a 1.0-kg steel ball, and (b) when the object is a raindrop of radius 2.0 mm.

30. (III) Consider the following model to describe the sequence of bounces that results when an object is released from rest above a solid horizontal surface (Fig. 2–9). Assume the object is released from an initial height h_1 and, after its first bounce from the surface, rises to a height of only $h_2 = \alpha h_1$, where α is a constant in the range $0 \leq \alpha < 1$ (because of dissipation that occurs during the bounce). Next, the object falls from height h_2 and, after its second bounce from the surface, rises to a height of $h_3 = \alpha h_2$, etc., such that the ratio of heights before and after a particular bounce always equals α:

$$\alpha = \frac{h_2}{h_1} = \frac{h_3}{h_2} = \frac{h_4}{h_3} = \cdots = \frac{h_{n+1}}{h_n} = \cdots$$

(a) Let the time interval Δt_n of the nth bounce be the elapsed time it takes for the object to fall from height h_n, bounce from the surface, and rise to the height h_{n+1}. Assuming the actual bounce takes negligible time, show that

$$\Delta t_n = \sqrt{\frac{2h_n}{g}} [1 + \sqrt{\alpha}] = \sqrt{\frac{2h_1}{g}} \left[\frac{1}{\sqrt{\alpha}} + 1\right] \alpha^{n/2}.$$

(b) Let T be the total elapsed time it takes for the object to make an infinite number of bounces. Then T is given by the following sum, which has an infinite number of terms:

$$T = \Delta t_1 + \Delta t_2 + \Delta t_3 + \cdots + \Delta t_n + \cdots$$

Using the expression you found in part (a), show that

$$T = \sqrt{\frac{2h_1}{g}} [1 + \sqrt{\alpha}][1 + \sqrt{\alpha} + (\sqrt{\alpha})^2 + (\sqrt{\alpha})^3 + \cdots]$$

and so

$$T = \sqrt{\frac{2h_1}{g}} \left[\frac{1 + \sqrt{\alpha}}{1 - \sqrt{\alpha}}\right].$$

[*Hint*: The sum of an infinite number of terms given by $1 + x + x^2 + x^3 + \cdots$ is called the geometric series. If $-1 < x < 1$, the geometric series equals $\frac{1}{1-x}$.]

(c) For a basketball bouncing on a gymnasium's hardwood floor, $\alpha = 0.65$. Use your result from part (b) to predict the time T, if the basketball is released from an initial height of 1.0 m. Is your prediction plausible?

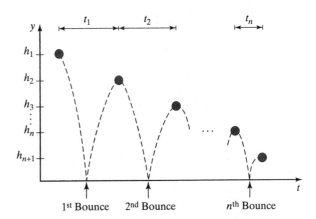

FIGURE 2–9

Kinematics in Two Dimensions; Vectors

Sections 3–1 to 3–5

1. (III) Vector **A** is horizontally directed and has a magnitude of 30, while vector **B** is directed at an angle of 30° above the horizontal and has a magnitude of 40. If **C** = **A** + **B**, find the magnitude and direction of **C** by two methods:
   (a) by adding the components of vectors **A** and **B**.
   (b) by drawing the graphical sum **C** = **A** + **B**, then using the law of cosines and the law of sines to find the magnitude and direction of **C**.

Section 3–6

2. (II) Let the time-dependent position vector of an object be given by $\mathbf{r} = (3.0\,t\,\mathbf{i} - 6.0\,t^3\,\mathbf{j})$ m.
   (a) Find the velocity **v** and acceleration **a** of the object.
   (b) Determine **r** and **v** at time $t = 1.50$ s.

3. (II) An object, which is at the origin at time $t = 0$, has initial velocity $\mathbf{v}_o = (-14\,\mathbf{i} - 7.0\,\mathbf{j})$ m/s and constant acceleration $\mathbf{a} = (6.0\,\mathbf{i} + 3.0\,\mathbf{j})$ m/s². Find the position **r** where the object's velocity **v** equals zero.

4. (II) A particle's position as a function of time t is given by $\mathbf{r} = (5.0\,t + 6.0\,t^2)\mathbf{i} + (7.0 - 3.0\,t^3)\mathbf{j}$, where all quantities have the appropriate SI units. At $t = 5.0$ s, find the magnitude and direction of the particle's displacement vector $\Delta\mathbf{r}$ relative to the point $\mathbf{r} = 5.0\,\mathbf{i} + 7.0\,\mathbf{j}$.

Sections 3–7 and 3–8

5. (II) (a) When competing in the long jump event, track-and-field athlete Rose leaves the ground at 45° above the horizontal and is able to long jump 8.0 m. What is her "take-off" speed v_o?
   (b) Rose is out on a hike and comes to the left bank of a river. There is no bridge across the river and the right bank is horizontally 9.5 m away and vertically 2.5 m below. If she long jumps from the edge of the left bank with the "take-off" speed v_o determined in part (a) directed at $\theta_o = 45°$ above the horizontal as in Fig. 3–1, will Rose make it across the river or will she get wet?

6. (II) Extreme sports enthusiasts have been known to jump off the top of El Capitan, a sheer granite cliff 910 m high in Yosemite National Park. Assume a jumper runs horizontally off the top of El Capitan with speed 3.0 m/s and enjoys a free fall until he is 150 m above the valley floor, at which time he opens his parachute (Fig. 3–2).
   (a) How long does the jumper get to enjoy his free fall?
   (b) It's important to be as far away from the cliff as possible before opening one's parachute on such a jump. How far from the cliff is this jumper when he opens his chute? Ignore air resistance.

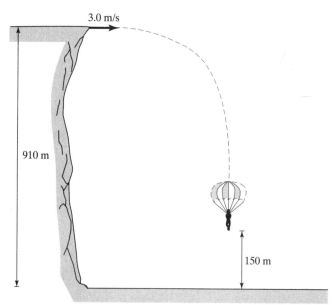

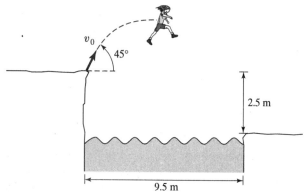

FIGURE 3–1

FIGURE 3–2

7. (II) Example 3–9 in the text is worked by arbitrarily defining the coordinate axes as follows: x-axis to the right and y-axis up. Rework this problem by defining the x-axis to the left and the y-axis down and show that the conclusion remains the same—the football lands on the ground 40.5 m to the right of where it departed the punter's foot.

8. (II) At a point that is 2.0 m above the ground, a basketball player releases a basketball with an initial speed $v_o = 12$ m/s directed at angle $\theta_o = 30°$ above the horizontal as shown in Fig. 3–3.

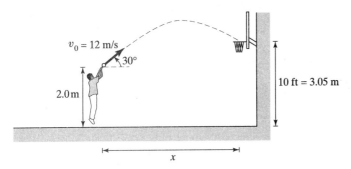

FIGURE 3–3

(a) How far from the basket should the player be in order to make a basket?
(b) At what angle relative to the horizontal will the ball enter the basket?

9. (II) A rock is thrown horizontally at 25 m/s from a hill with a 45° slope (see Fig. 3–4). How long does it take for the rock to hit the ground (i.e., land at a point on the 45° incline below that point from which it was thrown)?

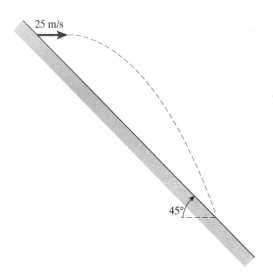

FIGURE 3–4

10. (II) In Example 3–5 of the text, the y-coordinate of the football at all times is given by the equation:

$$y = (12 \text{ m/s})t - \frac{1}{2}(9.8 \text{ m/s}^2)t^2.$$

(a) Using this equation, determine the two times t when the football passes through $y = -5.0$ m. Interpret the meaning of these two times. Are they relevant to the motion of the football?
(b) Using the above equation, determine the times t when the football passes through $y = +10.0$ m. Something unusual happens mathematically here. Interpret the physical meaning of this.

11. (II) Consider the trajectory of a projectile that returns to the same level as it had at launch. In Example 3–8 of the text, an expression for the projectile's range R is derived. The peak of the trajectory is defined by the condition $v_y = 0$. Use this condition to prove that the x-coordinate of the peak point is $R/2$.

12. (II) A daredevil wants you to estimate the absolute farthest distance one would expect a person on a motorcycle to be able to jump. What is your estimate, assuming the motorcycle returns to the same level from which it took off?

13. (II) The electric field produced in the region between two charged metal plates can be arranged to accelerate an electron with a constant upward-directed acceleration of 4.0×10^{13} m/s². This acceleration is present only during the time the electron is in the region between the plates. Assume an electron with horizontal speed 6.0×10^6 m/s enters the electric field region and that this region extends horizontally for 0.30 m as shown in Fig. 3–5. Ignore gravitational effects.
(a) Calculate the time it takes the electron to travel across the electric field region.
(b) Calculate the total vertical distance the electron will rise due to the field.
(c) At what angle will the electron emerge from the electric field region?
(d) Although gravity also acts on the electron as it moves, why is it unnecessary to take this influence into account in this calculation?

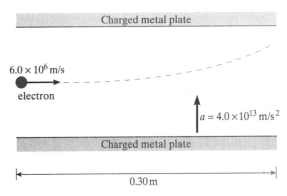

FIGURE 3–5

14. (II) Prove that when a projectile returns to the same level from which it was launched, its speed equals its initial launch speed v_o.

15. (III) A shot putter throws a shot (which is a small heavy ball) from a height of $h = 2.1$ m above the ground. Assume the shot putter is a world-class athlete who can release the shot with an initial speed of $v_o = 13.5$ m/s as shown in Fig. 3–6.
(a) Define θ_o to be the angle (above the horizontal) with which the shot leaves the shot putter's hand and R to be the horizontal distance from the shot putter's feet to the point on the ground where the shot lands. Derive a relation that describes how the range R depends on the release angle θ_o.
(b) Using the given values for v_o and h, use a graphing calculator or a computer to plot R vs. θ_o. According to your plot, what value of θ_o maximizes R?

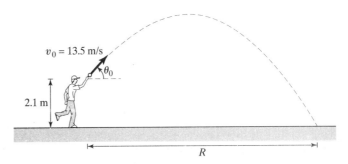

FIGURE 3–6

SECTIONS 3–7 and 3–8

16. (III) A boy is capable of throwing a ball with a maximum speed v_o of about 100 km/h. He also can run the 100 m dash in 15 s. This boy wants to throw a ball into the air with his maximum throwing speed v_o, then run over and catch the ball when it returns to the ground. What range of initial angles θ_o can the boy throw the ball and still be capable of catching it on its return flight? Assume the boy catches the ball at the same level at which it was thrown.

Section 3–9

17. (II) Professional skateboarder Tony Hawk can ride around a vertical circular track in a loop-the-loop trick.
    (a) At the bottom of the loop, Tony typically is moving at 8.0 m/s and, if standing upright, the center of his body would travel in a circle of radius 1.2 m. Find the centripetal acceleration a (in terms of g) in this situation.
    (b) In the actual trick, Tony crouches down at the bottom of the loop, so that the center of his body travels in a circle of radius 1.8 m, markedly reducing his centripetal acceleration (helping to prevent his legs from buckling). By what percent is a reduced via crouching (in comparison with the value for a when standing upright)?

18. (II) The centripetal acceleration equals g for a satellite in circular orbit near the surface of Earth (i.e., the orbit's radius is approximately Earth's radius). Determine the orbital speed v and period T for such a satellite. This approximation is valid for the Space Shuttle whose orbit is only about 250 km above Earth's surface.

Section 3–10

19. (II) City A is a distance x directly west of city B. A wind blows from west to east at a speed v_o relative to the ground. An airplane flies from city A to B (flying with the wind), then immediately returns from city B to A (flying into the wind). Show that the total round trip time between these cities is longer than if no wind were blowing.

20. (II) A person in the passenger basket of a hot-air balloon throws a ball horizontally outward from the basket with speed 10.0 m/s (Fig. 3–7).

    (a) If the hot-air balloon is rising upward at 5.0 m/s relative to the ground during this throw, what velocity does the ball have relative to a person standing on the ground?
    (b) If the hot-air balloon is descending downward at 5.0 m/s relative to the ground during this throw, what velocity does the ball have relative to a person standing on the ground?

21. (II) A child, who is 50 m from the bank of a river, is being carried helplessly downstream by the river's swift current of 1.0 m/s. As the child passes a lifeguard on the river's bank, the lifeguard starts swimming in a straight line until he reaches the child at a point downstream (Fig. 3–8). If the lifeguard can swim at a speed of 2.0 m/s relative to the water, how long does it take him to reach the child? How far downstream does the lifeguard intercept the child?

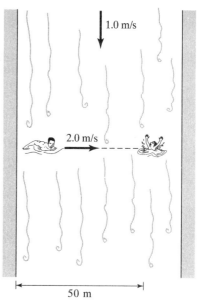

FIGURE 3–8

22. (III) A child, who is 50 m from the bank of a river, is being carried helplessly downstream by the river's swift current of 1.0 m/s. A lifeguard on the river's bank starts swimming with a speed of 2.0 m/s relative to the water directed at an angle θ upstream when the child is 20 m upstream (see Fig. 3–9).

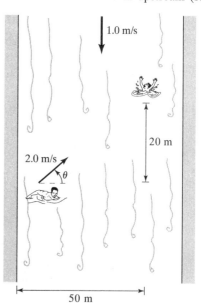

FIGURE 3–9

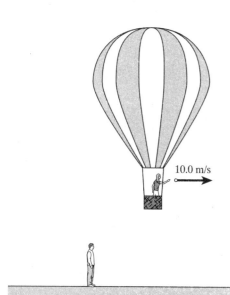

FIGURE 3–7

How long does it take the lifeguard to reach the child? How far downstream (relative to the lifeguard's starting point) does the lifeguard intercept the child?

General Problems

23. (II) Let objects A and B be on the x-axis and y-axis, respectively, and assume that the position of these objects changes linearly with time according to the following relations:

$$\mathbf{r}_A = \alpha t\, \mathbf{i}$$
$$\mathbf{r}_B = \alpha t\, \mathbf{j}$$

where α is a constant called the "expansion rate" (see Fig. 3–10). Show that the magnitude of the displacement vector $\Delta \mathbf{r} = \mathbf{r}_A - \mathbf{r}_B$, which equals the distance between these objects, also increases linearly with time at the rate $\sqrt{2}\alpha$. [Similarly, the expansion of space in the universe causes all galaxies to move away from each other at a constant rate.]

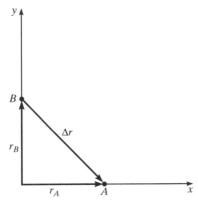

FIGURE 3–10

24. (II) Here's something fun to try at the next sporting event that you attend. Show that the maximum height H (m) attained by an object (e.g., a baseball, football, or soccer ball) projected into the air is approximately given by

$$H \approx 1.2 T^2$$

where T is the total time of flight for the object in seconds. Assume the object returns to the same level as that from which it was launched as in Fig. 3–11. For example, if you count to find that a baseball was in the air for $T = 5.0\,\text{s}$, the maximum height attained was $H \approx 1.2 \times (5.0)^2 = 30\,\text{m}$. The beauty of this relation is that H can be determined without knowledge of the launch speed v_o or launch angle θ_o.

25. (II) A grasshopper hops down a level road. On each hop, the grasshopper launches itself at angle $\theta_o = 45°$ and it achieves a range $R = 1.0\,\text{m}$. What is the average horizontal speed of the grasshopper as it progresses down the road? Assume that the time spent on the ground between hops is negligible.

26. (II) A traveler is carried down a moving walkway of length L in an airport in 60 s. The traveler can walk a distance L on stationary ground in 40 s. If the traveler walks on the moving walkway (in the same direction in which the walkway is moving), how long will it take to travel the distance L?

27. (II) In an atom trap, a laser slows atoms to very small velocities and confines them to a small spatial region. Assume all of the atoms in the trap have the same speed v_o and the confinement region is so small it can be considered to be a point. The atoms' speed v_o (and thus their temperature) can be determined by the following "time-of-flight" procedure: As shown in Fig. 3–12 a plane that is a distance d below the trap is the "detection plane." At time $t = 0$, the trap's laser is turned off and the atoms, each with initial speed v_o but with a range of possible launch angles $-90° \leq \theta_o \leq +90°$ (relative to the horizontal, positive is upward), follow projectile trajectories under the influence of gravity to the detection plane. Prove that the launch angles $\theta_o = -90°$ and $\theta_o = +90°$ produce the least and greatest fall times, respectively, to the detection plane. [Hint: This problem can be answered without the use of calculus.]

28. (II) You want to find out how fast you can throw a baseball, but only have a meter stick and a long piece of string. You climb to the roof of a small building and measure its height H as follows: Dangle the string from the roof until its lower end touches the ground, then mark the string's top. Pull up the string and measure from the mark to the string's lower end. Next, you throw a baseball with speed v_o horizontally as fast as possible and have your friend mark the point on the ground where the ball lands, then use the meter stick to measure the distance R from the base of the building to this landing point. Derive the equation that gives the speed v_o as a function of R and H. If, for your throw, $H = 10\,\text{m}$ and $R = 40\,\text{m}$, how fast (km/h) can you throw a baseball?

29. (II) (a) Austin Powers is forced into a north-facing car by Dr. Evil and then blindfolded. The car, which is a Mini Cooper, accelerates along a straight line for 8.0 s at its maximum rate. Austin knows that at maximum acceleration, a Mini Cooper can go from rest to 97 km/h (= 60 mph) in 6.9 s. What speed v does Austin have at the end of this straight-line acceleration?
(b) While at the speed v found in part (a), the car then turns right along a circular trajectory; Austin estimates from the tug that the car seat exerts on his velvet pants during the turn that he is being accelerated toward the center of the circle at about 30% of the acceleration due to gravity. Austin experiences this acceleration for 42 s. At the end of that time, in what direction can Austin Powers conclude that he is heading?

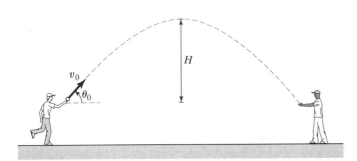

FIGURE 3–11

30. (III) In an atom trap, a laser slows atoms to very small velocities and confines them to a small spatial region. Assume all of the atoms in the trap have the same speed v_o and the confinement region is so small it can be considered to be a point. The atoms' speed v_o (and thus their temperature) can be determined by the following "time-of-flight" procedure: As shown in Fig. 3–12, a plane that is a distance d below the trap is the "detection plane." At time $t = 0$, the trap's laser is turned off and the atoms, each with initial speed v_o but with a range of possible launch angles $-90° \leq \theta_o \leq +90°$ (relative to the horizontal, positive is upward), follow projectile trajectories under the influence of gravity to the detection plane.

(a) The first and last atoms to arrive at the detection plane are the downward $(\theta_o = -90°)$ and upward $(\theta_o = +90°)$ directed atoms, respectively. In the time-of-flight experiment, the arrival times t_F and t_L of the (first) downward- and (last) upward-directed atoms are measured. Show that the speed of the atoms is given by

$$v_o = \frac{g(t_L - t_F)}{2}.$$

(b) Thermal physics theory suggests that the temperature of trapped atoms on the Kelvin scale (where absolute zero is 0 Kelvin) is given by

$$T = \frac{mv_o^2}{2k}$$

where m is the mass of an atom in the trap (commonly rubidium) and k is Boltzmann's constant, a fundamental constant of nature. In SI units, the approximate numerical values for m and k are 1.0×10^{-25} and 1.4×10^{-23}, respectively. If, in a time-of-flight experiment, one measures $t_L - t_F = 0.080\,\text{s}$ (this is a typical value), what is the temperature of the trapped atoms on the Kelvin scale?

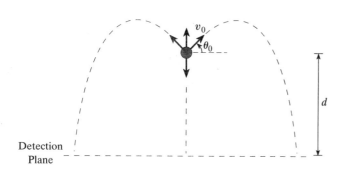

FIGURE 3–12

Dynamics: Newton's Laws of Motion

Sections 4–4 to 4–6

1. (I) An atomic force microscope can measure forces as small as a piconewton (pN). If a net force of this tiny magnitude were applied to a 50-kg person, how many years would it take to accelerate the person from rest to a speed of 1.0 m/s?

2. (II) The fan on a small fan cart (Fig. 4–1) commonly used in physics instructional laboratories can propel the 375-g cart forward with a force $F_P = 4.0\,\text{N}$. What is the steepest incline that this cart can climb? The friction produced by the cart's rolling wheels can be neglected in this situation.

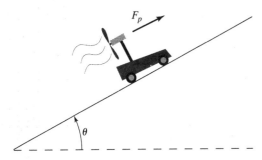

FIGURE 4–1

3. (II) The car is full, so a parent decides to buckle himself in with a seat belt and hold his 20-kg toddler on his lap during a short trip, which includes some highway driving at speed $v = 30\,\text{m/s}$.
   (a) If the maximum (magnitude) deceleration that the car's braking system can provide is about $10\,\text{m/s}^2$, show that the shortest distance over which the car can "emergency stop" while highway driving is about 50 m.
   (b) While traveling 30 m/s, assume the car's driver has to emergency stop the car over a distance of 50 m. In order to decelerate the toddler along with the car, the parent wraps his arms around the child. Assuming constant deceleration, how much force will the parent need to exert on the child during this deceleration period? Is this force achievable by an average parent?
   (c) Now assume the car is in an accident while highway driving and the car is brought to stop over a distance of 10 m. Assuming constant deceleration, how much force will the parent need to exert on the child during this deceleration period? Is this force achievable by an average parent?

4. (II) Modern skyscrapers would be impractical without high-speed elevators to whisk people to the upper floors. These elevators function under the following two limitations: the maximum magnitude of acceleration in a vertical plane that the human body can tolerate without discomfort is about $1.2\,\text{m/s}^2$ and the typical maximum speed attainable is about 9.0 m/s. Consider a person who boards an elevator on a skyscraper's ground floor and is transported to a floor 200 m above the ground level. Assume this vertical transport over 200 m occurs in a sequence of three steps: acceleration of magnitude $1.2\,\text{m/s}^2$ from rest to 9.0 m/s, followed by constant upward velocity of 9.0 m/s, and then deceleration of magnitude $1.2\,\text{m/s}^2$ from 9.0 m/s to rest. Determine the elapsed time and the percent deviation of the normal force magnitude exerted on the person from his or her weight during
   (a) the acceleration step.
   (b) the deceleration step.
   (c) the constant-velocity step.
   (d) During what percentage of the total transport time does the normal force magnitude not equal the person's weight?

5. (II) A 3.0-kg object has the following two forces acting on it:
$$\mathbf{F}_1 = (6.0\,\mathbf{i} + 12\,\mathbf{j})\,\text{N}$$
$$\mathbf{F}_2 = (-10\,\mathbf{i} + 22\,\mathbf{j})\,\text{N}$$
If the object is initially at rest, determine its velocity v at $t = 3.0\,\text{s}$.

6. (II) After work, a woman is commuting home in her 800-kg car at 24 m/s. She is 80 m from the beginning of an intersection when she sees the green traffic light change to yellow. The yellow light will last 4.0 s and the distance to the far side of the intersection is 100 m (Fig. 4–2).
   (a) If she chooses to accelerate, her car's engine will furnish a forward force of 1200 N. Will she make it completely through the intersection before the light turns red?
   (b) If she decides to panic stop, her brakes will provide a backward force of 2400 N. Will she stop before entering the intersection?

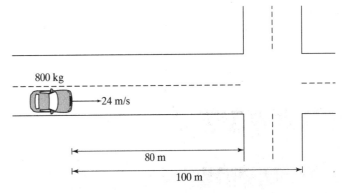

FIGURE 4–2

7. (II) In the game of paintball, players use "markers" (guns powered by pressurized gas such as air or CO_2) to propel 30-g spherical paintballs (gel capsules filled with paint) at opposing team members. For safety, game rules dictate that a paintball cannot leave the barrel of a marker with a speed greater than 92 m/s (= 300 ft/s). Model the firing of a paintball by assuming that pressurized gas applies a constant force F on this 30-g object over the length of the marker's barrel (typically 30 cm), accelerating it from rest to 92 m/s. Determine F and compare its magnitude with the weight of a paintball.

8. (III) You are stranded on the ice of a frozen pond with nothing except for the 540-g graphing calculator that you had hoped would help you get through your calculus-based physics course. The ice you stand on is a perfectly horizontal and frictionless surface and the nearest exit to safety is the edge of the pond 50 m away. You do one last calculation on your calculator to determine (correctly) the time T it is going take to reach safety, and then—doing what you have to do—throw the calculator horizontally away from you at a speed of 20 m/s. What value for T does your calculator display just before you toss it? Apply Newton's Laws to solve this problem, assuming that you provide a constant horizontal acceleration to the calculator, that the distance over which you are accelerated is negligible compared with 50 m, and that your mass is 75 kg. [Note: the time interval over which you exert a force on the calculator is equal to the time interval over which the calculator exerts a force on you.]

Section 4–7

9. (II) A skateboarder, with an initial speed of 2.0 m/s, rolls virtually friction-free down a straight incline of length 20 m in 3.3 s. At what angle θ is the incline oriented above the horizontal?

10. (II) Two 1.0-kg masses, connected by a string, hang vertically at rest, supported by another string attaching the upper mass to the ceiling of a room (Fig. 4–3). The string supporting the upper mass is then cut. At the instant that that string is severed, what is the downward acceleration of the upper mass?

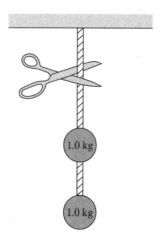

FIGURE 4–3

11. (II) Here is a new invention for the next X-Games—the Skateboard Launcher. A skateboarder of mass m standing at the bottom of a ramp is handed a rope, which passes over a frictionless pulley and is attached to a large mass M. The mass M, starting from rest, descends vertically and pulls the skateboarder up the incline, which is inclined at an angle $\theta = 30°$ above the horizontal (Fig. 4–4). At the top of the incline, the skateboarder, now with final speed v, releases the rope and is launched into the air where he can do his tricks. Through a proprietary mechanism, the rope always pulls on the skateboarder parallel to the incline and the pulley recesses out of the rider's way at the proper time. Because skateboard wheels roll with very little friction, it is plausible to assume that the skateboarder ascends the ramp in a friction-free manner. If $m = 80$ kg and $M = 160$ kg, find the skateboarder's (a) acceleration up the ramp, and (b) final speed v, if the acceleration takes place over a distance of 15 m.

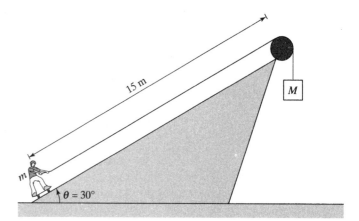

FIGURE 4–4

12. (II) On steep downhill stretches of mountain highways, there are commonly escape ramps for trucks with overheated brakes. In its simplest form (called a "gravity escape ramp"), the ramp is a straight, upwardly sloped roadway that uses only gravity to stop the runaway truck. A very long gravity escape ramp is required, if it is designed for all runaway trucks, including the fastest, which travel at 140 km/h. Assume that on a certain downhill stretch of highway there is funding only to build a gravity escape ramp of length 120 m inclined at about 11° above the horizontal.
(a) For a runaway truck, what is the maximum speed it can have and still be stopped by this ramp?
(b) If a brakeless truck entered the bottom of this ramp at a speed of 140 km/h, what speed (km/h) would it have at the top of the ramp?

13. (II) In rock climbing, a climber's safety rope is attached to a rock face with an "anchor" (e.g., a piece of metal wedged in a crack). If only a single anchor is used, the full tension F_{Tr} in the rope will pull on the anchor (Fig. 4–5a), possibly stressing it beyond its ultimate strength. This might occur when the rope stops the climber after an accidental fall. Thus, it is often essential to distribute the rope's tension between two anchors as shown in Fig. 4–5b. Assume the metal oval ring ("carabiner") that connects the rope to the two "anchor slings" (made of strong nylon webbing) remains stationary at all times.
(a) Prove that if two anchors are placed symmetrically (i.e., $\theta_1 = \theta_2 = \theta$), the tensions F_{Ta1} and F_{Ta2} on the two anchors are equal.
(b) If $\theta = 30°$, show that a force of only about 60% of the rope's tension will be applied to each anchor.

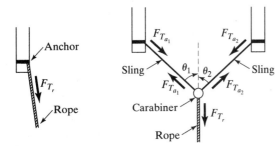

FIGURE 4–5

14. (II) In rock climbing, a climber's safety rope is commonly attached to a rock face with two "anchors." Consider two anchors symmetrically placed with $\theta_1 = \theta_2 = 20°$ (see Fig. 4–5b). Such anchors, when properly attached to a rock face, will stay in place unless subjected to a force greater than about $F_{Ta} = 10 \text{ kN}$. Assuming the metal oval ring ("carabiner") that connects the rope to the two "anchor slings" (made of strong nylon webbing) remains stationary at all times, find the greatest tension force F_{Tr} that can exist in the climbing rope (such as when a climber is being stopped during a fall) without the anchors "failing" (i.e., becoming detached from the rock face).

15. (II) In rock climbing, a climber's safety rope is commonly attached to a rock face with two "anchors." As a rule of thumb, the two anchors are always placed so that the total angle $\theta_1 + \theta_2$ between them is less than 90° (see Fig. 4–5b). Assume the metal oval ring ("carabiner") that connects the rope to the two "anchor slings" (made of strong nylon webbing) remains stationary at all times and that the two anchors are symmetrically placed (i.e., $\theta_1 = \theta_2 = \theta$).
    (a) Show that the force F_{Ta} applied to each anchor is only a fraction x (<1) of the rope's tension F_{Tr} when $\theta = 45°$. Determine $x \equiv \dfrac{F_{Ta}}{F_{Tr}}$ for this value of θ.
    (b) Show the basis of the rule of thumb by finding the threshold angle θ_o beyond which $x > 1$. That is, for angles greater than θ_o, the force applied to each anchor is greater than the rope's tension and so the benefit of "dual anchoring" has been lost.

16. (II) In rock climbing, a climber's safety rope is commonly attached to a rock face with two "anchors." If the two anchors are asymmetrically placed so that $\theta_1 \neq \theta_2$ (see Fig. 4–5b), the force applied to one anchor will be greater than the other. Let $\theta_1 = 20°$ and $\theta_2 = 40°$ and find how many times greater the force applied to the first anchor is in comparison with the force applied to the second anchor. That is, compute the ratio $\dfrac{F_{Ta1}}{F_{Ta2}}$. Assume the metal oval ring ("carabiner") that connects the rope to the two "anchor slings" (made of strong nylon webbing) remains stationary at all times.

17. (II) On steep downhill stretches of mountain highways, there are commonly escape ramps for trucks with overheated brakes. In its simplest form (called a "gravity escape ramp"), the ramp is a straight, upwardly sloped roadway that uses only gravity to stop the runaway truck. The steepest that a roadway can be inclined is about 11° above the horizontal and the fastest runaway truck travels at about 140 km/h. Using these values, determine the necessary length for a gravity escape ramp. [Note the large size of your calculated length. If sand is used for the bed of the escape ramp, rather than hard roadway, the ramp's length can be reduced by a factor of about two.]

18. (II) Two masses m_1 and m_2, connected by a string passing over a frictionless pulley, are supported on frictionless inclines with $\theta_1 = 50°$ and $\theta_2 = 40°$ as shown in Fig. 4–6. If it is found that these masses can be made to move at constant speed in either direction (i.e., the motion can be directed so that either m_1 or m_2 moves down its respective incline at constant speed), what is the ratio m_1/m_2?

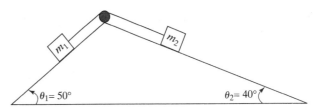

FIGURE 4–6

19. (III) Imagine a passenger seated in a sports car, which has a horizontal acceleration a. Assume that the seat that supports the passenger is rigid and slippery (i.e., essentially frictionless) so that its influence on this person can be modeled as two normal forces F_{N_1} and F_{N_2} due to the horizontal and inclined portions of the seat, respectively, as shown in Fig. 4–7. Also, assume the only other force that acts on the person is the weight force mg (i.e., there are no forces due to a seat belt, the car's floor, etc.).
    (a) Show that, as the car accelerates, the passenger's body will remain stationary with respect to the seat only if the incline angle θ of the seat's upper portion obeys the following relation
    $$\theta < \theta_o = \tan^{-1}\left(\frac{g}{a}\right)$$
    where θ_o is the threshold value for the body "slipping" relative to the seat.
    [Hint: The normal force due to a surface can only push outward from the surface.]
    (b) A typical sports car can accelerate at values up to $a = 7.0 \text{ m/s}^2$. For this maximum value of a, determine θ_o.

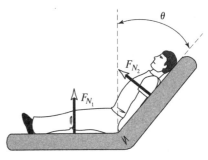

FIGURE 4–7

20. (III) In deciding upon the best choice for the Skateboard Launcher's descending mass M (see Problem 11), a designer would be confronted with the concept of "diminishing returns."
    (a) Use Newton's second law to prove that the acceleration a produced by the 30°-inclined Skateboard Launcher (compared with the acceleration due to gravity g) is given by
    $$\frac{a}{g} = \frac{x - 0.5}{1 + x}$$

where $x = \frac{M}{m}$ is the ratio of the descending mass compared with the skateboarder's mass.

(b) From your relation in part (a), show that the Skateboard Launcher is intrinsically limited in that it can, at most, provide an acceleration of g.

(c) Use a graphing calculator or a computer to plot the quantity $\frac{a}{g}$ (y-axis) versus x (x-axis). Based on this plot, along with the knowledge that the larger the value of the descending mass the more the Skateboard Launcher will cost to construct, explain why a designer would most likely choose x to be in the range of 1 to 5.

21. (III) For the situation described in Problem 19:
(a) taking the acceleration a and angle θ as givens, find expressions for the forces F_{N_1} and F_{N_2} in terms of multiples of the passenger's weight. That is, find expressions for F_{N_1}/mg and F_{N_2}/mg.
(b) with $a = 7.0 \text{ m/s}^2$, plot F_{N_1}/mg and F_{N_2}/mg versus angle θ in the range $0 \leq \theta < \theta_o$. At this acceleration, find the ratio of F_{N_2} at $\theta = 0°$ to F_{N_2} at $\theta = \theta_o$.

22. (III) Two masses $m_1 = 5 \text{ kg}$ and $m_2 = 10 \text{ kg}$, connected by a string passing over a frictionless pulley, are supported on frictionless inclines with $\theta_1 = 50°$ and $\theta_2 = 40°$. If a 100-N force, directed at an angle of 20° above the horizontal, acts continually on mass m_1 (Fig. 4–8), what is the resulting acceleration (magnitude and direction) of this system? Assume the string remains taut when the 100-N force acts.

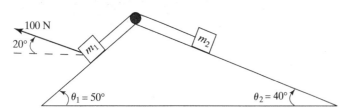

FIGURE 4–8

General Problems

23. (II) You are an 80-kg prisoner planning to escape from a jailhouse window 15 m above the ground. You have a rope made of bed sheets, but it can provide a maximum tension force of only 600 N. If you plan to leave the window with zero speed and want to hit the ground with no more than a speed of 7.0 m/s, should you go on a diet before escaping? If so, how many kilograms should you lose? Assume constant acceleration during the descent.

24. (II) A 200-m wide asteroid, composed of rock-like material and roughly spherical in shape, has a mass of 1.0×10^{10} kg. If such an object impacted Earth, the ecological changes following this collision could potentially put an end to humanity. Researchers have suggested that if a "doomsday" asteroid were identified a decade or so before it is scheduled to collide with Earth, a small "space tug" could be sent to its surface to deliver a gentle constant force of 2.5 N to the asteroid for an extended time period. If this force operated until it had changed the asteroid's orbital speed by 0.20 cm/s, it has been shown that the asteroid would then miss its future rendezvous with Earth. Determine how long the gentle force must act on a 1.0×10^{10} kg asteroid in order to avert its collision with Earth.

25. (II) A 55-kg swimmer is standing erect at the tip of a 3.0-m diving board. The swimmer then steps off the board and falls from rest for a vertical distance of 3.0 m, at which point her toes start to enter the swimming pool's water. As the swimmer falls 3.0 m downward through the air, she causes Earth to rise a distance x. Determine x, given that the mass of Earth is 6.0×10^{24} kg.

26. (II) Using focused laser light, optical tweezers can apply a force of about 10 pN to a 1.0-μm diameter polystyrene sphere. How many times greater is the bead's acceleration due to the optical tweezers than due to gravity? Polystyrene has a density about equal to that of water; i.e., a volume of 1.0 cm^3 has a mass of about 1.0 g. [After attaching the molecules to polystyrene beads, optical tweezers have been used to study the physical properties of biological molecules such as DNA.]

27. (II) The inexperienced captain of a small 2500-kg boat points the boat directly toward a dock and shuts off its engine when the boat is traveling at a speed of 0.50 m/s. When the boat is 1.0 m from the dock, a helpful person on the dock reaches out and starts pushing on the boat with a force of 200 N and continues to do so, trying to stop the boat. Will the person be able to stop the boat before it hits the dock? Ignore any resistive force on the boat due to the water.

28. (II) In a mountain-climbing technique called the "Tyrolean traverse," a rope is anchored on both ends (to rocks, strong trees, etc.) across a deep chasm and then a climber traverses the rope while attached by a sling as shown in Fig. 4–9. Because this technique can generate tremendous forces that stress the rope and anchors, a basic understanding of physics can be helpful in implementing it safely. A typical climbing rope can provide a tension force of up to 29 kN before breaking and, for safe use, a "safety factor" of 10 is usually recommended (i.e., the rope should only be required to supply a tension force of 2.9 kN). As we will demonstrate, in setting up the Tyrolean traverse, the length of rope used must allow for some "sag" in order for it to remain in its recommended safety range. Consider a 90-kg climber at the center of a Tyrolean traverse, which spans a 30-m chasm.
(a) In order for the rope to be within its recommended safety range, find the distance x that the rope sags in this situation.
(b) If the Tyrolean traverse is incorrectly set up so that the rope sags by only one-fourth the distance found in part (a), determine the tension force in the rope. Will the rope break?

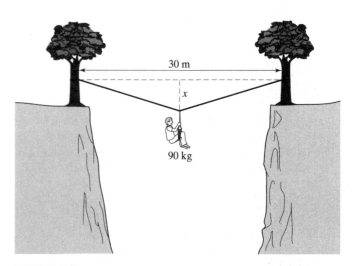

FIGURE 4–9

29. (II) Here's an invention to measure the upward acceleration of an elevator. As shown in Fig. 4–10, a sequence of masses—2.00, 2.05, 2.10, 2.15, and 2.20 kg—is hung from a cross bar, each mass supported by lightweight "5-pound test" fishing line. This type of fishing line will break when its tension force exceeds 22 N (= 5 pounds). Assume that this device is placed in an elevator and, when the elevator accelerates upward, it is observed that the 2.20, 2.15, and 2.10 kg masses each break their respective fishing lines, but the lines attached to the 2.05 and 2.00 kg masses remain intact. Within what range is the elevator's acceleration determined to be?

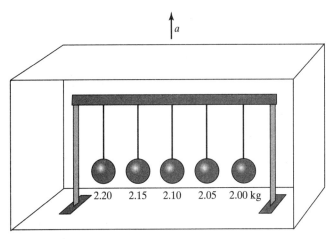

FIGURE 4–10

30. (III) An invention to measure the upward acceleration of an elevator consists of a sequence of masses each hung from "5-pound test" fishing line (see Fig. 4–10). This type of fishing line will break when its tension force exceeds 22 N (= 5 pounds). When placed in an elevator with acceleration a, all masses greater than a value m will break their attached fishing lines. Thus, for the given upward acceleration a, the tension in the line attached to mass m is exactly equal to the threshold "breaking" tension $F_{Tb} = 22$ N (or 5 pounds).

(a) Derive the relation between m and a and show that, if the elevator's acceleration is much less than the acceleration due to gravity,

$$m \approx \frac{F_{Tb}}{g} - \frac{F_{Tb}}{g^2} a$$

[*Hint*: The binomial expansion allows the following approximation $(1 + x)^{-1} \approx 1 - x$, if $x \ll 1$.]

(b) Show that if a sequence of hung masses is to resolve the elevator's acceleration to within a range of $\pm 0.10 \, \text{m/s}^2$, the difference between adjacent hung masses should be 23 g.

CHAPTER 5

Further Applications of Newton's Laws

Section 5–1

1. (II) Two masses $m_1 = 2.0$ kg and $m_2 = 50$ kg are on inclines and are connected together by a string passed over a frictionless and massless pulley as shown in Fig. 5–1. The coefficient of kinetic friction between each mass and its incline is $\mu_k = 0.30$. When the masses move such that m_1 moves up and m_2 moves down their respective inclines, find the masses' acceleration.

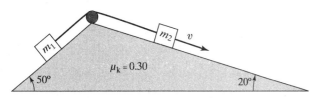

FIGURE 5–1

2. (II) The Sheriff has bad-guy Bob pinned against a vertical wall by exerting a force of magnitude F_P directed at an angle of 50° above the horizontal as shown in Fig. 5–2. Over what range of values for F_P can the Sheriff keep Bob at rest against the wall, if the coefficient of static friction between Bob and the wall is $\mu_s = 0.60$? Bob's mass is 70 kg and, impressively, the Sheriff has pinned Bob against the wall so that Bob's feet do not touch the ground.

FIGURE 5–2

3. (II) Two masses ($m_1 = 10$ kg and $m_2 = 20$ kg) are connected by a rope and are being pulled up a 20° incline by a force F_P as shown in Fig. 5–3. The coefficient of kinetic friction between each mass and the inclined surface is $\mu_k = 0.20$. If the maximum tension force that the rope can provide without breaking is 200 N, what is the maximum value allowable for the force F_P?

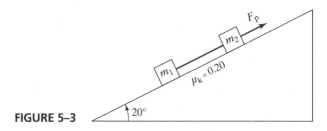

FIGURE 5–3

4. (II) A package of mass m is dropped onto a horizontal conveyor belt. The speed of the conveyor belt is $v = 1.5$ m/s and the coefficient of kinetic friction between the package and the belt is $\mu_k = 0.70$.
(a) For what length of time will the package slide on the belt (i.e., how long is it until the package is at rest relative to the belt)?
(b) How far will the package move during the time interval that it is being accelerated up to the speed of the belt?

5. (II) A heavy piano, without wheels and of mass m, is being slowly slid down an inclined ramp of angle $\theta = 30°$ at the back of a mover's truck (Fig. 5–4). The coefficient of kinetic friction between the piano and the incline is $\mu_k = 0.13$. The mover exerts a horizontal force F_P on the piano to retard its slide in such a way that the piano moves with constant velocity. Determine the horizontal force's magnitude F_P, expressing it as a fraction of the piano's weight.

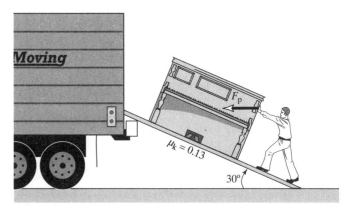

FIGURE 5–4

6. (II) A 70-kg water skier is being accelerated by a boat on a flat ("glassy") lake. The coefficient of kinetic friction between the skis and the water surface is $\mu_k = 0.25$ (see Fig. 5–5).
(a) Assume the rope that attaches the skier to the boat applies a completely horizontal tension force (i.e., $\theta = 0°$) of magnitude $F_T = 200$ N to the skier. Determine the horizontal acceleration of the skier.
(b) Again, assume the rope that attaches the skier to the boat applies a tension force of magnitude $F_T = 200$ N to the skier, but now this force is directed upward at an angle $\theta = 10°$ relative to the horizontal. Given that the skier only moves horizontally (i.e., does not move up or down from the water surface), determine the horizontal acceleration of the skier.
(c) Explain why the skier's acceleration in part (b) is greater than that in part (a).

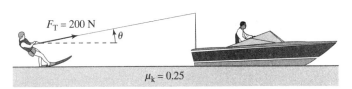

FIGURE 5–5

7. (II) It's moving day and you are dragging a heavy dresser at constant velocity across the horizontal floor of your new apartment. The force you are applying to the dresser has magnitude F_P and it is directed at an angle θ above the horizontal as shown in Fig. 5–6. The weight of the dresser is $mg = 400$ N and the coefficient of kinetic friction between it and the floor is $\mu = 0.40$. Determine the required value of F_P if
(a) you apply your force completely horizontally; that is, $\theta = 0°$.
(b) you apply your force at an angle $\theta = 20°$.
(c) Explain qualitatively why the required force at $\theta = 20°$ (somewhat surprisingly) turns out to be less than that required at $\theta = 0°$.

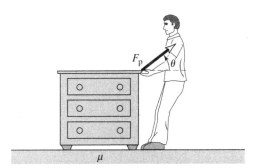

FIGURE 5–6

8. (II) On level hard-packed snow, Billy (mass $m_B = 40$ kg), while seated on a sled (mass $m_S = 5.0$ kg) as shown in Fig. 5–7, holds onto a rope and is towed at constant velocity by Susie. The rope is horizontally directed and the coefficient of kinetic friction between the sled and the snow is $\mu_k = 0.15$. Only the static friction force is available to hold Billy in place on the sled (i.e., no other force attaches him to the sled).
(a) Assuming that Billy remains at rest with respect to the sled, determine the rope's tension force.

FIGURE 5–7

(b) Determine the static friction force and the normal force that act on Billy due to the sled.
(c) If the coefficient of static friction between Billy and the sled is $\mu_s = 0.30$, will the static friction actually be capable of holding Billy in place on the sled?

9. (II) On level hard-packed snow, Billy (mass m_B), while seated on a sled (mass m_S) as shown in Fig. 5–7, holds onto a rope and is towed at constant velocity by Susie. The rope is horizontally directed and the coefficient of kinetic friction between the sled and the snow is $\mu_k = 0.15$. Only the static friction force is available to hold Billy in place on the sled (i.e., no other force attaches him to the sled). Let μ_s be the coefficient of static friction between Billy and the sled.
(a) Show that the "no-slip" criterion for Billy remaining at rest with respect to the sled, while being towed, is

$$\mu_s \geq \left(1 + \frac{m_S}{m_B}\right)\mu_k.$$

(b) In the limit where Billy is much more massive than the sled, what does the minimum "no-slip" value for μ_s approximately equal?

10. (III) On level hard-packed snow, Billy (mass $m_B = 40$ kg), while lying on a sled (mass $m_S = 5.0$ kg) as shown in Fig. 5–8, holds onto a rope and is towed at constant velocity by Susie. The angle of the rope is $\theta = 30°$ above the horizontal and the coefficient of kinetic friction between the sled and the snow is $\mu_k = 0.15$. Only the static friction force is available to hold Billy in place on the sled (i.e., no other force attaches him to the sled).
(a) Assuming that Billy remains at rest with respect to the sled, determine the magnitude of the rope's tension force.
(b) Determine the magnitudes of the static friction force and the normal force that act on Billy due to the sled.
(c) If the coefficient of static friction between Billy and the sled is $\mu_s = 0.30$, will the static friction force actually be capable of holding Billy in place on the sled?

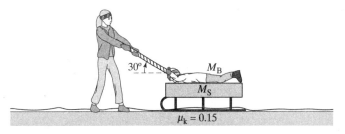

FIGURE 5–8

11. (III) (a) It's moving day and you are dragging a heavy dresser at constant velocity across the horizontal floor of your new apartment. The force you are applying to the dresser has

magnitude F_P and it is directed at an angle θ above the horizontal as shown in Fig. 5–6. The weight of the dresser is mg and the coefficient of kinetic friction between it and the floor is μ. Show that the required F_P at an angle θ is given by

$$F_P = \frac{\mu mg}{\cos\theta + \mu \sin\theta}.$$

(b) For most combinations of surfaces, the coefficient of kinetic friction falls in the range $0 < \mu < 1$. Taking $mg = 400$ N, plot F_P vs. θ for the following three choices of μ: $\mu = 0.10$, $\mu = 0.40$, and $\mu = 1.00$. On each plot, identify the angle θ_{min} at which F_P is smallest. θ_{min} is the angle at which the minimum magnitude force F_P is required to drag the dresser at constant velocity for a given μ. As μ is increased, what happens to the value of θ_{min}?

12. (III) It's moving day and you are dragging a heavy dresser at constant velocity across the horizontal floor of your new apartment. The force you are applying to the dresser has magnitude F_P and it is directed at an angle θ above the horizontal as shown in Fig. 5–6. The weight of the dresser is mg and the coefficient of kinetic friction between it and the floor is μ. Identify generic trends in this situation by using "dimensionless quantities" to perform the analysis (rather than taking a particular choice for mg).
(a) Show that the required F_P at a given angle θ is given by

$$F_P = \frac{\mu mg}{\cos\theta + \mu \sin\theta}$$

or, equivalently,

$$\frac{F_P}{mg} = \frac{\mu}{\cos\theta + \mu \sin\theta}.$$

In the second relation, the force F_P is conveniently described in comparison with the dresser's weight mg. For example, if $F_P/mg = 0.10$, that means that F_P is 10% as large as mg.
(b) For most combinations of surfaces, the coefficient of kinetic friction falls in the range $0 < \mu < 1$. Plot (F_P/mg) vs. θ for the following three choices of μ: $\mu = 0.10$, $\mu = 0.40$, and $\mu = 1.00$. On each plot, identify the angle θ_{min} at which the quantity (F_P/mg) is smallest. Since mg is a constant, θ_{min} is the angle at which the minimum magnitude force is required to drag the dresser at constant velocity for a given μ. As μ is increased, what happens to the value of θ_{min}?

13. (III) (a) It's moving day and you are dragging a heavy dresser at constant velocity across the horizontal floor of your new apartment. The force you are applying to the dresser has magnitude F_P and it is directed at an angle θ above the horizontal as shown in Fig. 5–6. The weight of the dresser is mg and the coefficient of kinetic friction between it and the floor is μ. Show that the required F_P at an angle θ is given by

$$F_P = \frac{\mu mg}{\cos\theta + \mu \sin\theta}.$$

(b) Show that the derivative of F_P with respect to θ is

$$\frac{dF_P}{d\theta} = -\frac{\mu mg}{(\cos\theta + \mu \sin\theta)^2}(-\sin\theta + \mu \cos\theta).$$

[*Hint*: For help in evaluating derivatives, see Appendix B in the text.]

(c) Setting $\dfrac{dF_P}{d\theta} = 0$, show that, for a given value of μ, the force's required magnitude is smallest when its applied angle is $\theta_{min} = \tan^{-1}(\mu)$. Determine θ_{min} when $\mu = 0.10$, 0.40, and 1.00.

14. (III) Consider two masses m_1 and m_2 on a rough horizontal surface, connected by a taut string. A force F_P directed at an angle θ above the horizontal acts on mass m_1, moving it to the right while towing mass m_2 behind as in Fig. 5–9. The coefficient of kinetic friction between each mass and the surface is μ_k. Show that in this situation the string's tension force depends on the value of μ_k and is given by

$$F_T = F_P \frac{m_2(\cos\theta + \mu_k \sin\theta)}{(m_1 + m_2)}.$$

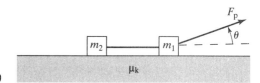

FIGURE 5–9

Sections 5–2 and 5–3

15. (II) You are designing a carnival ride. A passenger "bucket" is attached by a 10.0-m cable to a 6.0-m crossarm (Fig. 5–10). When this device is made to rotate about the mast, the bucket will swing out to an angle θ with respect to the vertical. For maximum thrills, you wish this angle θ to be 60°. How many revolutions about the mast per second will be required?

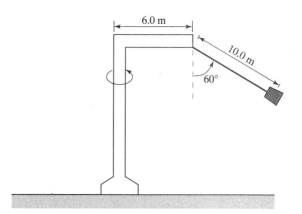

FIGURE 5–10

16. (II) Except for the transportation required, here is a painless diet plan. Assume that you first stand on a bathroom scale at the North Pole and the scale reports your weight to be 650 N. Next you travel to the equator and, once there, again stand on your scale. Because of Earth's rotation, at the equator you are traveling in a circular path of radius $R = 6.38 \times 10^6$ m ($=$ Earth's radius) with a period of 1 day. What value does the scale now report for your weight? Take $g = 9.80$ m/s².

17. (II) In a jungle mishap, Tarzan, whose mass is 80 kg, finds himself swinging around a tree in a horizontal circle at one end of a vine 10 m long that makes an angle of θ with the vertical (Fig. 5–11). Tarzan travels once around the circle every

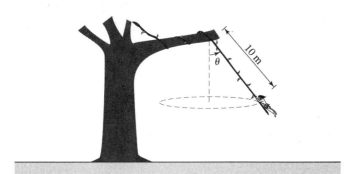

FIGURE 5–11

5.0 s. If this vine can supply a tension force of up to 2000 N without breaking, will Tarzan continue circling or will he break the vine?

18. (III) In a jungle mishap, Tarzan has a vine tangled around his foot while he is swinging on an overhead vine. Thus, he finds himself attached to two vines and swinging around in a horizontal circle as shown in Fig. 5–12. If 80-kg Tarzan rotates around the circumference of the horizontal circle at constant speed of 8.5 m/s, find the tensions in both the upper and lower vines.

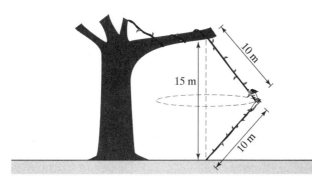

FIGURE 5–12

19. (III) Consider a roller-coaster car traveling around the roller coaster's vertical loop, which has a radius of curvature R at its top (see Fig. 5–13). Assume that, for a person of mass m seated in the car, the only forces acting on the rider are the weight force mg and the normal force due to the car's seat. The normal force is variable in magnitude.
    (a) Show that when the car travels at critical speed $v_c = \sqrt{gR}$ at the loop's top, the normal force acting on the rider is zero.
    (b) If the speed of the car at the top of the loop exceeds the critical value by 10%, i.e., $v_t = 1.10\sqrt{gR}$, determine the magnitude (expressed as a multiple of the rider's weight) and direction of the resulting normal force on the rider. Briefly explain why a simple seat without even lap restraints could provide this force.
    (c) If the speed of the car at the top of the loop falls short of the critical value by 10%, i.e., $v_t = 0.90\sqrt{gR}$, determine the magnitude (expressed as a multiple of the rider's weight) and direction of the required normal force on the rider if the person still moves on the loop's curved path of radius R. Explain why the car's seat must now be equipped with lap-bar and/or shoulder-harness restraints.
    (d) If a person was seated in the car with a speed of $v_t = 0.90\sqrt{gR}$ at the top of the loop and the seat was unable to provide any normal force (i.e., there were no restraints holding the person in the seat), on what radius circle would the person travel? Assuming the car continued on the loop's path (because it was somehow attached to the roller-coaster's track), qualitatively sketch the subsequent trajectory of the person relative to the car.

20. (III) (a) In older, "looping" roller coasters, the roller coaster's vertical loop is a perfect circle of radius R as in Fig. 5–14a. With this design, it can be shown (see Problem 85 in Chapter 8 of the text) that the apparent weight of a rider (i.e., the normal force on the rider due to the roller-coaster car) must be at least six times the rider's weight at the bottom of the loop. When the normal force takes on the minimum value of $6.00\,mg$ at the loop's bottom, what is the rider's speed v_b?
    (b) In modern roller coaster design, loops are teardrop-shaped, so that if the loop's total height is $2R$, the top part of the arc has, e.g., half the radius of the bottom arc. Consider a teardrop loop such that its top radius is $\frac{2}{3}R$ and its bottom radius is $\frac{4}{3}R$ as in the Fig. 5–14b. Show that if a rider at the bottom of this teardrop loop has the same speed v_b found in

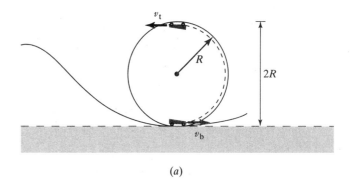

(a)

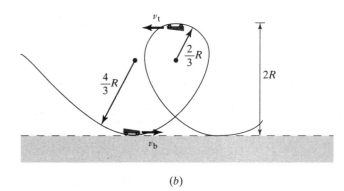

(b)

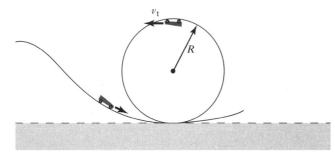

FIGURE 5–13

FIGURE 5–14

part (a), the rider's apparent weight is only 4.75 mg (rather than the value of 6.00 mg for a circular-shaped loop). [Teardrop shapes are used in roller-coaster design to reduce the "g-forces" (which can produce ill effects) experienced by riders.]

21. (III) Teardrop-shaped roller-coaster loops, when compared with circular-shaped loops, reduce the "g-forces" experienced by riders at the loop's bottom. Let's explore another advantage of the teardrop shape in roller-coaster design.
(a) Assuming a perfectly circular loop of radius R (Fig. 5–14a), show that when a roller-coaster car travels at critical speed $v_c = \sqrt{gR}$ at the loop's top, the normal force acting on the rider is zero. Speeds slower than v_c then are to be avoided because they would require the track to exert a normal force on the car that is directed away from the center of the circle, necessitating some complicated attachment mechanism between track and car (including riders).
(b) Now assume a teardrop-shaped loop of equal height ($=2R$) with a top radius of $\frac{2}{3}R$ (Fig. 5–14b). Show that if a car travels at the speed $v = \sqrt{gR}$ at the top of this loop, an (substantially nonzero) inwardly directed normal force equal to half its weight is exerted by the track on the car.
(c) Show that, in comparison with the circular loop, the teardrop design reduces the critical speed v_c by 18%.

Section 5–4

22. (III) A race car, starting at rest, travels continually on a horizontal circular track of radius $R = 50$ m. If the car is constantly accelerating along the direction of the track (i.e., tangent to the circle of radius R) at $a = 5.0 \text{ m/s}^2$, at what speed will the car's tires lose grip with the track and start skidding? Assume the coefficient of static friction between the tires and track is $\mu_s = 0.80$.

23. (III) You suddenly apply the brakes while driving your car at speed v around a curve on a level unbanked road, intending to decelerate the car with an acceleration of magnitude a_o along the road's path (i.e., tangential to the road's radius of curvature $R = 100$ m). Assume that the coefficient of static friction between your car's tires and the road is $\mu_s = 0.80$.
(a) Show that the car will not start to skid (i.e., the tires will not lose grip with the road) if

$$a_o \leq \sqrt{(\mu_s g)^2 - \left(\frac{v^2}{R}\right)^2}.$$

(b) If $v = 100$ km/h, what is the maximum allowed value for a_o if you wish to avoid skidding?
(c) At a certain critical speed v_c, you will be able to negotiate this curve at constant speed, but if you apply the car's brakes in any fashion, your car will start skidding. Determine v_c (km/h).

General Problems

24. (II) One rainy night, Wade is driving his car at speed $v_o = 65$ km/h along an unfamiliar back road. Suddenly, he realizes that the road dead-ends straight ahead and that there is a long brick wall at the road's end blocking his path (Fig. 5–15). When he is a distance $x = 30$ m from the wall, Wade will either brake or turn his car to avoid the wall. Assume that the road is flat (i.e., not inclined with respect to the horizontal) and that the coefficient of static friction between the car's tires and the wet road surface is $\mu_s = 0.70$.

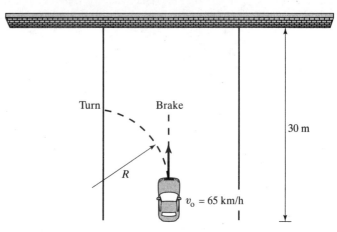

FIGURE 5–15

(a) If Wade decides to brake his car (along a straight line) and is able to do so using the maximum static friction force available, will his car stop before colliding with the wall?
(b) If Wade decides to turn his car and is able to use the maximum static friction force available to move his car on a circular path of radius R, will his car avoid colliding with the wall?

25. (II) A mass $m = 2.0$ kg is on an incline that makes an angle θ with the horizontal. The coefficients of static and kinetic friction are $\mu_s = 0.50$ and $\mu_k = 0.20$, respectively. A physicist uses her hand to exert an additional constant force F_P, pressing on the mass in the direction normal to the surface of the plane, as shown in Fig. 5–16.
(a) For the special case $F_P = 0.0$ N and the mass initially at rest, the mass will have a nonzero acceleration down the incline only if θ exceeds a threshold value θ_{min}. Determine θ_{min}.
(b) If $\theta = 40°$ and $F_P = 8.0$ N, with what acceleration will the mass slide down the incline?
(c) If $\theta = 40°$ and the mass is initially at rest, for what range of values for F_P will the mass not be able to slide down the incline?

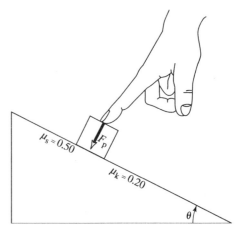

FIGURE 5–16

26. (II) A ski resort owner hires a consultant to determine the size of the tension force that the rope tow on the resort's "bunny" (beginner) hill must provide. When a skier grabs the rope, it pulls him or her up a 20° incline at constant velocity (Fig. 5–17). The consultant decides that the heaviest skier who is ever likely to use the rope tow has a weight of 1400 N (about 315 lbs) and assumes that the coefficient of kinetic friction between the skier's skis and the snow is $\mu_k = 0.10$. With these assumptions, what tension force will the consultant determine is required of the rope tow?

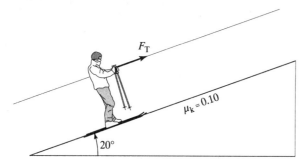

FIGURE 5–17

27. (II) With her dog, Hershey, lying asleep in the bed of her pickup truck, Devon is driving around town. An intersection is up ahead with a red light. If the truck is moving at 55 km/h initially, and Devon doesn't want Hershey to start sliding across the bed of the pickup while she is braking to a stop, at what minimum distance from the edge of the intersection should Devon start applying the pickup's brakes? Assume constant deceleration of the truck during the braking process, that the road is flat, and that the coefficient of static friction between Hershey and the metal pickup bed is $\mu_s = 0.30$.

28. (II) Consider two masses m_1 and m_2 on a rough horizontal surface, connected by a taut string. A horizontal force F_P acts on mass m_1, moving it to the right while towing mass m_2 behind as in Fig. 5–18. The coefficient of kinetic friction between each mass and the surface is μ_k. Show that the string's tension force in this situation is independent of μ_k and is the same value as when the surface is frictionless.

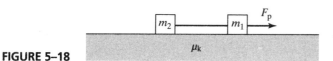

FIGURE 5–18

29. (II) After performing a quick calculation to determine the required rotational period T for its scheme, a malevolent extraterrestrial grabs the North Pole region and with a flick of the wrist, speeds up Earth's rotation so that people on the equator become "weightless" and are on the verge of flying off into space. What value of T (in minutes) is required for the extraterrestrial's evil plan?

30. (III) A 70-kg snowboarder has an initial velocity of 5.0 m/s at the top of a 30° incline (see Fig. 5–19). After sliding down the 100-m long incline (on which the coefficient of kinetic friction is $\mu = 0.10$), the snowboarder has attained a velocity v. The snowboarder then slides along a flat surface (on which $\mu = 0.20$) and comes to rest after a distance x. Use Newton's second law to find the snowboarder's accelerations while on the incline and while on the flat surface. Then use these accelerations to determine x.

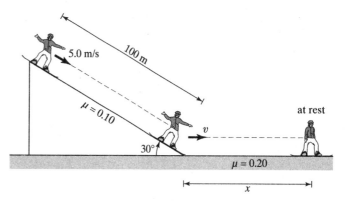

FIGURE 5–19

CHAPTER 6

Gravitation and Newton's Synthesis

Sections 6–1 to 6–3

1. (II) Imagine that you are the first experimental physicist to attempt a determination of the fundamental constant G that appears in Newton's law of universal gravitation. In order to design your experiment, you try to make an educated guess about the size of this quantity. You know the following facts: the acceleration due to gravity at Earth's surface is about 10 m/s^2; Earth is approximately spherical in shape and, from the original definition of the meter, its circumference is about 40×10^6 m; and rocks found on Earth's surface typically have densities of about 3000 kg/m^3. You know also that Earth possibly changes in composition as one descends into its interior, but accounting for this effect will make your estimation job difficult. If you make the simplifying assumption that Earth has constant density, what value do you estimate for G from these known facts?

2. (II) Determine how large the force of attraction is between two lovers. Assume the attractive force is gravitational. Also assume that the two people have masses of 75 kg and 50 kg, respectively, and that they can be assumed to be point masses separated by a distance of 20 cm. What two things would Newton suggest the lovers do to make themselves more attractive?

3. (II) Consider a pair of identical point masses, each of mass M, that are securely held in place so that they always remain separated by a distance of $2R$. A third mass m is then placed a distance x along the perpendicular bisector of the original two masses, as shown in Fig. 6–1. Show that the gravitational force on the third mass is directed inward along the perpendicular bisector and has a magnitude of
$$F = \frac{2GMmx}{(x^2 + R^2)^{3/2}}.$$

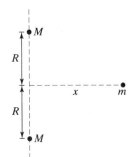

FIGURE 6–1

4. (II) A sphere of uniform construction has a mass M and radius R. A spherical cavity (which contains no mass) of radius $R/2$ is then carved within this sphere as shown in Fig. 6–2 (the cavity's surface passes through the sphere's center and just touches the sphere's outer surface). The centers of the original sphere and the cavity lie on a straight line, which defines the x-axis. With what gravitational force will the hollowed-out sphere attract a point mass m, which lies on the x-axis a distance d from the sphere's center?

[*Hint*: For computational purposes, think of the hollowed-out sphere as a composite object, which consists of a uniform sphere of positive mass M plus a uniform sphere with an appropriate amount of negative mass.]

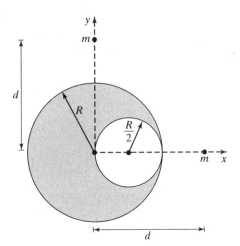

FIGURE 6–2

5. (III) A sphere of uniform construction has a mass M and radius R. A spherical cavity (which contains no mass) of radius $R/2$ is then carved within this sphere as shown in Fig. 6–2 (the cavity's surface passes through the sphere's center and just touches the sphere's outer surface). The centers of the original sphere and the cavity lie on a straight line, which defines the x-axis. With what gravitational force (expressed in terms of its x- and y-components) will this sphere attract a point mass m, if m lies on the y-axis a distance d from the sphere's center?

[*Hint*: For computational purposes, think of the hollowed-out sphere as a composite object, which consists of a uniform sphere of positive mass M plus a uniform sphere with an appropriate amount of negative mass.]

6. (III) A sphere of uniform construction has a mass M and radius R. Two spherical cavities (which contain no mass) of radius $R/2$ are then carved within this sphere as shown in Fig. 6–3 (each cavity's surface passes through the sphere's center and just touches the sphere's outer surface). The centers of the original sphere and two cavities all lie on a straight line, which defines the y-axis. With what gravitational force will the hollowed-out sphere attract a point mass m, which lies on the x-axis a distance d from the sphere's center?

[*Hint*: For computational purposes, think of the hollowed-out sphere as a composite object, which consists of a uniform sphere of positive mass M plus two uniform spheres with appropriate amounts of negative mass.]

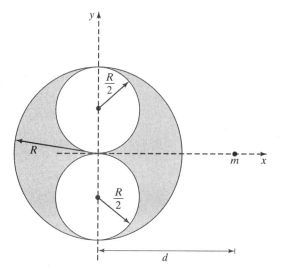

FIGURE 6–3

7. (III) Experiments to determine an accurate value for the fundamental constant G are notoriously difficult. An experimental physicist colleague comes to you with the following scheme to measure G. A steel sphere of mass $M_1 = 10\,\text{kg}$ and radius $R_1 = 6.7\,\text{cm}$ is to be hung from a cable of length L, which, when undisturbed, hangs motionless along the vertical direction (Fig. 6–4a). A second steel sphere of mass $M_2 = 10\,\text{kg}$ and radius $R_2 = 6.7\,\text{cm}$ is then brought alongside the first, gravitationally attracting it. When the two spheres come to rest, the cable will be deflected by an angle θ from the vertical. In this "equilibrium state," the centers of the spheres will be a distance $D = 14\,\text{cm}$ apart and aligned horizontally (Fig. 6–4b).
(*a*) Show that the gravitational constant can be determined in this experiment through the relation

$$G = \frac{gD^2 \tan\theta}{M_2}.$$

(*b*) In order to determine θ accurately, your colleague wants the first sphere, when in the equilibrium state, to be deflected by a horizontal distance $x = 1\,\text{cm}$ from the vertical. You estimate the value of G will be on the order of $1 \times 10^{-10}\,\text{N}\cdot\text{m}^2/\text{kg}^2$ (see Problem 1 for one estimation method). Assuming $G = 1 \times 10^{-10}\,\text{N}\cdot\text{m}^2/\text{kg}^2$, to obtain the desired deflection distance x, what length L is required for the cable? Express your answer as a fraction of Earth's radius.
(*c*) Sobered by your finding in part (*b*), your colleague decides that an optical technique involving lasers may be able to

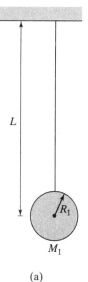

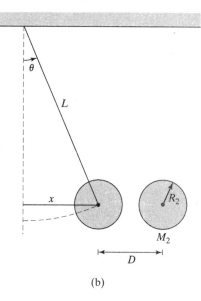

FIGURE 6–4

measure a horizontal deflection distance x for the first sphere on the order of the wavelength of light ($\approx 1 \times 10^{-6}\,\text{m}$). If $x = 1 \times 10^{-6}\,\text{m}$, what length L is required for the cable?

8. (III) Analyze the experiment described in Problem 7 more deeply. Assume the first and second spheres have densities and radii of ρ_1, R_1, ρ_2, and R_2, respectively. Assume that in the equilibrium state, the spheres are almost touching so that the distance D between their centers is about equal to $R_1 + R_2$. Remember an object's mass $M = \rho V$, where ρ and V are the object's density and volume, respectively.
(*a*) Show that in the equilibrium state, the cable's deflection angle θ is given by

$$\theta = \tan^{-1}\left[\frac{4\pi G \rho_2 R_2^3}{3g(R_1 + R_2)^2}\right].$$

(*b*) In Problem 7, the radii of the spheres were chosen as $R_1 = R_2 = 6.7\,\text{cm}$, and each was made of steel so that $\rho_1 = \rho_2 = 7800\,\text{kg}/\text{m}^3$. If, instead, the radius of the first object is decreased by a factor of ten and the radius of the second object is increased by a factor of ten, i.e., $R_1 = 0.67\,\text{cm}$ and $R_2 = 67\,\text{cm}$, and the second object is made of lead ($\rho_2 = 11{,}300\,\text{kg}/\text{m}^3$), by what factor will the deflection angle θ increase?

9. (III) No experimentally determined quantity x can be measured with perfect accuracy; there is always some uncertainty Δx (called the *absolute uncertainty*) in any measurement. The fractional uncertainty in x is defined as $\Delta x/x$. Among all of Nature's fundamental constants (e.g., speed of light, charge of the electron), the gravitational constant G is, by far, the least accurately known. One major problem in many of the experiments designed to measure G is precisely determining the distance between the centers of two attracting spherically shaped objects. Consider an experiment where two spherical masses m_1 and m_2, whose centers are separated by a distance r, attract each other with a force F. Assume that m_1, m_2, and F can be determined with negligible uncertainty, but that r is uncertain by an amount Δr.

(a) Using Newton's law of universal gravitation, show that

$$\frac{dG}{dr} = \frac{2Fr}{m_1 m_2}$$

if m_1, m_2, and F are considered known constants.

(b) Starting with

$$\Delta G \approx \frac{dG}{dr} \Delta r$$

show that the fractional uncertainty in G that results from a fractional uncertainty in r is given by

$$\frac{\Delta G}{G} \approx 2 \frac{\Delta r}{r}.$$

(c) In all laboratory experiments to measure G, the distance between the center of masses of attracting masses must be measured. Typically this distance is of the order $r = 10$ cm and, in an extremely accurate experiment, can be measured to an uncertainty of $\Delta r = 0.5\ \mu$m. Show that this limitation in precision for the distance measurement alone will cause a determination of G to be uncertain by about $\Delta G = 0.00007 \times 10^{-11}\ \text{N} \cdot \text{m}^2/\text{kg}^2$. [The best current value for G, which has more contributing uncertainties than just the "center-to-center" distance measurement, has an uncertainty of $\Delta G = 0.00085 \times 10^{-11}\ \text{N} \cdot \text{m}^2/\text{kg}^2$.]

10. (III) A mass M is ring-shaped with radius R. A small mass m is then placed at a distance along the ring's axis as shown in Fig. 6–5a. Show that the gravitational force on the mass m due to the ring is directed inward along the axis and has a magnitude of

$$F = \frac{GMmx}{(x^2 + R^2)^{3/2}}.$$

[*Hint*: Think of the ring as a collection of very small point masses ΔM, where the sum of all these ΔM gives the total mass of the ring, i.e., $\sum_{\text{entire ring}} \Delta M = M$ (see Fig. 6–5b). Then determine the force ΔF that a particular pair of ΔM, located on the opposite sides of the ring, produce on the mass m. Finally, sum the forces ΔF of all possible pairs of ΔM on the ring.]

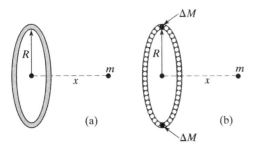

FIGURE 6–5

Section 6–4

11. (II) A doomsday asteroid is on collision course with Earth. If the asteroid's arrival time at Earth's orbit is delayed by the time τ it takes Earth to travel a distance equal to its own diameter D, the collision will be avoided. First, determine the orbital speed of Earth, assuming it travels in a perfectly circular path about the Sun; then use this speed to find τ.

12. (II) In 1922, John Plaskett discovered two nearly equal-mass stars that orbit about each other. He measured their separation and orbital period to be about 90,000,000 km and 14.4 days, respectively. Let's analyze his data with the simplifying assumption that the stars have exactly the same mass M and that they revolve in a circular orbit about a point midway between them (Fig. 6–6). In this model, find the mass M of each star, expressing M in units of the solar mass (1 solar mass equals the mass of our Sun). This pair, now called Plaskett's star, is one of the most massive binary star systems ever found.

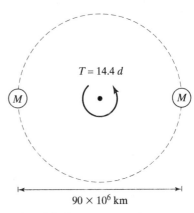

FIGURE 6–6

13. (II) Determine the length of one month (i.e., the time between successive new moons) with the following model (see Fig. 6–7).

(a) Assume Earth is solely attracted by the Sun of mass M_S and as a result moves in a perfectly circular orbit of radius r_{ES}. Additionally, assume the Moon is solely attracted by Earth of mass M_E and, as a result, moves in a perfectly circular orbit of radius r_{ME}. Show that Earth's orbital speed v_E about the Sun and that the Moon's orbital speed v_M about Earth are given by

$$v_E = \sqrt{\frac{GM_S}{r_{ES}}} \quad \text{and} \quad v_M = \sqrt{\frac{GM_E}{r_{ME}}}.$$

(b) Show that, because of Earth's orbital motion, the time τ between successive new moons is given by

$$\tau = \frac{2\pi}{\dfrac{v_M}{r_{ME}} - \dfrac{v_E}{r_{ES}}};$$

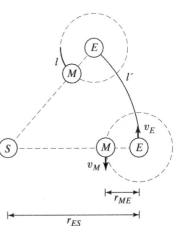

FIGURE 6–7

then use the known values for M_S, M_E, r_{ES}, and r_{ME} to predict τ.

[*Hint*: The Sun, the Moon, and Earth align at each new moon. The Moon and Earth travel distances $(2\pi r_{ME} + l)$ and l', respectively, during the time τ.]

14. (II) In between the orbits of Mars and Jupiter, an assortment of several thousand small objects called asteroids move in nearly circular orbits around the Sun. Consider a rocky asteroid that is spherically shaped with radius R and has a density $\rho = 2700 \text{ kg/m}^3$ (which is typical of solid rock). Also, assume that the maximum speed at which you can throw a baseball is 20 m/s (about 45 mph).
    (*a*) You find yourself on the surface of a spherical rocky asteroid and, to have some fun, you decide to throw a baseball so that it travels around the asteroid in a circular orbit. Assuming the baseball's orbital radius can be taken to be R, what is the largest radius asteroid on which you are capable of accomplishing this feat?
    (*b*) On the largest radius asteroid you determined in part (*a*), after you throw the baseball properly in one direction, you will be able to turn around and face the opposite direction, wait a time T, and then catch the baseball. How much time T will elapse between your throw and your catch?

15. (III) The mathematician Joseph-Louis Lagrange discovered five special points in the vicinity of Earth's orbit about the Sun, where a small satellite (mass m) will orbit the Sun with the same period T as Earth's period ($= 1$ year). Consider one of these "Lagrange Points" called L1, which lies between Earth (mass M_E) and the Sun (mass M_S) on the line connecting these large masses. Assume that Earth and the small satellite at L1 both orbit the Sun with period T and that Earth and the satellite are always separated by a distance d (see Fig. 6–8). Then, if Earth's orbital radius is r, the satellite's orbital radius is $(r - d)$.

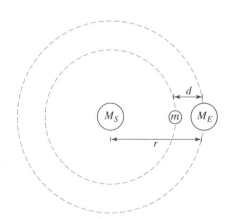

FIGURE 6–8

(*a*) Because the satellite is assumed to have minuscule mass in comparison with the Sun, to an excellent approximation, Earth's orbit will be determined solely by the Sun. Apply Newton's second law to Earth and show

$$\frac{GM_S}{r^2} = \frac{4\pi^2 r}{T^2}.$$

(*b*) Apply Newton's second law to the satellite and show

$$\frac{GM_S}{r^2}\left(1 - \frac{d}{r}\right)^{-2} - \frac{GM_E}{d^2} = \frac{4\pi^2 r}{T^2}\left(1 - \frac{d}{r}\right).$$

(*c*) The binomial expansion allows the following approximation $(1 + x)^n \approx 1 + nx$, if $x \ll 1$. Using the binomial expansion, show that if $d \ll r$ (or, equivalently, $d/r \ll 1$), the expression from part (*b*) can be approximated as

$$\frac{GM_S}{r^2}\left(1 + 2\frac{d}{r}\right) - \frac{GM_E}{d^2} = \frac{4\pi^2 r}{T^2}\left(1 - \frac{d}{r}\right).$$

(*d*) Combining your results from parts (*a*) and (*c*), show that the distance d is given by

$$d = \left(\frac{M_E}{3M_S}\right)^{1/3} r.$$

Using the known values for M_E, M_S, and r, show that d is about four times the distance from Earth to the Moon.
(*e*) Now that you have determined d, check to see whether the approximation $d \ll r$ that you used during the above derivation is valid.
[Placing a satellite at L1 has two advantages: the satellite's view of the Sun is never eclipsed, and it is always close enough to Earth to transmit data easily. The L1 point of the Earth-Sun system is currently home to the Solar and Heliospheric Observatory (SOHO) satellite.]

Section 6–5

16. (II) From astronomical data of the time, Newton had the information given in Table 6–1 about the orbit of Venus about the Sun, as well as the orbit of Jupiter's moon, Callisto, about Jupiter, and the orbit of Earth's Moon about Earth. One astronomical unit (AU) is the average radius of Earth's orbit about the Sun. Newton also had the following values for the relative size of these objects: in terms of the Sun's radius R_S, the radii of Jupiter and Earth were thought to be 0.0997 R_S and 0.0109 R_S, respectively.
Newton used this information to determine that the average density ρ of Jupiter is slightly less than that of the Sun, while the average density of Earth is four times that of the Sun. Thus, without leaving his home planet, Newton was able to predict that the compositions of the Sun and Jupiter are markedly different than that of Earth. Reproduce Newton's calculation and find his values for the ratios ρ_J/ρ_S and ρ_E/ρ_S (the modern values for these ratios are 0.93 and 3.91, respectively). Remember that for an object of mass M and volume V, its average density $\rho = M/V$.

TABLE 6–1 Newton's Astronomical Data (circa 1726)		
	Orbital Radius (units of AU)	Orbital Period T (units of Earth days)
Venus about Sun	0.724	224.70
Callisto about Jupiter	0.01253	16.69
Moon about Earth	0.003069	27.32

17. (II) From astronomical data of the time, Newton had the information given in Table 6–1 about the orbit of Venus about the Sun, as well as the orbit of Jupiter's moon, Callisto, about Jupiter, and the orbit of Earth's Moon about Earth. One astronomical unit (AU) is the average radius of Earth's orbit about the Sun. Newton also had the following values for the

relative size of these objects: in terms of the Sun's radius R_S, the radii of Jupiter and Earth were thought to be $0.0997\, R_S$ and $0.0109\, R_S$, respectively. Newton used this information to determine, in comparison with the Sun's values, the acceleration due to gravity g at the surface of Jupiter and Earth, as well the total mass M of Jupiter and Earth. Reproduce Newton's calculation and find his values for the ratios g_J/g_S, g_E/g_S, M_J/M_S, and M_E/M_S. The modern values for these ratios are 0.084, 0.036, 9.6×10^{-4}, and 3.0×10^{-6}, respectively.

18. (II) Newton's laws tell us that the orbital period T of the Moon about Earth is related to its orbital radius r by (see Eq. 6–6 in the text)

$$T = \frac{2\pi}{\sqrt{GM_E}} r^{3/2}$$

where M_E is Earth's mass. Let T_o and r_o be the current values for the Moon's orbital period and radius, respectively. If somehow the Moon's orbital radius increased to $r = r_o + \Delta r$, then its orbital period would increase to $T = T_o + \Delta T$.
(a) Starting with

$$T_o + \Delta T = \frac{2\pi}{\sqrt{GM_E}} (r_o + \Delta r)^{3/2}$$

and assuming that the radial increase is slight, i.e., $\Delta r \ll r_o$, use the binomial expansion to show that

$$\Delta T \approx 3\pi \sqrt{\frac{r_o}{GM_E}} \Delta r.$$

The binomial expansion allows the following approximation $(1 + x)^n \approx 1 + nx$, if $x \ll 1$.
(b) Due to tidal friction effects, the Moon's orbital radius about Earth increases by 3.8 cm/y. By how much does the Moon's orbital period change per year due to tidal friction?

19. (II) For convenience, let's use the period and semimajor axis of Earth's orbit as units for measuring T and s, respectively. The period of Earth is, of course, 1 year and the semimajor axis of Earth's orbit is called 1 astronomical unit (AU). 1 AU = 149.6×10^9 m.
(a) From Kepler's third law, show that the orbit of any satellite orbiting about the Sun obeys

$$s = T^{2/3}$$

if s and T are measured in AU and Earth-years, respectively.
(b) Halley's Comet travels about the Sun with an orbital period of 76 years. Its elliptical orbit has a large eccentricity (i.e., $e \approx 1$). If, at its closest approach, Halley's Comet is within 0.587 AU of the Sun, what is the farthest distance (in AU) it will travel from the Sun in its orbit? Does Halley's Comet ever travel beyond the orbit of Pluto? For Pluto, $s = 39$ AU.

20. (II) After ten years of painstaking measurements, an international team of astronomers recently announced it has directly observed an otherwise normal star (called S2) closely orbiting an extremely massive, but small, object at the center of the Milky Way Galaxy. The midpoint of our galaxy is located in the southern constellation Sagittarius and is centered on an object called SgrA, which is an intense emitter of radio waves and X-rays. The new data show that S2 is moving along an elliptical orbit with SgrA at one focus and an orbital period of 15.2 years. The orbit is rather elongated with an eccentricity $e = 0.87$. In the spring of 2002, S2 reached its closest approach to SgrA, a distance of only 123 AU, or just 3 times the Sun-Pluto distance. One astronomical unit (AU) is defined to be the semimajor axis of Earth's orbit about the Sun, which equals 1.496×10^{11} m. Determine the mass M of SgrA, the compact object at the center of our galaxy, and compare it with the mass of our Sun. [It is believed that SgrA is a supermassive black hole.]

21. (III) Newton's laws tell us that the orbital period T of the Moon about Earth is related to its orbital radius r by (see Eq. 6–6 in the text)

$$T = \frac{2\pi}{\sqrt{GM_E}} r^{3/2}$$

where M_E is Earth's mass. If the Moon's current orbital radius r_o were to increase by an amount Δr, use calculus to show that the resulting increase in its orbital period would be

$$\Delta T \approx 3\pi \sqrt{\frac{r_o}{GM_E}} \Delta r.$$

[*Hint*: For a small increase Δr in orbital radius from an initial value of r_o, the resulting increase in orbital period ΔT will approximately be $\Delta T \approx \left.\frac{dT}{dr}\right|_{r=r_o} \Delta r$, where $\left.\frac{dT}{dr}\right|_{r=r_o}$ means the derivative of T with respect to r evaluated at $r = r_o$.]

22. (III) The orbital periods T and mean orbital distances r for Jupiter's four largest moons are given in Table 6–2.

TABLE 6–2

Moon	Period T (Earth Days)	Mean Distance r (km)
Io	1.77	422,000
Europa	3.55	671,000
Ganymede	7.16	1,070,000
Callisto	16.7	1,883,000

(a) Starting with Kepler's third law in the form

$$T^2 = \left(\frac{4\pi^2}{GM_J}\right) r^3$$

where M_J is the mass of Jupiter, show that this relation implies that a plot of $\log(T)$ vs. $\log(r)$ will yield a straight line. What are the predicted slope and y-intercept of this straight-line plot?
(b) Using the data for Jupiter's four moons, plot $\log(T)$ vs. $\log(r)$ and show that you get a straight line. Determine the slope of this plot and compare it with the value you expect if the data are consistent with Kepler's third law. Determine the y-intercept of the plot and use it to compute the mass of Jupiter.

General Problems

23. (II) Is gravity strong enough to hold an atom together? Consider the simplest atom, hydrogen, in which an electron orbits a stationary proton (mass $m_p = 1.7 \times 10^{-27}$ kg). It is known

from experiment that the radius r of the electron's orbit is about 0.5×10^{-10} m. A famous relation from quantum theory called the Heisenberg Uncertainty Principle tells us that, when confined to a spatial region of 1×10^{-10} m, an electron must have a speed v of at least 6×10^5 m/s. Assume that an electron with speed v is held in a circular orbit about a stationary proton by the gravitational attraction between these two masses. Show that this assumed model for the hydrogen atom requires that the gravitational constant have the following value:

$$G = \frac{rv^2}{m_p}.$$

Taking the experimental value of $r = 0.5 \times 10^{-10}$ m, and the minimum value of $v = 6 \times 10^5$ m/s dictated by quantum theory, determine the required value for G if the atom is held together by gravity. Compare this required value with the actual value for G. Is gravity strong enough to hold an atom together?

24. (II) The variation in gravitational forces across Earth's volume slightly distorts our planet's shape, which we experience as tides. Explore this idea using the known values for r_{ME}, r_{ES}, M_M, M_S, and r_E, which are the average distance between the centers of Earth and the Moon, the average distance between the centers of Earth and the Sun, the Moon's mass, the Sun's mass, and Earth's radius, respectively.
(a) Consider two small pieces of Earth, each of mass m, one on the side of Earth nearest the Moon, the other on the side farthest from the Moon. Show that the ratio of the Moon's gravitational forces on these two masses is

$$\left(\frac{F_{near}}{F_{far}}\right)_{Moon} = 1.0687.$$

(b) Consider two small pieces of Earth, each of mass m, one on the side of Earth nearest the Sun, the other on the side farthest from the Sun. Show that the ratio of the Sun's gravitational forces on these two masses is

$$\left(\frac{F_{near}}{F_{far}}\right)_{Sun} = 1.000171.$$

(c) Now consider the actual size of the Sun's and the Moon's gravitational forces on Earth. Show that the ratio of the Sun's average gravitational force on Earth compared with that of the Moon's is

$$\left(\frac{F_{Sun}}{F_{Moon}}\right)_{ave} = 178.$$

(d) From above we see that, in the case of the Moon, its relatively small force varies by a large percentage across Earth's diameter, while for the Sun, its relatively large force varies by only a small percentage. For each case, we can estimate the resulting "force differential" ΔF (which is the cause of tides) by

$$\Delta F \equiv F_{near} - F_{far} = F_{far}\left(\frac{F_{near}}{F_{far}} - 1\right) \approx F_{ave}\left(\frac{F_{near}}{F_{far}} - 1\right).$$

Show that the ratio of the tide-causing force differential due to the Moon compared with that due to the Sun is

$$\frac{\Delta F_{Moon}}{\Delta F_{Sun}} \approx 2.3.$$

Thus, while the force differentials of the Moon and the Sun are of the same order-of-magnitude, the Moon's influence on tide-production is over two times as great.

25. (II) In between the orbits of Mars and Jupiter, a distributed assortment of several thousand small, rocky objects called asteroids move in nearly circular orbits around the Sun. However, within this distribution of asteroids, several "gaps" are observed, i.e., there are several orbital radii at which few asteroids are found. These asteroid-free zones are called "Kirkwood gaps" and are located at radii r from the Sun where the ratio α of the orbital period of an asteroid at r compared with the orbital period of Jupiter is equal to a ratio of integers. The most prominent gaps occur when $\alpha = \frac{1}{2}, \frac{3}{7}, \frac{2}{5}, \frac{1}{3}$. In a Kirkwood gap, an asteroid's motion is synchronized with the motion of Jupiter so that it feels a Jupiter-induced tug at exactly the same position every orbit. The result of these tugs is to pull the asteroid out of that orbit; hence, no asteroids are observed at that distance from the Sun. Find the orbital radii r (in astronomical units) for the four prominent Kirkwood gaps. One astronomical unit (AU) equals the average distance between Earth and the Sun. The radius of Jupiter's orbit is 5.20 AU.

26. (II) A pulsar is a compact, rotating stellar object that emits a radio-wave beam that sweeps through space as the star rotates, much like a revolving lighthouse beacon. We receive one radio pulse each time the beacon sweeps past Earth, and so the observed time between pulses is equal to the rotational period T of the pulsar. Since the mid-1980s, astronomers have observed "millisecond pulsars," pulsars with rotational periods of only about 0.001 seconds. This rotation rate is so rapid that millisecond pulsars are believed to be on the verge of break up, and are only prevented from flying apart due to intense gravitational forces. Consider the following model for a millisecond pulsar: A sphere of radius R and mass M is rotating about an axis with period T. A small mass m is held on the sphere's equator solely by its gravitational attraction to the sphere and so rotates around a circle of radius R with period T (see Fig. 6–9).
(a) Assuming the sphere is of constant density, show that the mass m will be held in place if the density is at least

$$\rho_o = \frac{3\pi}{GT^2}.$$

In our model, if a millisecond pulsar has this "threshold" density ρ_o, it is on the verge of flying apart at its surface.
[Hint: Density equals mass per volume, i.e., $\rho = M/V$.]

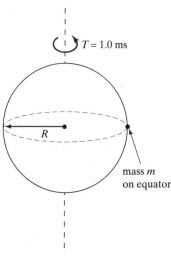

FIGURE 6–9

(b) Taking $T = 0.0010\,\text{s}$, determine the "threshold" density for a stable millisecond pulsar.

(c) For stability, a millisecond pulsar's density ρ must equal or exceed the threshold density ρ_o you determined in part (b). Is such a density possible in nature? Compare the value for ρ_o with a rough estimate of the density of a neutron (or proton). A neutron can be thought of as a sphere with a radius of about 1.0×10^{-15} m. What does this comparison suggest about a possible composition of a millisecond pulsar?

(d) A pulsar is the imploded core of a massive star that underwent a supernova explosion at the end of its lifespan as a normal star. If a millisecond pulsar has twice the mass of the Sun and has the "threshold" density determined in part (b), what is its radius (km)?

27. (II) Imagine you are a theoretical physicist trying to invent the law of universal gravitation between two point masses M and m separated by a distance r. You decide the magnitude of the attractive gravitational force F should double if only one mass is doubled and quadruple if both masses are doubled. So, you make F proportional to the product of the masses. Additionally, you decide that the masses should influence each other less when they are farther apart. The simplest way to do this mathematically is to put r in the denominator of your force relation as follows:

$$F \propto \frac{Mm}{r}.$$

Of course, Nature may be more complicated than this—any increasing function of r, for example, r^n, $r + r^2 + r^3$, or $\log(r)$, in the denominator also makes F get weaker with increasing distance. To cover more of these possibilities, you decide to modify the above relation by raising r to the power n in the denominator (your "simplest" relation is the special case of $n = 1$). After including a constant of proportionality G, your postulated force law is

$$F = G\frac{Mm}{r^n}.$$

Johannes Kepler then tells you that by analyzing experimental data he has found, for any planet going around the Sun, its orbital period T and average orbital distance r are related by

$$\frac{T^2}{r^3} = \text{constant}.$$

Use your postulated force relation to predict the period of a planet orbiting about the Sun, assuming a circular orbit of radius r. Then compare your prediction with Kepler's empirical relation. What does this comparison tell you about your postulated force law? That is, is r^n the correct function to put in the denominator? If so, what is the correct value for n?

28. (II) Imagine you are a theoretical physicist trying to invent the law of universal gravitation between two point masses M and m separated by a distance r and have postulated that the gravitational force obeys the relation $F = G\dfrac{Mm}{r^2}$, where you have established that r in the denominator of your force law should be squared with reasoning based on the motion of the Moon (see Section 6–1 in the text). You now become concerned that the product of masses in the numerator possibly ought to be raised to some power p as well. So, you propose the following force law:

$$F = G\frac{(Mm)^p}{r^2}.$$

Your friend Galileo Galilei claims that, near Earth's surface, the acceleration due to gravity g is the same for all objects, regardless of their mass m. Use your proposed force law to predict the acceleration g of a mass m near Earth's surface due to Earth's gravitational attraction. Show that your predicted relation for g is consistent with Galileo's claim only if $p = 1$.

29. (III) From our vantage point on its exterior, Earth appears to be a mass M with spherical shape of radius b. Thus, it attracts other pieces of matter outside of itself (e.g., us) toward its center. Consider the possibility that Earth is actually a hollow sphere of mass M with inner radius a and outer radius b. By observing from our vantage point how it attracts things gravitationally, could we discern that Earth does in fact have a hollow structure? Answer this question as follows. Place a point mass m at a distance $d > b$ from the center of a spherical mass M of outer radius b. Determine the gravitational force on m, first supposing that the sphere is solid (Fig. 6–10a), then supposing that it is hollow (Fig. 6–10b). Take the mass as evenly distributed in each assumed spherical structure.

[*Hint*: For computational purposes, think of the hollow sphere as a composite object, which consists of a larger uniform sphere of positive mass plus a nested smaller uniform sphere of negative mass. Also, an object's mass $M = \rho V$, where ρ and V are the object's density and volume, respectively.]

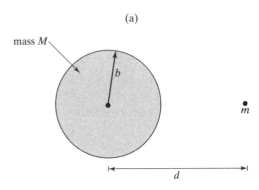

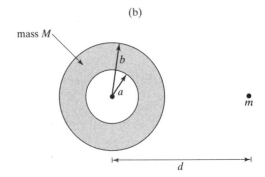

FIGURE 6–10

30. (III) Analyze the tide-related situation described in Problem 24 using calculus.
    (a) Consider a small mass m that can be a variety of distances r away from the center of a gravitationally attracting sphere of mass M. Show that the change in gravitational force F on m with respect to changes in r is
    $$\frac{dF}{dr} = -\frac{2GMm}{r^3}.$$
    (b) For a small change Δr from a given value of $r = d$ (where d is a constant such as r_{ME} or r_{ES}), the resulting change in gravitational force will approximately be $\left.\frac{dF}{dr}\right|_{r=d} \Delta r$, where $\left.\frac{dF}{dr}\right|_{r=d}$ means the derivative of F with respect to r evaluated at $r = d$. Then the gravitational force differential ΔF on mass m on opposite sides of Earth is found by
    $$\Delta F = F_{near} - F_{far} \approx \left.\frac{dF}{dr}\right|_{r=d}(-2r_E).$$
    Starting with this expression, show that
    $$\frac{\Delta F_{Moon}}{\Delta F_{Sun}} \approx \frac{M_M}{M_S}\left(\frac{r_{ES}}{r_{ME}}\right)^3.$$
    Using the known values of M_M, M_S, r_{ME}, and r_{ES}, show that $\frac{\Delta F_{Moon}}{\Delta F_{Sun}} \approx 2.2.$

CHAPTER 7

Work and Energy

Section 7–1

1. (I) A farmer lifts a 25-kg bag of potatoes off of the ground and places it onto his shoulder, which is 1.6 m above the ground. He next carries the bag 100 m across a level field and then lifts it down from his shoulder to the bed of a pickup truck, which is 1.0 m above the ground. According to the definition of work in physics, how much work did the farmer do on the bag of potatoes during this process?

Section 7–2

2. (II) Given $\mathbf{A} = 1.0\,\mathbf{i} + 1.0\,\mathbf{j} - 2.0\,\mathbf{k}$ and $\mathbf{B} = -1.0\,\mathbf{i} + 1.0\,\mathbf{j} + 2.0\,\mathbf{k}$,
   (a) what is the cosine of the angle between these two vectors?
   (b) explain the significance of the sign in part (a), and then determine the angle between the two vectors.
   (c) what is the smallest angle between $\mathbf{B}$ and its projection on the x-y plane?

3. (II) Find a vector of unit length in the x-y plane that is perpendicular to $3.0\,\mathbf{i} + 4.0\,\mathbf{j}$.

4. (II) The velocity and acceleration of an object are given by $\mathbf{v}$ and $\mathbf{a}$, respectively. Prove $\dfrac{d(v^2)}{dt} = 2\,\mathbf{a}\cdot\mathbf{v}$, where $v^2 = v_x^2 + v_y^2 + v_z^2$.

5. (II) The time-dependent position of an object which moves along the circumference of a circle (radius R) with constant speed v is given by

$$\mathbf{r} = R\cos(\omega t)\,\mathbf{i} + R\sin(\omega t)\,\mathbf{j}$$

where the constant $\omega = v/R$. Show that the velocity $\mathbf{v}$ of this object is perpendicular to $\mathbf{r}$ at all times.

Section 7–3

6. (II) As an object moves along the x-axis from $x_a = 0.0\,\text{m}$ to $x_b = 20.0\,\text{m}$ it is acted upon by a force given by $F = \sqrt{100 - (x-10)^2}\,\text{N}$. Determine the work done by the force on the object,
   (a) by first sketching F vs. x and then determining the area under this curve.
   (b) by evaluating the integral $\displaystyle\int_{x_a}^{x_b} F\,dx$.

   [Hint: $\displaystyle\int \sqrt{a^2 - x^2}\,dx = \dfrac{x}{2}\sqrt{a^2 - x^2} + \dfrac{a^2}{2}\sin^{-1}\!\left(\dfrac{x}{a}\right).$]

7. (II) (a) Consider the force $F_1 = \dfrac{A}{\sqrt{x}}$, which acts on an object during its journey along the x-axis from $x = 0.0\,\text{m}$ to $x = 1.0\,\text{m}$, where $A = 2.0\,\text{N}\cdot\text{m}^{1/2}$. Show that during this journey, even though F_1 is infinite at $x = 0.0\,\text{m}$, the work done on the object by this force is finite.
   (b) Next consider the force $F_2 = \dfrac{B}{x}$, which acts on an object during its journey along the x-axis from $x = 0.0\,\text{m}$ to $x = 1.0\,\text{m}$, where $B = 2.0\,\text{N}\cdot\text{m}$. Show that during this journey the work done on the object by this force (which is infinite at $x = 0.0\,\text{m}$) is infinite.

8. (II) Assume that a force acting on an object is given by $\mathbf{F} = ax\,\mathbf{i} + by\,\mathbf{j}$, where the constants $a = 3.0\,\text{N}\cdot\text{m}^{-1}$ and $b = 4.0\,\text{N}\cdot\text{m}^{-1}$. Determine the work done on the object by this force as it moves from the origin to $\mathbf{r} = (10.0\,\mathbf{i} + 20.0\,\mathbf{j})\,\text{m}$.

9. (II) A force given by $\mathbf{F} = (2.0\,\mathbf{i} + 5.0\,\mathbf{j})\,\text{N}$ acts on an object that can only move in the upper half of the x-y plane, i.e., $-\infty \le x \le +\infty$ and $y \ge 0$. If the object starts at the origin and moves a distance of 10.0 m, determine which path the object moves along if
   (a) $\mathbf{F}$ does zero work.
   (b) $\mathbf{F}$ does the maximum possible work. How many joules of work does $\mathbf{F}$ do in this case?

10. (II) An object, moving along the circumference of a circle with radius R, is acted upon by a force of constant magnitude F that is directed at an angle of 30° with respect to the tangent of the circle at all times as shown in Fig. 7–1. Determine

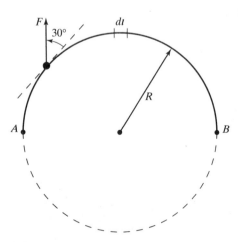

FIGURE 7–1

the work done by this force when the object moves from point A to point B, which are directly opposite each other on the circle.

11. (III) An object moves from $\mathbf{r}_1 = (1.0\,\mathbf{i} + 3.0\,\mathbf{j})$ m to $\mathbf{r}_2 = (2.0\,\mathbf{i} + 12.0\,\mathbf{j})$ m along a parabolic-shaped path described by $y = Ax^2$, while acted upon by a force given by $\mathbf{F} = Bxy\,\mathbf{i} + Bxy\,\mathbf{j}$, where the constants $A = 3.0\,\text{m}^{-1}$ and $B = 10.0\,\text{N}\cdot\text{m}^{-2}$. Determine the work done on the object by this force along the parabolic path.

[*Hint*: Write F_x and F_y as functions of only the variables x and y, respectively.]

12. (III) A small mass m hangs at rest, supported at angle $\theta = 0°$ with respect to the vertical by a rope of length L, which is fixed to the ceiling. A force **F** then pushes on the mass, always directed perpendicular to the taut rope, until the string is oriented at an angle $\theta = \theta_o$ and the mass has been raised by a vertical distance h (see Fig. 7–2). Assume that during this process the force's magnitude F is adjusted so that the mass moves at constant speed along its circular trajectory. Show that the work done by **F** during this process equals mgh, which is equivalent to the amount of work it takes to simply lift a mass m straight upwards at constant velocity to a height h above its initial level.

[*Hint*: When the angle is increased by $d\theta$ (in radians), the mass moves along an arclength $ds = L\,d\theta$.]

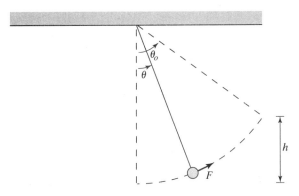

FIGURE 7–2

13. (III) An object moves from $\mathbf{r}_1 = (1.0\,\mathbf{i} + 3.0\,\mathbf{j})$ m to $\mathbf{r}_2 = (4.0\,\mathbf{i} + 48.0\,\mathbf{j})$ m along a path, while acted upon by a force given by $\mathbf{F} = Bxy\,\mathbf{i} + Bxy\,\mathbf{j}$, where the constant $B = 10.0\,\text{N}\cdot\text{m}^{-2}$. Determine the work done on the object by this force,
(a) assuming that the path has the parabolic shape $y = Ax^2$, where $A = 3.0\,\text{m}^{-1}$.
(b) assuming that the path has the straight-line shape $y = Cx + D$, where $C = 15.0$ and $D = -12.0$ m.
(c) For the given force **F**, does the amount of work done in moving from $\mathbf{r}_1$ to $\mathbf{r}_2$ depend on the particular path taken?

[*Hint*: Write F_x and F_y as functions of only the variables x and y, respectively.]

Section 7–4

14. (II) An object (mass $m = 2.50$ kg) moving in two dimensions initially has a velocity $\mathbf{v}_1 = (10.0\,\mathbf{i} + 20.0\,\mathbf{j})$ m/s. A net force **F** then acts on the object for a certain time interval, at the end of which the object's velocity is $\mathbf{v}_2 = (15.0\,\mathbf{i} + 30.0\,\mathbf{j})$ m/s. Determine the work done by **F** on the object.

15. (II) Stretchable ropes are used to safely arrest the fall of rock climbers. Consider the following model of a rock-climbing fall. One end of a rope with unstretched length L is anchored to a rock cliff and a climber of mass m is attached to the other end. When the climber is a height L above the anchor point, he loses contact with the cliff and falls only under the influence of gravity for a distance $2L$, after which the rope becomes taut and stretches a distance x as it stops the climber (see Fig. 7–3). Assume that, as it stretches, the rope behaves as a spring with spring constant k.
(a) Applying the work-energy principle, show that
$$x = \frac{mg}{k}\left[1 + \sqrt{1 + \frac{4kL}{mg}}\right].$$
(b) Assuming typical values of $m = 75$ kg, $L = 5.0$ m and $k = 2200$ N/m, determine x/L and kx/mg, the fractional stretch of the rope and the force that the rope exerts on the climber compared with his own weight, respectively, at the moment the climber's fall has been stopped.

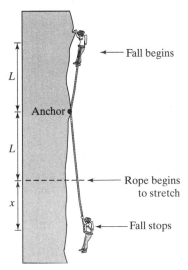

FIGURE 7–3

16. (II) On steep downhill stretches of mountain highways, there are commonly escape ramps provided for runaway trucks with overheated brakes. The ramps are straight, upwardly sloped beds of loosely packed gravel. Assume that the steepest such ramp is inclined at 11° above the horizontal and that the coefficient of friction for a truck while moving on the ramp is $\mu = 0.25$. If an escape ramp is designed to stop all possible runaway trucks (the fastest of which travels at 140 km/h), use the work-energy principle to determine the necessary length for the ramp.

17. (II) A train is moving along a track with constant speed V relative to the ground. A person standing in a car of the train and holding a ball of mass m throws it toward the front of the train so that it moves with a speed v relative to the train.
(a) Calculate the change in kinetic energy of the ball in the ground frame of reference, as well as in the train frame of reference.
(b) Relative to each frame of reference, how much work was done on the ball?
(c) Explain why the results in part (b) are not the same for the two frames.

18. (II) The car is full, so a parent decides to buckle himself in with a seat belt and hold his 20-kg toddler on his lap during a short trip, which includes some highway driving at speed $v = 30$ m/s. Use the work-energy principle to answer the following questions:
(a) While traveling 30 m/s, assume the car's driver has to emergency stop the car over a distance of 50 m. In order to decelerate the toddler along with the car, the parent wraps his arms around the child. Assuming constant deceleration, how much force will the parent need to exert on the child during this deceleration period? Is this force achievable by an average parent?
(b) Now assume that the car is in an accident while highway driving and the car is brought to stop over a distance of 10 m. Assuming constant deceleration, how much force will the parent need to exert on the child during this deceleration period? Is this force achievable by an average parent?

19. (II) The effects of an automobile collision can be studied by dropping a car from a height and allowing it to strike the ground. Use the work-energy principle to determine from what height h a car would have to be released in order to simulate a collision at 160 km/h.

20. (II) An object of mass m_1 sits on a horizontal surface and is attached on one side to a spring (spring constant k) and on the other side to a massless rope. The rope passes over a pulley and supports a second object of mass m_2 as shown in Fig. 7–4. The coefficient of kinetic friction between m_1 and the horizontal surface is μ.
(a) Assume m_1 and m_2 are initially at rest with the spring at its unstretched length. If m_2 then falls a distance y and attains a speed v, determine the work done by gravity, friction, and the spring; then use the work-energy principle to find v.
(b) How far will m_2 fall before being brought to rest?

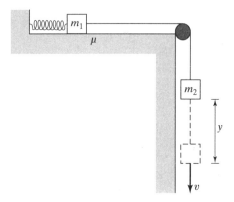

FIGURE 7–4

Section 7–5

21. (II) Compare the relativistic kinetic energy K_R with the classical kinetic energy K_C by constructing the ratio $R = K_R/K_C$. Write R as a function of $\beta = v/c$, where β represents an object's velocity v as a fraction of the speed of light c.
(a) Use a graphing calculator or a computer program to plot R vs. β in the range $0 \leq \beta \leq 0.99$. [Hint: $R=1$ for $\beta=0$.]
(b) One might plausibly define an object to be "relativistic" when K_R deviates from K_C by more than 1%. Using this criterion and your plot, determine the value for β (to one significant figure) at the threshold of the "relativistic regime" (i.e., find the value of β for which $R = 1.01$).

22. (II) In Section 7–5 of the text the binomial expansion is used to show that the term $\gamma = 1 \bigg/ \sqrt{1 - \dfrac{v^2}{c^2}}$ in the expression for relativistic kinetic energy (Eq. 7–12) can be approximated as $\gamma' = \left(1 + \dfrac{1}{2}\dfrac{v^2}{c^2}\right)$ when an object's speed v is much smaller than c.
(a) Determine the speed v_R (as a multiple of c) for which the percent difference between the true γ and the approximate γ' is 1.0%. For speeds $v > v_R$ one might plausibly conclude that the above approximation is not valid.
(b) To demonstrate the validity of the above approximation, determine the percent difference between the true γ and the approximate γ' for the speed $v = \dfrac{v_R}{100}$.

23. (II) When fully loaded, the mass of the Space Shuttle is 103,000 kg. If, in the future, a spacecraft with the same mass as the Shuttle is used to travel to nearby stars at a speed $v = \dfrac{3}{4}c$ (i.e., its speed equals three-fourths the speed of light), determine the kinetic energy K of this spacecraft. This kinetic energy would have to be given to the spacecraft during its launch from Earth. Compare (as a ratio) K with the amount of energy presently consumed worldwide in one year ($\approx 5 \times 10^{20}$ J).

General Problems

24. (II) In the game of paintball, players use "markers" (guns powered by pressurized gas such as air or CO_2) to propel 30-g spherical paintballs (gel capsules filled with paint) at opposing team members. For safety, game rules dictate that a paintball cannot leave the barrel of a marker with a speed greater than 92 m/s ($= 300$ ft/s). Model the firing of a paintball by assuming that pressurized gas applies a constant force F on this 30-g object over the length of the marker's barrel (typically 30 cm), accelerating it from rest to 92 m/s. Determine F
(a) using the work-energy principle.
(b) using the kinematic equations and Newton's second law. You should, of course, get the same answer for F using the two different methods.

25. (II) The car is full, so a parent decides to buckle himself in with a seat belt and hold his 20-kg toddler on his lap during a short trip, which includes some highway driving. While traveling at a speed of 30 m/s, assume the car's driver has to emergency stop the car over a distance of 50 m. In order to decelerate the toddler along with the car, the parent wraps his arms around the child. Assuming constant deceleration, how much force F will the parent need to exert on the child during this deceleration period?
(a) Answer the above question using the work-energy principle.
(b) Answer the above question using Newton's method (i.e., using the kinematic equations and Newton's second law). You should, of course, get the same answer for F using the two different methods.

26. (II) An escape lane for runaway trucks with overheated brakes consists of a horizontal bed of loosely packed sand. Assume that the sand is very deep so that as a truck moves down the ramp, it sinks into the sand; also assume that this process can be modeled as a position-dependent coefficient of friction given by $\mu(x) = \mu_o + \alpha x$, where x is the distance from the lane's entrance and μ_o and α are constants. If a runaway truck enters the lane with speed v_o, determine the distance D it will travel before being brought to a stop by using the work-energy principle.

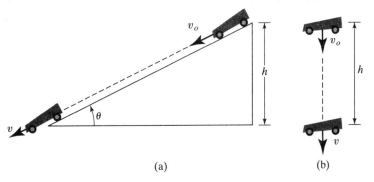

FIGURE 7–5

27. (II) A section of a roller-coaster track is a straight incline oriented at angle θ to the horizontal, which drops a vertical height h (Fig. 7–5a). A car, with an initial speed v_o, moves friction-free down the incline, attaining a final speed v at the bottom. (a) Determine the work done by gravity on the car as it moves down the incline. Then use the work-energy principle to find v in terms of v_o and h.
(b) Assume that a roller-coaster car, disconnected from the track and with initial speed v_o, simply free falls over a vertical distance h (Fig. 7–5b). Use the work-energy principle to show that the final speed v the car attains in this situation is the same as for the situation in part (a).

28. (II) A constant force $\mathbf{F} = (2.0\,\mathbf{i} + 4.0\,\mathbf{j})$ N acts on an object as it moves along a straight-line path. If the object's displacement is $\mathbf{d} = (1.0\,\mathbf{i} + 5.0\,\mathbf{j})$ m, calculate the work done by $\mathbf{F}$ using the following two alternate ways of writing the dot product:
(a) $W = Fd\cos\theta$ (Eq. 7–3 in the text).
(b) $W = F_x d_x + F_y d_y$ (Eq. 7–4 in the text).

29. (II) You are a farmer in ancient Egypt and Pharaoh has "volunteered" you to help build a new pyramid. Each day, you are made to push many 50.0-kg stones from the bottom to the top of a 7.00° incline at constant velocity. The incline is 120 m long. You push on each stone with a force that is directed parallel to the incline and the coefficient of kinetic friction between the stone and the incline is $\mu = 0.400$ (see Fig. 7–6).
(a) As it is moved from the bottom to the top of the incline, how much work is done on the stone by gravity, by friction, and by you?

(b) Positive and negative work can be interpreted as energy added to and taken away from an object, respectively. When the stone arrives at the incline's top, what percentage of the energy that you have added to the stone has been taken away by the gravitational force and what percentage has been taken away by the frictional force?
(c) If you push three 50.0-kg stones to the top of the incline per hour and work 8.0 hours per day, how many Food Calories are equivalent to your daily work done on the stones? Note: 1 Food Calorie = 4186 J.

30. (II) A net force $\mathbf{F}_{net}$ acts on an object of mass m whose velocity and acceleration are given by $\mathbf{v}$ and $\mathbf{a}$, respectively.
(a) Prove $\dfrac{d(v^2)}{dt} = 2\left(\dfrac{\mathbf{F}_{net}}{m}\right)\cdot\mathbf{v}$, where $v^2 = v_x^2 + v_y^2 + v_z^2$.
(b) Starting with the work-energy principle, use the result from part (a), to prove $dW_{net} = \mathbf{F}_{net}\cdot\mathbf{v}\,dt$, where dW_{net} is the work done by $\mathbf{F}_{net}$ on the object during the time interval dt.

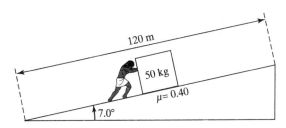

FIGURE 7–6

CHAPTER 8

Conservation of Energy

Sections 8–1 and 8–2

1. (II) A particle is constrained to move in one dimension along the x-axis and is acted upon by a force given by

$$\mathbf{F}(x) = -\frac{k}{x^3}\mathbf{i}$$

where k is a constant with units appropriate to the SI system. Find the potential energy function $U(x)$, if U is arbitrarily defined to be zero at $x = 2.0\,\text{m}$, i.e., $U(2.0\,\text{m}) = 0$.

2. (II) A particle is constrained to move in one dimension and is subject to a force $F(x)$, which varies with position x as

$$\mathbf{F}(x) = A\sin(kx)\mathbf{i}$$

where A and k are constants. What is the potential energy function $U(x)$, if $U = 0$ at the point $x = 0$?

Sections 8–3 and 8–4

3. (II) An object of mass m is held at rest at a height h above a tabletop surface. The tabletop surface is itself a height H above the ground. If the object is dropped, use conservation of energy to predict the velocity v with which the object strikes the tabletop surface, by defining the zero-level for gravitational potential energy at
   (a) the tabletop surface.
   (b) the ground.
   (c) Both definitions for the zero-level yield the same result for v. Practically speaking, though, which definition for the zero-level is the wisest (i.e., yields the simplest equations to solve)?

4. (II) A roller-coaster car on a frictionless track passes point A with a speed of 10 m/s. Will it make it over a hill that is a vertical height of 6.0 m above point A (Fig. 8–1)? If not, to what vertical height on the hill will the car rise?

5. (II) Two masses are connected by a string that hangs over a massless pulley as shown in Fig. 8–2. Mass $m_1 = 4.0\,\text{kg}$ rests

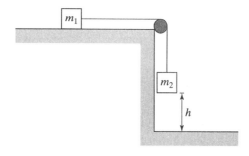

FIGURE 8–2

on a frictionless horizontal surface, while mass $m_2 = 5.0\,\text{kg}$ is initially held at a height $h = 0.75\,\text{m}$ above the floor.
(a) If m_2 is allowed to fall, what will be the resulting acceleration of the masses?
(b) If the masses were initially at rest, use the kinematic equations to find their speed just before m_2 hits the floor.
(c) Use conservation of energy to find the speed of the masses just before m_2 hits the floor. You should get the same answer as in part (b).

6. (II) When a mass m sits continually at rest on a spring, the spring is compressed by a distance d from its undeformed length (Fig. 8–3a). Suppose, instead, that the mass is released from rest when it barely touches the undeformed spring

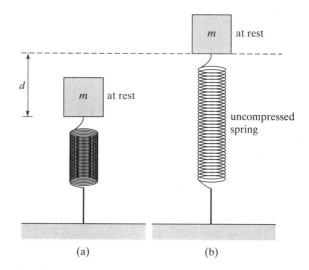

FIGURE 8–3

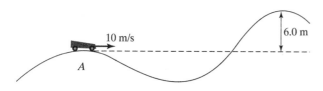

FIGURE 8–1

(Fig. 8–3b). Find the distance D (expressed as a multiple of d) that the spring is compressed before it is able to stop the mass.

7. (II) Tarzan, seeing Jane at the edge of the alligator pond, lets out a tremendous macho yell, grabs a 7.0-m long (horizontally aligned) jungle vine at point A, and swings down to rescue her. This is followed by an even louder yell as he discovers that the ground is 10.0 m below point A (see Fig. 8–4). If he releases the (vertically aligned) vine at point B, does Tarzan make it to the other side of the pond or is he alligator bait?

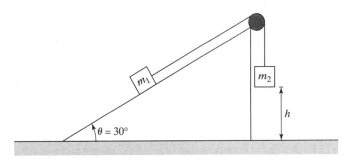

FIGURE 8–5

11. (III) Conservation of energy provides a method for predicting the velocity of an object at later times. Once the object's velocity v is determined, the time t to journey from position $x_o = 0$ to position $x = L$ can be found from the following integral

$$t = \int_0^t dt = \int_0^L \frac{dx}{v}$$

where, during each incremental increase of position dx throughout its journey, the object's instantaneous velocity is given by the position-dependent function v.
Consider a frictionless plane inclined at angle θ with a total length L. Assume an object of mass m starts from rest and slides from the top to the bottom of the incline (Fig. 8–6).
(a) Defining the top of the incline to be $x_o = 0$, use conservation of energy to show that when the object has traveled a distance x down the incline, its velocity is given by

$$v = \sqrt{2gx \sin \theta};$$

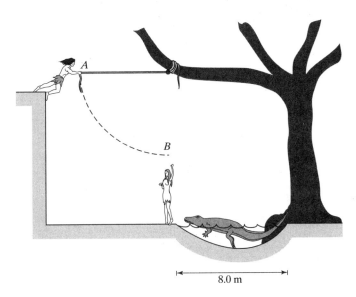

FIGURE 8–4

8. (II) Assume an object of mass m moves in one dimension under the influence of only conservative forces. Then the total energy E of the object can be written as

$$E = \frac{1}{2}mv^2 + U$$

where the potential energy U accounts for all of the influences of conservative forces. Now assume the object's energy is constant, i.e., $\frac{dE}{dt} = 0$. Show that these relationships can then be used to derive Newton's second law.

9. (II) Human beings are limited in their ability to toss objects to heights above Earth's surface. Make an educated guess as to the largest speed that a person is capable of throwing an object upward; then, based on this estimate, use conservation of energy to determine the maximum height h for such a toss.

10. (III) Two masses are connected by a string as shown in Fig. 8–5. Mass $m_1 = 4.0$ kg rests on a frictionless inclined plane, while mass $m_2 = 5.0$ kg is initially held at a height $h = 0.75$ m above the floor.
(a) If m_2 is allowed to fall, what will be the resulting acceleration of the masses?
(b) If the masses were initially at rest, use the kinematic equations to find their speed just before m_2 hits the floor.
(c) Use conservation of energy to find the speed of the masses just before m_2 hits the floor. You should get the same answer as in part (b).

then use this expression to predict the time it takes for the object to slide from the top to the bottom of the incline.
(b) Implement Newton's second law to determine the sliding object's acceleration; then use the kinematic equations to predict the time t it takes for the object to slide from the top to the bottom of the incline. This alternate approach should yield the same answer for t as the conservation of energy method.

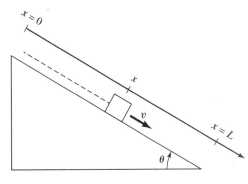

FIGURE 8–6

Sections 8–5 and 8–6

12. (II) A rule of thumb in roller-coaster design is that the total length L of the ride should be 35 times the height H of the first hill. If $L < 35H$, the car will end the ride with excess

kinetic energy that will then have to be wasted in braking the car to a stop.

(a) Based on this rule of thumb, estimate the average frictional force F_{fr} on a roller-coaster car of mass m during the ride. Assume that, when $L = 35H$, the car ends the ride with negligible kinetic energy.

(b) Assuming the first hill that a roller-coaster car descends is a straight incline at angle 70° above the horizontal and that a constant frictional force F_{fr} as found in part (a) acts on the car during its descent down this incline, what percentage of the car's initial potential energy at the hill's top is converted to thermal energy during this descent? Assume the car starts at rest at the top of the hill.

13. (II) A 70-kg snowboarder has a velocity of 5.0 m/s at the top of a 30° incline (see Fig. 8–7). After sliding down the 100-m long incline (on which the coefficient of kinetic friction is $\mu = 0.10$), the snowboarder then slides along a flat surface (on which $\mu = 0.20$) and comes to rest after a distance x.

(a) Implement Newton's method (find acceleration from second law, then use the kinematic equations) to determine x.

(b) Find x using conservation of energy (you'll need to use the magnitude of the friction force found from Newton's method).

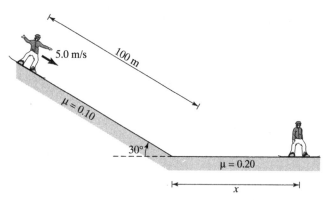

FIGURE 8–7

14. (II) A child is initially swinging back and forth on a playground swing, coming to rest at a vertical height of 4.0 m above the ground at the end point of each arc of the swing. At the low point of the swing's arc, the child is 0.75 m above the ground (see Fig. 8–8). The child's parent says it is time to leave the park and go home so, at the bottom of each arc of the swing, the child drags her feet on the ground for a distance of 2.0 m. If she has to drag her feet seven times in order to stop the swing completely, how large is the friction force that the ground exerts while she drags her shoes? Express your answer in a multiple of the child's weight.

15. (II) Two masses m_1 and m_2 are connected by a string that passes over a frictionless pulley. The mass m_1 is attached to a spring with spring constant k and lies on a horizontal surface (see Fig. 8–9). The coefficient of kinetic friction between m_1 and the horizontal surface is μ. The system is initially held at rest and the spring is undeformed. If the masses are released, how far will mass m_2 fall before it is brought to rest?

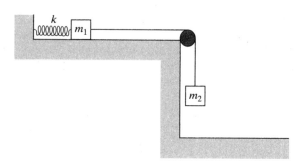

FIGURE 8–9

Section 8–7

16. (II) Find the velocity with which an object has to be projected vertically upward from Earth's surface so that it reaches a height above the surface equal to Earth's radius R.

17. (II) An astronaut is stranded on a spherical chunk of solid (water) ice with a radius of 1000 m. How fast must the astronaut jump away from the surface in order to completely escape the chunk's gravitational pull? Assume the density of water ice is 917 kg/m³.

18. (III) The potential energy between two attracting masses m and M, separated by a distance r, is

$$U(r) = -\frac{GmM}{r} + C$$

where it is conventional to arbitrarily define $U(r = \infty) = 0$ so that the constant $C = 0$.

(a) If, instead, we arbitrarily define $U(r = R) = 0$, where R is finite, determine the resulting form of $U(r)$, i.e., find the resulting value of the constant C.

(b) If $U(r = R) = 0$, what is the value of U at $r = \infty$?

(c) Let R be the radius of Earth and define $U(r = R) = 0$. With this definition for the location of $U = 0$, derive an expression for an object's escape velocity v_{esc} from Earth. How does your expression for v_{esc} compare with that of Eq. 8–20 in the text, which was derived under the assumption of $U(r = \infty) = 0$?

19. (III) To illustrate how tiny the gravitational attraction is between everyday-sized objects, Newton suggested in his book *Principia* that two homogenous 1-foot (=0.3 m) diameter spheres made of rock, if placed in an isolated space with their surfaces 1/4 inch (=0.6 cm) apart, would take over a month to come into contact. Let's analyze Newton's thought experiment.

(a) Consider two identical spherical objects of radius R and mass M in a region of space far removed from any other objects that might influence their motion. The spheres are initially at

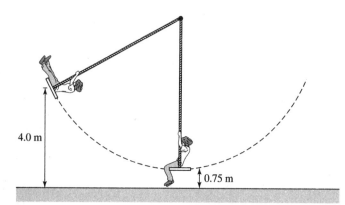

FIGURE 8–8

rest and their centers are initially separated by a distance d (Fig. 8–10a). Due to their mutual gravitational attraction, at some later time, each sphere will have moved toward the other by a distance x and have acquired a speed v, as shown in Fig. 8–10b. Use conservation of energy to show

$$v = \sqrt{\frac{2GM}{d}} \sqrt{\frac{x}{d - 2x}}.$$

(b) Show that, if $d \gg 2x$, (as in Newton's imagined experiment) the expression in part (a) can be approximated as

$$v \approx \frac{\sqrt{2GM}}{d} \sqrt{x}.$$

(c) Consider now the motion of the left-hand sphere in Fig. 8–10, which initially has its center at $x = 0$. When the spheres finally come in contact, we see from Fig. 8–10c that $d = 2R + 2x$, so the center of the left-hand sphere will be at $x = \left(\frac{d}{2} - R\right)$. The time for this sphere to travel from $x = 0$ to $x = \left(\frac{d}{2} - R\right)$ can be found by evaluating the integral

$$t = \int_{x=0}^{x=d/2-R} \frac{dx}{v}.$$

Using the expression for v found in part (b), evaluate this integral to show that the time that elapses until these two objects collide is given by

$$t = d\sqrt{\frac{(d - 2R)}{GM}}.$$

(d) For Newton's thought experiment, $R = 0.500\,\text{ft} = 0.152\,\text{m}$ and $d = (1.00\,\text{ft} + .250\,\text{in}) = 0.311\,\text{m}$. Also, assuming rock has a density of $2700\,\text{kg/m}^3$, $M = 40\,\text{kg}$. Use these values to show that Newton wildly overestimated the elapsed time t until the spheres collide (due to his lack of knowledge of the value for G).

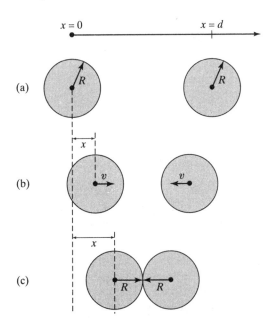

FIGURE 8–10

Section 8–8

20. (II) A 100-kg skier grips a moving rope, which is powered by an engine, and is pulled at constant speed to the top of a hill that is inclined at 20° to the horizontal. The skier is pulled a distance of $x = 200\,\text{m}$ along the incline and it takes 1.0 minute to reach the top of the hill. Assuming the coefficient of kinetic friction between the snow and the skis is $\mu = 0.10$, what horsepower engine is required to do this job?

21. (II) A world-class dragster accelerates from rest to a speed of 150 m/s in 4.5 s. The mass of a dragster is 980 kg.
(a) What power (hp) is required to accomplish this large change in speed in such a short time?
(b) The engine power of a real dragster is typically 6000 hp. Is this consistent with the value you calculated in part (a)? If not, explain the discrepancy.

22. (II) The drag force F_D (N) on a typical car is given by (see Problem 61 in Chapter 5 of the text for a description of the two terms in this equation)

$$F_D = 200\,\text{N} + \left(0.80\,\frac{\text{N}\cdot\text{s}^2}{\text{m}^2}\right)v^2$$

where the car's speed v is in m/s and F_D opposes the motion. In order to maintain the car at constant velocity, the engine must provide a power $P = F_D v$.
(a) Determine P (hp) for the two freeway drivers—the law abider traveling at 100 km/h and the speeder cruising at 130 km/h.
(b) Assuming the engines in their cars have the same efficiency, how much more gasoline (expressed as a percent) will the speeder burn than the law abider on a 100-km trip?

23. (III) Rather than describing a race car's motion using a "constant-acceleration" model, consider the following "constant-power" model: Assume that a race car of mass m, starting from rest, operates its engine at constant power P and that all of the energy added to the car from the engine goes into accelerating the car down a straight race track.
(a) If the car starts from rest at $t = 0$, determine an expression for $v(t)$, the car's speed as a function of time in this model.
(b) Using your expression for $v(t)$, find $a(t)$, the car's (non-constant) acceleration as a function of time. At what time is a greatest?
(c) Starting with your expression for $v(t)$, determine $x(t)$, the car's position as a function of time. Take $x(0) = 0$.
(d) A world-class 980-kg dragster, starting from rest, travels a distance of 402 m ($=0.25$ mile) in 4.5 s. According to this model, what power P (hp) is required for this feat?
(e) The engine power of a real dragster is typically 6000 hp. Is this consistent with the value you calculated in part (d)? If not, explain the discrepancy.

Section 8–9

24. (II) A particle of mass m moves under the influence of a potential energy $U(x)$ given by

$$U(x) = \frac{a}{x} + bx$$

where a and b are positive constants and the particle is restricted to the region $x > 0$. Find a point of equilibrium for the particle and demonstrate that it is stable.

25. (III) A 2.0-kg particle moving in one dimension has potential energy

$$U(x) = \begin{cases} A\dfrac{\sin(bx)}{bx} & |x| \le 3.0\,\text{m} \\ 0 & |x| > 3.0\,\text{m} \end{cases}$$

where the constants $A = 100\,\text{J}$ and $b = \pi\,\text{m}^{-1}$. The argument of the sine function is in radians. $U(x)$ is plotted in Fig. 8–11.
(a) Assume the particle starts at $x = +\infty$ with kinetic energy K and a velocity directed toward the origin. Qualitatively describe the particle's motion if $K > 100\,\text{J}$, $K = 30\,\text{J}$, and $K = 5.0\,\text{J}$, respectively?
(b) Assume that at some instant the particle is located at $x = 2.0\,\text{m}$ and has kinetic energy $K = 5.0\,\text{J}$ with its velocity directed toward the origin. Give a qualitative description of the future motion of the particle and determine its maximum speed during this motion.
(c) Calculate the force on the particle as a function of x.

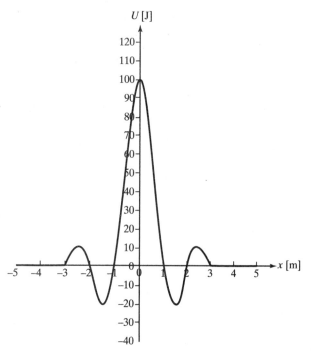

FIGURE 8–11

General Problems

26. (II) A portion of an amusement-park slide is shown in Fig. 8–12. At point A, a 50-kg child is a height of 30 m above the ground and is moving with a speed of 5.0 m/s. No frictional force acts as the child moves from point A to point B, but an average frictional force of 30 N acts as the child moves from point B to point C. The distance from B to C is 70 m.
(a) What is the child's speed at point B where the height is 8.0 m?
(b) Point B is at the bottom of a semicircular curve with a radius of 10 m. How large is the normal force that acts on the child at this point? State your answer as a multiple of the child's weight.
(c) What is the child's speed at point C where the height is 14 m?

(d) Point C is at the top of a semicircular curve of radius R. What minimum value must R have if the child does not lose contact with the slide at point C?

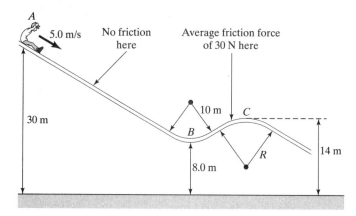

FIGURE 8–12

27. (II) You are a farmer in ancient Egypt and Pharaoh has 'volunteered' you to help build a new pyramid. You are made to push a 70-kg stone from the bottom to the top of a 7.0° incline. The incline is 120 m long. You push on the stone with a force that is directed parallel to the incline and has a magnitude of 500 N and, as you push, a constant frictional force of 200 N acts on the stone (see Fig. 8–13).
(a) How much work do you do on the stone as you push it from the bottom to the top of the incline?
(b) Assume the stone has zero velocity at the bottom of the incline. When the stone arrives at the incline's top, the work you have done on the stone will have been transformed into frictional heat as well as gravitational potential energy and kinetic energy of the stone. What percentage of your work ends up in each of these three energy types?
(c) It takes you 9.5 min to push the stone to the top of the incline. How can you convince Pharaoh to replace you with a horse? Assume the power output of a horse is 1 hp.

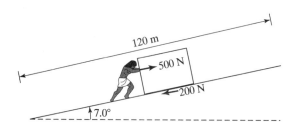

FIGURE 8–13

28. (II) Suppose the gravitational potential energy of an object of mass m at a distance r from the center of Earth were given by

$$U(r) = -\frac{GMm}{r}e^{-\alpha r}$$

where α is a constant and e is the exponential function. (Newton's Universal Gravitation formula, of course, is the special case of $\alpha = 0$.)
(a) What would be the force on the object as a function of r?
(b) What would be the object's escape velocity, if Earth's radius is R?

29. (II) Sally has recently purchased a set of delicate hemispherically shaped wine goblets. Each goblet has a radius of curvature $R = 6.0$ cm and is made from delicate glass that can withstand only a radially directed (i.e., directed perpendicular to its surface) force of up to 2.0 N without breaking. When Sally is not looking, her two-year old son releases a small 100-g steel ball from rest at the rim of one of the goblets (Fig. 8–14). As the ball slides down the inside of the goblet, at what vertical distance above the hemisphere's bottom will the glass break? Assume the goblet exerts a negligible friction force on the steel ball.

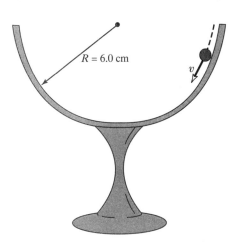

FIGURE 8–14

30. (III) A particle, constrained to move only along the x-axis, is subject to a force given by

$$\mathbf{F}(x) = (ax - bx^3)\mathbf{i}$$

where a and b are positive constants appropriate for SI units.
(a) What is the potential energy $U(x)$ of the particle if $U(0) = 4.0$ J?
(b) At what values of x is the particle in equilibrium?
(c) Which equilibrium points are stable?

31. (III) Problem 85 of the text focuses on a rider's apparent weight when traveling around a circular roller-coaster loop of height $2R$. Consider, instead, a roller-coaster loop that is teardrop-shaped where its total height is still $2R$, but the top part of the arc has half the radius of the bottom arc. Hence, assume the top radius is $2R/3$ and the bottom radius is $4R/3$ as shown in Fig. 8–15. Show that, at the minimum speed needed for the roller-coaster car to make it around this loop, the rider's apparent weight at the bottom of the loop is $4.5\,mg$ (rather than the value of $6.0\,mg$ for a circular-shaped loop), where mg is the weight of the rider. [Teardrop shapes are used in roller-coaster design to reduce the g-forces (which can produce ill effects) experienced by riders.]

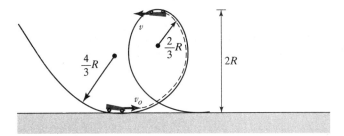

FIGURE 8–15

32. (III) Suppose the attractive gravitational force between two masses m and M were given by

$$\mathbf{F} = -\frac{GmM}{r^n}\hat{\mathbf{r}}$$

where n is an integer. For Newton's law of universal gravitation, of course, $n = 2$.
(a) Define $U(r = \infty) = 0$; then show that the potential energy of this system when the masses are separated by a distance r is

$$U(r) = -\frac{1}{n-1}\frac{GmM}{r^{n-1}}.$$

(b) Consider the situation of mass m with speed v orbiting a stationary mass M in a circle of radius R. Assuming m is held in its orbit by the gravitational force given above, apply Newton's second law to determine its orbital speed v. Then show that the ratio of the potential energy to kinetic energy for this system is given by

$$\frac{U}{K} = \frac{-2}{n-1}.$$

This result is a special case of a famous theorem called the Virial Theorem. What is $\frac{U}{K}$ when $n = 2$?

CHAPTER 9

Linear Momentum and Collisions

Section 9–1

1. (II) The car is full, so a parent decides to buckle himself in with a seat belt and hold his 20-kg toddler on his lap during a short trip that includes some highway driving at speed $v = 30$ m/s.
(a) If the coefficient of static friction between the car's tires and the road is $\mu_s = 1.0$ and the car is moving at 30 m/s, find the minimum time interval t_{min} over which the car can be stopped in case of an emergency.
(b) While traveling 30 m/s, assume the car's driver has to emergency stop the car over the time interval t_{min}. In order to decelerate the toddler along with the car, the parent wraps his arms around the child. Determine the toddler's change in momentum Δp during the stop, and then use Δp to determine how much (assumed constant) force the parent needs to exert on the child during this deceleration period? Is this force achievable by an average parent?
(c) Now assume the car is in an accident while highway driving and the car is brought to stop over a distance of 10 m. Find the stopping time t in this case and then use Δp to determine how much force the parent will need to exert on the child during this deceleration period. Is this force achievable by an average parent?

Section 9–2

2. (II) Two identical objects of mass m, each with the same initial speed v_o, approach each other with an angle $\alpha = \theta + \phi$ between their directions of travel. The objects collide, stick together, and move away at a speed of $\frac{1}{2}v_o$, as shown in Fig. 9–1. Assume that the two objects are isolated (i.e., no net outside force acts on them). Determine α.

3. (II) A mass $m_1 = 2.0$ kg, moving with velocity $\mathbf{v}_1 = (4\mathbf{i} + 5\mathbf{j} - 2\mathbf{k})$ m/s, collides with a mass $m_2 = 3.0$ kg, which is initially at rest. Immediately after the collision, mass m_1 is observed traveling at velocity $\mathbf{v}_1' = (-2\mathbf{i} + 3\mathbf{k})$ m/s. Find the velocity of mass m_2 after the collision. Assume no net outside force acts on the two masses during the collision.

4. (II) You are stranded on the ice of a frozen pond with nothing except for the 540-g graphing calculator that you had hoped would help you get through your calculus-based physics course. The ice you stand on is a perfectly horizontal and frictionless surface and the nearest exit to safety is the edge of the pond 50 m away. You do one last calculation on your calculator to find out the time T it is going take you to reach safety, and then—doing what you have to do—throw

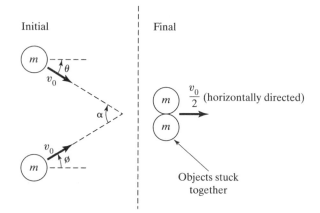

FIGURE 9–1

the calculator horizontally away from you at a speed of 20 m/s. What value for T does your calculator display just before you toss it? Apply conservation of momentum to solve this problem, assuming that the distance over which you are accelerated is negligible compared with 50 m and that your mass is 75 kg.

5. (II) Two starry-eyed ice skaters, one of mass 50 kg with speed 4 m/s and the other of mass 80 kg with speed 3 m/s, glide straight toward each other and, upon meeting, passionately embrace. A physicist watching this scene is, of course, very interested in determining the pair's final velocity during the embrace. What magnitude and direction does the physicist determine for this final velocity?

6. (II) You, too, can move Earth.
(a) Assume that you are a person with mass $m = 80$ kg initially standing at rest upon the surface of Earth. For simplicity, also assume that Earth is a motionless slab of mass $M = 6 \times 10^{24}$ kg. Being an Olympic-caliber sprinter, you then start running to the east with a speed $v_1' = 10$ m/s. Determine the speed v_2' with which Earth must recoil in response to your motion. In which direction does Earth recoil?
(b) To get a feel for the size of the recoil speed v_2' determined in part (a), find the number of days it would take Earth to traverse a distance equivalent to the size of a single proton, if traveling at that speed. The spatial extent of a proton is about 2×10^{-15} m.
(c) If all 6 billion people (assume each has mass of 80 kg) on Earth simultaneously started running east at 10 m/s, find Earth's recoil speed. Would this recoil speed be noticeable to the runners?

Section 9–3

7. (II) Light of a given color consists of a collection of moving identical particles called photons, each of which carries a certain amount of momentum. For the light used in a laser cooling experiment, the momentum of each photon is $p = 8.5 \times 10^{-28}$ kg·m/s. Consider the absorption of such photons by a rubidium atom (atomic mass 85), which is initially at room temperature and has a velocity of $v = 300$ m/s due to its thermal motion. Assume the following: (1) the rubidium atom's initial velocity v is directed into the laser beam, e.g., the photons are moving right and the atom is moving left, (2) when a particular photon collides with the atom, the photon ceases to exist and (by conservation of momentum) imparts its momentum to the atom, and (3) the atom can collide with a new photon every 25 ns. How long will it take for this process to completely stop ("cool") the rubidium atom? [Note: A more detailed analysis predicts that the atom can be slowed to about 1 cm/s by this light absorption process, but that it cannot be completely stopped.]

Sections 9–4 and 9–5

8. (II) Between the "Initial" and "Final" situations shown in Fig. 9–2, the two objects undergo an elastic collision. Determine the velocities of the 4.0-kg and 5.0-kg objects (i.e., v_1' and v_2', respectively) after the collision.

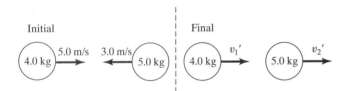

FIGURE 9–2

9. (II) Two balls with unequal masses m and M, moving with the same speed v in opposite directions, collide elastically. The ball of mass M is stationary after the collision. What is the ratio m/M of the balls' masses?

10. (III) Consider the elastic collision in one dimension between two masses m_1 and m_2 with initial velocities v_1 and v_2, respectively. After the collision, the velocities of these masses are v_1' and v_2', respectively.
(a) The change in kinetic energy of mass m_2 as a result of the collision is
$$\Delta K_2 = \tfrac{1}{2} m_2 v_2'^2 - \tfrac{1}{2} m_2 v_2^2.$$
Show that
$$\Delta K_2 = \frac{2 m_1 m_2}{(m_1 + m_2)^2} (m_1 v_1 + m_2 v_2)(v_1 - v_2).$$
[Hint: Use the result of Problem 40 in the text.]
(b) Consider the "rear" head-on collision shown in Fig. 9–3a where, in order for the masses to collide, $v_1 > v_2 > 0$. Use your result from part (a) to show that, as a result of this type of collision, mass m_2 will gain kinetic energy.
(c) Consider the "front" head-on collision shown in Fig. 9–3b where $v_1 > 0$ and $v_2 < 0$. Use your result from part (a) to show that, as a result of this type of collision, mass m_2 will gain kinetic energy only if $v_1 > -\dfrac{m_2}{m_1} v_2$.

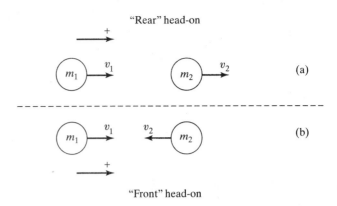

FIGURE 9–3

[Note that for equal masses $(m_1 = m_2)$, both rear and front head-on collisions will speed up the initially slowest mass (and thus slow down the initially fastest mass).]

Section 9–6

11. (II) A car and a truck approach each other as shown in Fig. 9–4.
(a) If they collide and stick together, find the velocity (magnitude and direction) of the tangled wreckage.
(b) How much of the vehicles' initial kinetic energy is dissipated (i.e., used to bend fenders, break glass, cause temperature rises, etc.) as a result of this collision? Express your answer as a percentage.

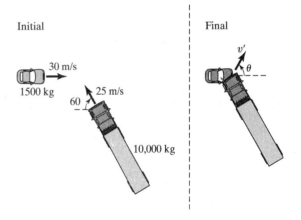

FIGURE 9–4

12. (II) Astronomers estimate that a 2.0-km wide asteroid composed of rock collides with Earth once every million years. Such collisions are highly destructive and pose a threat to life on Earth, as we will show in this problem.
(a) The density of rock is about 3000 kg/m³. What is the mass of a spherical asteroid with a diameter of 2.0 km?
(b) Asteroids, when crossing Earth's orbit about the Sun, typically are traveling at about 10 to 20 km/s relative to Earth. Model a collision between these two objects by assuming an asteroid moving at 15 km/s collides with an at-rest Earth and these objects stick together after colliding. How much destructive energy will be released by this collision?
(c) A large nuclear weapon releases about 4.0×10^{16} J of energy. How many such nuclear bombs would have to be exploded simultaneously to release an equivalent amount of

destructive energy to that found in part (*b*)? [Two kilometers is thought to be the threshold asteroid diameter for causing global catastrophe upon impact.]

13. (II) A pendulum consists of a mass M hanging at the bottom end of a massless rod of length L, which has a frictionless pivot at its top end. A mass m, moving as shown in Fig. 9–5 with the velocity v, impacts M, where it becomes embedded. What is the minimum value of v sufficient to cause the pendulum (with embedded mass m) to swing clear over the top of its arc?

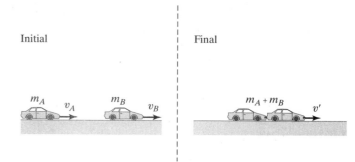

FIGURE 9–6

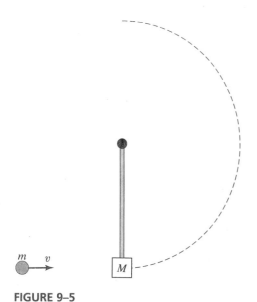

FIGURE 9–5

in comparison with $\frac{1}{2}(m_A + m_B)v'^2$; this is an invalid assumption in this situation, as we will now demonstrate.
(*b*) Momentum is conserved during any collision, as long as the only forces on the colliding objects are due to each other (i.e., no net external force acts). Apply conservation of momentum to the cars' collision and determine the final velocity v'.
(*c*) Now that v' has been determined by momentum conservation, go back to the energy conservation equation in part (*a*) and find E_d.
(*d*) As a result of the collision, the cars' initial kinetic energy ends up being distributed between the wreckage's kinetic energy and the dissipated energy. To demonstrate that it is invalid to assume E_d is negligibly small in comparison with $\frac{1}{2}(m_A + m_B)v'^2$, determine the percentage of the cars' initial kinetic energy that ends up in the wreckage's kinetic energy and the percentage that ends up in the dissipated energy.

Section 9–7

14. (II) A bullet of mass $m = 0.0010$ kg embeds itself in a wooden block with mass $M = 0.999$ kg, which then compresses a spring by a distance $x = 5.0$ cm before coming to rest. The spring's spring constant is $k = 100$ N/m and the coefficient of kinetic friction between the block and the table is $\mu = 0.50$.
(*a*) What is the initial speed of the bullet?
(*b*) What fraction of the bullet's initial kinetic energy is dissipated (i.e., causes damage to the wooden block, temperature rises, etc.) in the collision between the bullet and the block?

15. (II) If unaided by momentum conservation, conservation of energy has limited predictive powers in situations involving colliding (or exploding) objects. To demonstrate this fact, consider the following collision between two cars: On a level highway, initially car A of mass $m_A = 800$ kg has velocity $v_A = +40.0$ m/s and is behind car B, which has mass $m_B = 1000$ kg and velocity $v_B = +10.0$ m/s (positive is defined to the right). Car A plows into the rear end of car B, they stick together, and the tangled wreckage moves off with velocity v' (see Fig. 9–6). Our goal is to predict v'.
(*a*) Define E_d to be the energy dissipated in the cars' collision. That is, E_d equals the total energy that goes into bending fenders, making sound, creating temperature rises, etc. Explain how conservation of energy, when applied to this situation, yields

$$\frac{1}{2}m_A v_A^2 + \frac{1}{2}m_B v_B^2 = \frac{1}{2}(m_A + m_B)v'^2 + E_d.$$

This single equation involves the two unknowns v' and E_d. It can only be solved for v', if we assume E_d is negligibly small

16. (II) A lit firecracker of mass M is thrown horizontally at 30 m/s. It then explodes into two fragments. Just after the explosion, the larger fragment (whose mass is $\frac{3}{4}M$) is observed to fly at an angle of 30° above the horizontal with a speed of 50 m/s, as shown in Fig. 9–7.
(*a*) What is the velocity (magnitude and direction) of the other fragment whose mass is $\frac{1}{4}M$?
(*b*) If the mass of the firecracker is $M = 0.050$ kg, how much energy is released by the explosion?

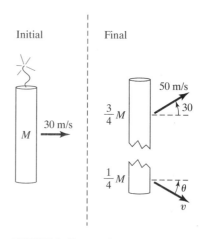

FIGURE 9–7

17. (II) Car 1 ($m_1 = 1000$ kg) collides with car 2 ($m_2 = 1200$ kg) in an intersection as shown in Fig. 9–8. The tangled wreckage moves off at an angle of 30° with speed v'. The wreckage is brought to rest by frictional forces ($\mu = 0.50$) after traveling a distance $x = 20$ m.
(a) Show that $v' = 14$ m/s.
(b) Find the speed of each car before the collision (v_1 and v_2).
(c) If the speed limit is 50 km/h, was either car speeding?

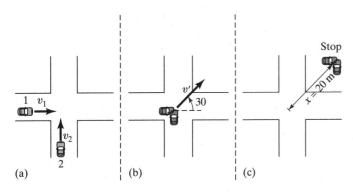

FIGURE 9–8

18. (III) Show that for an elastic collision between two particles of equal mass when one of the particles is initially at rest, the energy transferred to the originally stationary particle is $E_1 \sin^2 \theta_1'$, where E_1 is the initial energy of the incoming particle and θ_1' is its angle of deflection.

Section 9–8

19. (II) Define a coordinate system so that the acceleration due to gravity g near the surface of Earth is directed in the negative z-direction and the zero-level for the gravitational potential energy is defined as the x-y plane (see Fig. 9–9). Also assume the acceleration due to gravity has a constant magnitude g at all locations. Consider an extended object of total mass M as a collection of infinitesimal masses dM at various (x, y, z) points. Show that the gravitational potential energy U of this extended object is given by $U = Mgz_{CM}$, where z_{CM} is the z-coordinate of the object's center of mass.

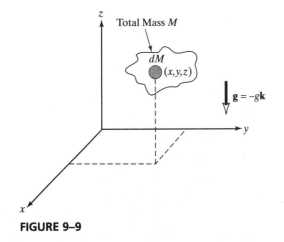

FIGURE 9–9

20. (III) One technique for detecting nearby extrasolar planetary systems is to observe the wobble a large planet produces on its central star. To understand this technique, consider a system consisting of two gravitationally attracting masses m_1 and m_2. If no external forces act on this system, then its center of mass (CM) is stationary. Thus, if the two objects are moving, they must move in such a way that their CM remains at a fixed location (e.g., a straight line drawn between the objects will always pass through the CM).
(a) Assume the two masses m_1 and m_2 are in circular orbits with radii r_1 and r_2, respectively, about the system's CM. Show that

$$r_1 = \frac{m_2}{m_1} r_2.$$

(b) Take the following model for the "generic" solar system: a Sun-like star and a single massive planet with the same characteristics as Jupiter. That is, the planet is one-thousandth as massive as the star and it has an orbital radius of 8.0×10^{11} m. For this generic solar system, determine the radius r_1 of the star's orbit about the system's CM.
(c) When viewed from a distance d away, the generic solar system's star will appear to "wobble" over a distance of $2r_1$. Currently, astronomers are able to detect angular displacements θ of about 1 milliarcsec. From what distance d (in light-years) can the star's wobble be detected? 1 arcsec = 1/3600 deg and 1 light-year = 9.46×10^{15} m.
(d) The nearest-neighbor star to our Sun is about 4.0 light-years away. Assuming stars are uniformly distributed throughout our neighboring region of the Milky Way galaxy, to about how many stars can this detection technique be applied in the search for extrasolar planetary systems?

21. (III) A rope of uniform linear density and total length L is placed on a frictionless tabletop, whose height from the ground is coincidentally L, such that exactly one-half of the rope hangs over the tabletop's edge. The rope is released from rest and slips under the influence of gravity. Show that, at the moment that the rope first touches the ground, the rope is moving at a speed $v = \sqrt{\frac{3}{4} gL}$.

Section 9–9

22. (III) Consider a system consisting of two particles $m_1 = 1.0$ kg and $m_2 = 2.0$ kg, which always remains in the x-y plane. The system is acted upon by a constant external force **F** (i.e., a force with fixed direction and constant magnitude F throughout the x-y plane) and the particles exert forces on each other in a manner consistent with Newton's third law. If the particles have positions at various times as given in Table 9–1, determine **F** (i.e., the magnitude and direction of the external force).

TABLE 9–1

Time (s)	(x, y) Position of Mass 1 (m)	(x, y) Position of Mass 2 (m)
0.0	(2.0, 20.0)	(3.0, 4.0)
1.0	(14.0, 10.0)	(3.0, 7.0)
2.0	(22.0, 2.0)	(15.0, 1.0)

General Problems

23. (II) In an introductory physics lab experiment, a 1.0-kg cart, which rolls along a track subject to very little friction, is given a velocity v and made to rebound off of a stationary vertical surface ("wall"). The portion of the cart that actually comes in contact with the wall is equipped with a computer-interfaced force probe, which measures the force F acting on the cart at a sequence of times during the collision. Suppose that the force probe can accurately measure the magnitude of forces only within the range from 0 to 50 N. Assuming that F is constant and that a typical collision lasts for about 1/50 of a second, determine the maximum speed at which the cart can hit the wall and still have the collision yield useful data. In working this problem, assume that the cart rebounds off the wall with the maximum possible speed (i.e., with the same speed with which it is incident upon the wall).

24. (II) A rider of mass M, standing erect on his skateboard of mass m, travels with a horizontal speed v_o along a horizontal surface. While moving this way on his skateboard, he suddenly bends his knees to assume a crouch position, lowering the vertical height of his center of mass by an amount Δy. Assume that friction forces acting on the rider and his skateboard while traveling on this surface are negligible and that the rider maintains rigid contact with his skateboard at all times.
    (a) What is the rider's horizontal speed after assuming the crouch position on his skateboard?
    (b) What happens to the potential energy that the rider's body loses when going from standing erect to the crouch position?

25. (II) Consider a system of two gravitationally attracting spheres: one of mass $m_1 = 13$ kg and radius $R_1 = 10$ cm and the other of mass $m_2 = 42$ kg and radius $R_2 = 15$ cm. In a region of otherwise empty space, where no external forces act on this system, assume the two spheres are initially at rest with their centers separated by a distance $d = 100$ cm. At some later time, the spheres will come in contact. Determine the speed of each sphere at that moment.

26. (II) Consider the following two possible scenarios when a karate expert attempts to break a wooden board (see Problem 29). Assume the fist and board masses are $M = 0.70$ kg and $m = 0.15$ kg, respectively. Also assume, as is commonly the case, that each edge of the board lies on a support and that the karate expert is striking the board at its center.
    (a) In the first scenario, the karate expert successfully breaks the board, with the initial fist velocity slightly above v, the threshold velocity for board-breaking. The fist's final velocity will then be approximately equal to the final velocity v' given by the model in Problem 29. Show that the momentum change of the fist due to its collision with the board then is

$$\Delta p_{break} \approx Mv\left(\frac{m}{M+m}\right).$$

    (b) In the second scenario, the karate expert has a failure of nerves and so slows his hand down to a speed just below v as his fist approaches the board. In this case, due to the board's edge support, the fist is brought to a complete stop as a result of the collision. Show that the momentum change of the fist due to its collision with the board then is

$$\Delta p_{fail} \approx Mv.$$

    (c) Assuming both collisions described in parts (a) and (b) take about the same amount of time, how many times larger will the force be on the karate expert's fist if he has a failure of nerves rather than if he confidently breaks the board?

27. (III) Imagine throwing a ball in the horizontal direction, so that it rebounds off a sturdy vertical wall. Let's prove that, to a very good approximation, the wall remains stationary during this process. Consider a ball of mass $m = 0.5$ kg with speed $v = 15$ m/s colliding head-on and rebounding off of a second much-larger mass M, which is initially at rest. Let v' and V' be the speeds of m and M after the collision, respectively.
    (a) Provide a brief argument that proves that the rebound speed of mass m can, at most, be equal to its initial speed before the collision; i.e., explain why $v' \leq v$.
    (b) Given the allowed range for v', show that the velocity V' of mass M after the collision obeys the following inequality:

$$V' \leq \frac{2m}{M}v.$$

    (c) Let mass M be a wall, which is rigidly attached to Earth so that it does not move relative to Earth during its collision with the ball. Then the wall's mass is effectively the entire mass of the Earth, so $M = M_{earth} = 6 \times 10^{24}$ kg. Determine the maximum speed V' imparted to such a wall when a ball rebounds off it. Is it valid to assume that, to a good approximation, the wall remains at rest (i.e., stationary) after the collision?

28. (III) Two balls, each of mass m, are suspended from cords as shown in Fig. 9–10. Ball B is initially at rest, hanging vertically down from a string of length L. Ball A, attached to the end of a string of length $2L$, is released from rest at an angle θ relative to the vertical. If the collision between balls A and B is perfectly elastic, what must be the value of θ for ball B to rise just to a horizontal position after the collision? Assume the centers of the balls align horizontally during the collision.

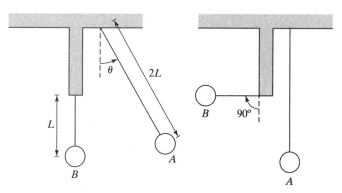

FIGURE 9–10

29. (III) The energy exchange involved during the collision between a karate expert's fast-moving fist and a wooden board can be used to break the board.
    (a) How much energy E_b does it take to break a board? By anchoring the edges of a horizontal, 30-cm long pine board between two sawhorses and by applying a sequence of increasing downward forces F (via hanging weights) to its center, it is found that the board's displacement x is approximately proportional to the applied force (i.e., Hooke's law is obeyed) until $F = 700$ N and $x = 2.0$ cm, when the board is deformed to its breaking point. In this situation, the work done by the applied force F is stored as elastic deformation energy

E_d in the board. Determine E_b, the threshold deformation energy at which fracture occurs.

(b) What fist-velocity is needed to break a board? Consider the following model, which allows us to neglect the complicating effects of any supports at the edge of the board. Imagine a pine board of mass m floating in outer space unsupported. As depicted in Fig. 9–11a, a fist of mass M initially has velocity v, which is the threshold velocity needed to deform the board to its breaking point, but not actually break it. The fist then collides completely inelastically with the board, deforming it to its breaking point (i.e., the board's deformation energy equals E_b) as shown in Fig. 9–11b. The fist and the deformed board move off with a common velocity v' after the collision. Using this model, show that the threshold fist-velocity for breaking the board is

$$v = \sqrt{\frac{2(M+m)}{Mm} E_b}.$$

(c) For typical values of $M = 0.70$ kg and $m = 0.15$ kg, determine v. [When supported at its edges, a board is actually easier to break than in our model.]

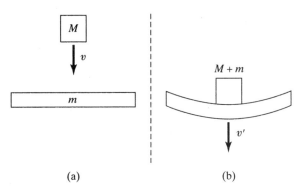

(a) (b)

FIGURE 9–11

30. (III) The collision between rigid objects is a fast event, typically occurring on time scales of several milliseconds; hence, the internal forces that the colliding objects exert on each other (called "contact forces") are large. Conservation of momentum predicts the effect of these large contact forces on the motion of the colliding objects. Let's investigate these ideas quantitatively for the following two-object system: On a horizontal surface, initially object A of mass $m_A = 10$ kg has velocity $v_A = +10$ m/s and is behind object B, whose mass $m_B = 15$ kg and velocity $v_B = +5$ m/s (where positive is defined to the right). Object A collides with object B, they stick together, and this composite mass $(m_A + m_B)$ moves off with velocity v' just after the collision (see Fig. 9–12). Our goal is to predict v'.

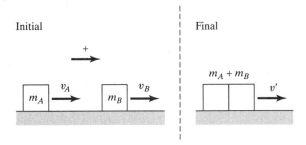

FIGURE 9–12

(a) Assume the horizontal surface is frictionless and no net external force acts on the system; thus, momentum will be conserved during the collision. Use conservation of momentum to show that $v' = +7$ m/s.

(b) From the result in part (a) we can conclude that, over the duration of the collision, the contact forces exerted by the objects on each other slow object A down from 10 m/s to 7 m/s and speed object B up from 5 m/s to 7 m/s. Assume that the collision lasts for a time period of $\Delta t = 5$ ms. Determine the (average) contact force F_C that object A exerts on object B during the collision. (By Newton's third law, the value you determine for F_C is also the magnitude of the contact force that object B exerts on object A during the collision.)

(c) Suppose now that the horizontal surface is not frictionless, but instead that the coefficient of kinetic friction between the surface and object B is $\mu_k = 0.3$. Show that the resulting frictional force F_{fr} on object B is very small in comparison with the contact force F_C determined in part (b). Since $F_{fr} \ll F_C$, is it valid to use conservation of momentum to predict v' at the conclusion of this collision on a frictional surface?

(d) To show that the conclusions drawn in part (c) are not always true, rework this problem with $m_A = 1$ kg, $v_A = +1$ m/s, $m_B = 100$ kg, $v_B = +0.5$ m/s, $\Delta t = 5$ ms, and $\mu_k = 0.3$. Show that for this particular collision, $F_{fr} > F_C$ and so it would be completely invalid to use momentum conservation to predict motion in this situation.

Rotational Motion about a Fixed Axis

Sections 10–1 to 10–3

1. (II) On a 12.0-cm diameter audio compact disc (CD), digital bits of information are encoded sequentially along an outward spiraling path. The spiral starts at radius $R_1 = 2.5$ cm and winds its way out to radius $R_2 = 5.8$ cm. To read the digital information, a CD player rotates the CD so that the player's readout-laser scans along the spiral's sequence of bits at a constant linear speed of 1.20 m/s. Thus, the player must accurately adjust the rotational frequency f of the CD when the laser is located at differing radii. Determine the two extreme values for f (rpm); i.e., find the required f when the laser is located at R_1 and then when it is at R_2.

2. (II) The Moon's gravitational attraction on the tidal bulges on opposite sides of Earth causes a torque that decelerates Earth's rotational motion about its polar axis. As a result of this "tidal torque," astronomers have determined observationally that the length of the day T is increasing at a rate of ΔT per century, where ΔT is currently equal to 2.3 ms.
   (a) At a particular moment, Earth's angular speed in rad/s is given by $\omega = 2\pi/T$, where T is Earth's rotational period (i.e., time for one day). Using the definition of instantaneous angular acceleration $\alpha = d\omega/dt$, show that
   $$\alpha = -\frac{2\pi}{T^2}\frac{dT}{dt}$$
   Determine the current value of α (in units of rad/s per century), given $dT/dt = 2.3$ ms/century.
   (b) Assuming constant angular acceleration, use the rotational kinematic equations to predict the length of the day 500 million years ago.

3. (II) From the study of fossil corals, researchers have discovered that Earth's rotation about its polar axis has slowed over time. Because Earth spun faster in the past, the number of days per year was greater. Table 10–1 shows the history of Earth's rotation over the past 600 million years, where time t is measured in million of years (My) before the present. The numerical value for the "Number of Days per Year" at a particular moment in history is equivalent to the numerical value for Earth's rotational frequency f in units of rev/year at that moment.

TABLE 10–1									
Time Before Present (My)	65	136	180	230	280	345	405	500	600
Number of Days per Year	371	377	381	385	390	396	402	412	424

(a) Explain why, if one graphs the given data as a plot of Rotational Frequency f (y-axis) vs. Time Before Present t (x-axis), a straight line will result, provided Earth's angular acceleration α has been constant over the past 600 My. How can the constant α be determined from the slope of such a straight-line plot? What value would you expect for the y-intercept of such a straight-line plot?
(b) Using the data in the table, graph f (y-axis) vs. t (x-axis) and show that a straight-line plot does indeed result. Determine Earth's constant angular acceleration α in units of rad/s^2 from the slope of this graph. Does the straight line's y-intercept have the value you expect?

4. (II) When a compact disc (CD) player is turned on, it requires a "spin-up" time of about 2.0 s to bring a disc from rest to the rotational frequency $f = 500$ rpm used during data readout near the center of the CD (where its information-containing spiral begins). Assuming constant angular acceleration, how many revolutions does the disc rotate through during the spin-up time period?

5. (II) Earth's orbital speed about the Sun is $v_E = 29.8$ km/s and the Moon's orbital speed about Earth is $v_M = 1.02$ km/s, assuming both orbits are perfectly circular. Let r_M and r_E be the orbital radii of the Moon about Earth and Earth about the Sun, respectively.
(a) A sidereal month is defined as the time it takes the Moon to complete one revolution of its orbit. Determine the number of days in a sidereal month.
(b) A synodic month is defined as the time between successive new moons (the Sun, the Moon, and Earth are aligned at new moon; see Fig. 10–1). Show that, because of Earth's

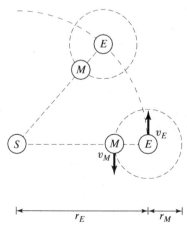

FIGURE 10–1

orbital motion about the Sun, the time τ in a synodic month is given by

$$\tau = \frac{2\pi}{\dfrac{v_M}{r_M} - \dfrac{v_E}{r_E}};$$

then determine τ in days.

6. (III) An automatic swing door in a commercial building (such as a supermarket) rotates open from the plane of the doorway after it detects the presence of an approaching person. Assume the door is set to detect a person when 1.2 m away and that the person walks toward the door at the moderate (constant) speed of 1.5 m/s. The door must rotate from its fully closed to its fully open position (a rotation angle of 90°) during the time interval from when the approaching person is first detected until he or she passes through the plane of the doorway.
(a) Assume, during the opening process, the door is accelerated from rest at constant angular acceleration α over the first half of its angular rotation and then decelerated to rest at constant acceleration $-\alpha$ over the second half of its travel. This scenario will require the least powerful driving motor. Find the required value for α.
(b) What maximum angular speed does the door attain during its opening motion?
(c) If a person walks toward the door at a speed $v > 1.5$ m/s, the door will not be fully open when he or she arrives at the plane of the doorway. If the person needs the door to have swung open at least 60° so as not to collide with the moving door, what is the maximum value for v so that the person can pass unimpeded through the doorway?

7. (III) The Moon's gravitational attraction on the tidal bulges on opposite sides of Earth causes a torque that decelerates Earth's rotational motion about its polar axis. As a result of this "tidal torque," astronomers have determined observationally that the length of the day T is increasing at a rate of ΔT per century, where ΔT is currently equal to 2.3 ms.
(a) Let T and $(T + \Delta T)$ be the length of the day at times $t_o = 0$ and $t = 1$ century, respectively. Assume Earth's angular acceleration α is constant over this time period and show that

$$\alpha = -\frac{2\pi}{Tt}\left[1 - \left(1 + \frac{\Delta t}{T}\right)^{-1}\right];$$

then go on to show that, if $\Delta T \ll T$, the above expression can be approximated as

$$\alpha \approx -\frac{2\pi}{T^2}\frac{\Delta T}{t}.$$

[*Hint*: The binomial expansion allows the quantity $(1 + x)^n$ to be approximated as $(1 + nx)$, if $x \ll 1$.]
(b) Using your relation from part (a), determine the current value for Earth's angular acceleration (in units of rad/s per century) due to tidal torque. Then, assuming α has remained constant over the past one-half billion years, predict what the length of the day (in hours) was 500 million years ago.
(c) From the study of fossil corals, researchers have concluded that 500 million years ago there were 412 days per year on Earth. How does this experimental result compare with your prediction in part (b), which was based on the assumption of constant α?

Sections 10–6 and 10–7

8. (II) A compact disc (CD) is a thin cylinder of mass $m = 15$ g with a hole of radius $R_1 = 0.75$ cm in its center and an outer radius $R_2 = 6.0$ cm. When mounted in a CD player, the CD is attached on top of the post-like spindle of an electric motor. In a typical player, the moment of inertia of the motor's spindle about its axis of rotation is 4.0×10^{-6} kg·m².
(a) The spindle and the attached CD form a composite object that the player's electric motor rotates during data readout. Determine the moment of inertia of this composite object.
(b) When a CD player is turned on, a short "spin-up" time period is required to bring the disc from rest to the rotational frequency $f = 500$ rpm used during data readout near the center of the CD (where its information-containing spiral begins). A typical player's motor applies a constant "starting torque" of 8.0×10^{-4} N·m to the composite object of part (a) during spin-up. After it is turned on, how long does the player have to wait until data readout can begin (i.e., how long does the spin-up take)?

9. (II) A couple of night owls, who like to party after the Sun goes down, hatch the following plan on a cocktail napkin one evening so that, from now on, they can party all the time (Fig. 10–2). They propose a device (as seen from the napkin, the details of its construction are a little sketchy) that can apply a constant tangential force of 6.4×10^{21} N on Earth's equator.
(a) Once the device is turned on, how many radians will Earth spin through before it stops?
(b) If the bar in which the night owls are seated is to be facing away from the Sun when Earth stops, should the force be turned on during their daytime or their nighttime?

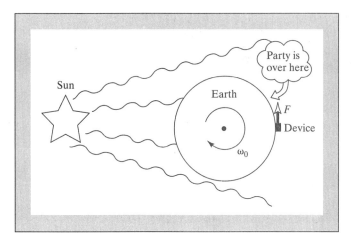

FIGURE 10–2

10. (II) The official disc of Ultimate Frisbee has a mass of 175 g and a diameter of 27 cm. Assuming that about half of its mass is spread uniformly in a solid, cylinder-like shape, while the other half is concentrated in its outer rim, this disc can be modeled as a composite object consisting of a 27-cm diameter cylinder and a 27-cm diameter hoop, each with 87.5 g of mass. When throwing such a disc, an Ultimate Frisbee player typically rotates it a quarter-turn while accelerating it from rest to 10 rev/s. For this typical throw, determine the

magnitude of the tangential force that the player applies at the disc's rim.

11. (II) A motionless gate is open at 90° to a fence. A farmer then closes the gate by pushing on it with a perpendicular force (i.e., the force is perpendicular to the gate at all times) of 20 N at a point that is 3.0 m from the gate's hinge. Assume the gate to be a long uniform rod of mass $M = 35$ kg and length $L = 4.0$ m with an axis at one end. How long does it take the farmer to shut the gate?

12. (II) A spherical asteroid with a radius of 100 m and a uniformly distributed mass of 1×10^{10} kg has been determined to be on a collision course with Earth. A "tug" spaceship is assigned the mission of landing on the asteroid's surface where it is to fire an engine, applying a straight-line push in order to shift the asteroid to an Earth-friendly orbit. Unfortunately, when the tug arrives at the asteroid, it is discovered that the asteroid spins about an axis at four revolutions per day, complicating the tug's original plan of applying a straight-line force. So the tug's first order of business is to stop the asteroid's rotational motion. The tug lands on the asteroid's equator (as defined by its axis of rotation), points its engine horizontally along the equator and fires it until the rotation is brought to a halt. If the tug's engine applies a small force of 2.5 N tangentially at the asteroid's equator, how long must the engine be fired in order to completely stop the asteroid's spinning motion? [Researchers have proposed this scenario as a viable method for "spinning down" a rotating asteroid.]

Section 10–8

13. (II) To perform a single axle (1.5 revolutions about vertical axis while airborne) and a triple axle (3.5 revolutions) jump, an ice skater must be able to reduce his body's moment of inertia about the vertical axis by factors of about 2 and 5, respectively. Moving one's outstretched arms to a position much closer to the body's center provides one simple rotational-inertia "reduction" method. Model this action as shown in Fig. 10–3, which makes use of the fact that for a person of mass M (in kg), each of the person's arms typically contains 5% of the person's total mass. In our crude model, each arm of mass $0.05M$ is taken to be a thin uniform rod of length 60 cm and the rest of the body is taken to be a uniform cylinder of mass $0.90M$ and radius 12 cm. Within this model, determine I_o and I, the body's moment of inertia with arms outstretched (Fig. 10–3a) and with arms at the side (Fig. 10–3b), respectively, in terms of M. Then show that the ratio I_o/I suggests that this action provides a more than adequate reduction method for performing the single axle jump, but is inadequate for the triple axle. [*Hint:* When at the side, assume all points on each arm are 12 cm from the axis.]

14. (II) To perform a triple axle jump (3.5 revolutions about vertical axis while airborne), an ice skater must re-position both her arms and legs in order to reduce her body's moment of inertia about the vertical axis by a factor of about five. Demonstrate that this level of "inertia reduction" is within the range of possibility for the human body by considering the following model for an action that produces maximal inertia reduction. This model makes use of the fact that for a typical person of mass M (in kg), each arm and each leg contains 5% and 16%, respectively, of the person's total mass, while the trunk and head contain the remaining 58%.

(*a*) In Fig. 10–4a, our model for the skater with both arms and both legs outstretched is shown. The body's trunk and head are taken to be a cylinder of mass $0.58M$ and radius 12 cm with two horizontally aligned arms (each modeled as a thin uniform rod of mass $0.05\,M$ and length 60 cm) attached on both sides. Model each leg as a thin uniform rod of mass $0.16M$ and length 60 cm that is horizontally aligned with the axis at one of its ends. Determine I_o (in terms of M), the body's moment of inertia about the vertical axis in this orientation.

(*b*) In Fig. 10–4b, the skater's arms are at her sides and each leg is vertically aligned with the axis along one of its curved sides. Model each leg now as a cylinder of radius 6.0 cm; also assume all points on each arm are 12 cm from the axis. Determine I (in terms of M), the body's moment of inertia about the vertical axis in this orientation.

(*c*) Show that the ratio I_o/I suggests that this action provides a more than adequate reduction method for performing the triple axle jump.

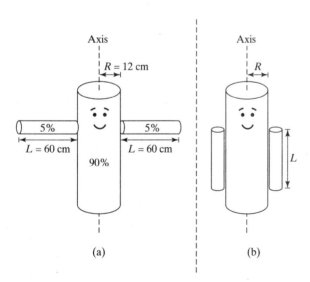

FIGURE 10–3

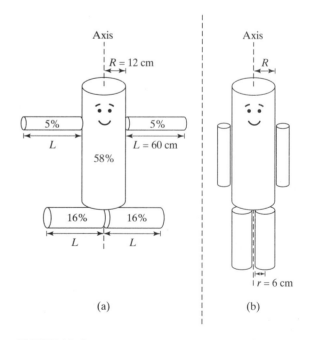

FIGURE 10–4

15. (III) When an extended object rotates about a distant axis, one can approximate the rotational inertia of the object as that of a point mass. Let's explore the validity of this approximation.
(a) For the sake of argument, consider an extended object that is a uniform sphere with mass M and radius r_o. We know that, if this object rotates about an axis through its center of mass (CM), its rotational inertia is $I_{CM} = \frac{2}{5} M r_o^2$. Now assume the sphere rotates about a new axis that is parallel to, but a distance h away from, the CM axis. Show that, if $h \gg r_o$ (i.e., the new axis is "distant"), then the rotational inertia of the sphere about this distant axis is approximately $I \approx Mh^2$, which is the moment of inertia for a point mass.
(b) Show that, if $h > 10 r_o$, the point-mass approximation in part (a) will deviate from the true value of I by, at most, 0.4%.
(c) To make this reasoning general, consider an arbitrarily shaped object that is rotating about an axis that is a distance h away from its CM. Starting with the definition $I = \int R^2 \, dm$, prove that, regardless of the object's shape, its moment of inertia about this axis is approximately that of a point mass as long as the distance h is much larger than the spatial extent of the object.

16. (III) A thin spherical shell has its mass M uniformly distributed over a spherical surface of radius R. The shell has negligible thickness in comparison with its radius so that all of its mass may be assumed to be a distance R from the shell's geometric center.
(a) Define area mass density σ to be the amount of infinitesimal mass dM per infinitesimal surface area dA, i.e., $\sigma = dM/dA$. In our given situation, the shell has a uniform distribution of mass; thus, σ has the same value at every location on the shell. Determine the constant value of σ for the uniform shell.
(b) In Fig. 10–5, all of the mass within the infinitesimal slice of mass dM shown is a distance $r = R \sin \theta$ from the axis. The infinitesimal area of this slice is $dA = (2\pi r)(R d\theta)$. Use these facts to show that the moment of inertia of this spherical shell about an axis through its center equals $\frac{2}{3} MR^2$.

[Hint: $\int \sin^3 \theta \, d\theta = -\cos \theta + \frac{1}{3} \cos^3 \theta$]

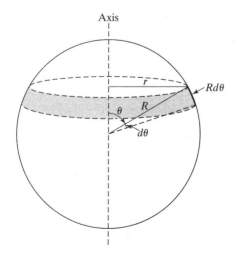

FIGURE 10–5

17. (III) Often one can find the moment of inertia of a rigid object by assuming it is composed of a collection of simpler objects whose moments of inertia are known. Consider a solid sphere of uniform construction with mass M and radius R to be a collection of nested thin spherical shells. The shells each have an infinitesimal thickness dr and, within the collection, there is a shell with every radius between $r = 0$ and $r = R$, so that when nested together they form a solid sphere. In Fig. 10–6, one of the infinitesimal shells is shown, which has radius r and infinitesimal mass dM. Since this shell has a surface area of $4\pi r^2$ and thickness dr, its infinitesimal volume $dV = (4\pi r^2) dr$. Assume that we know (see Problem 16) that the moment of inertia of this infinitesimally thin spherical shell about an axis through its center equals $\frac{2}{3} r^2 \, dM$. Summing (via integration) the moments of inertia of all the nested shells with radii from $r = 0$ to $r = R$, show that the moment of inertia of a uniform solid sphere about an axis through its center equals $\frac{2}{5} MR^2$.

[Hint: The mass dM of the infinitesimal shell can be found using $\rho = dM/dV$, where ρ is the volume mass density of the uniform solid sphere.]

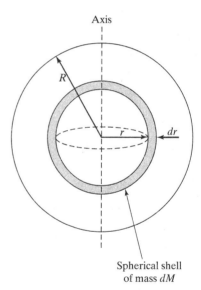

FIGURE 10–6

Section 10–9

18. (II) You, too, can move Earth.
(a) Assume that you are a person with mass $m = 80$ kg initially standing at rest upon the surface of Earth at its equator. Assume that Earth is a motionless uniform sphere of mass $M = 6.0 \times 10^{24}$ kg and radius $R = 6.4 \times 10^6$ m. Being an Olympic-caliber sprinter, you then start running to the east with a speed $v = 10$ m/s relative to Earth's surface. Determine the angular speed ω with which Earth must recoil in response to your motion.
(b) To get a feel for the size of the recoil angular speed ω determined in part (a), find the number of days it would take a point on Earth's surface to traverse a distance equivalent to the size of a single proton, if Earth rotates at speed ω. The spatial extent of a proton is about 2×10^{-15} m.

(c) If all 6 billion of Earth's people were on the equator and simultaneously started running east at 10 m/s, find Earth's recoil angular speed. Would this recoil angular speed be noticeable to the runners? Assume each person has a mass of 80 kg.

19. (III) A space pirate with mass $m = 80$ kg has been marooned on a very small planet, which is a hollow spherical shell of mass $M = 1000$ kg and radius $R = 100$ m (Fig. 10–7). If left unperturbed, the planet does not rotate about any axis. The pirate, standing at the "top" of the planet, decides to walk to the planet's "middle," where it is warmer (because the Sun's rays strike the planet's surface from directly above there). The pirate figures that the distance traveled during his walk will be $(2\pi R)/4$; so, if he walks at a speed v relative to the planet's surface, his walk should take a time $t = \pi R/2v$. Show that the pirate's walk will actually take about 14% longer than he figures. The moment of inertia of a hollow spherical shell with mass M and radius R about an axis through its center is $I = \frac{2}{3}MR^2$. Assume that the pirate's height is small in comparison with R, so that he can be modeled as a point mass.

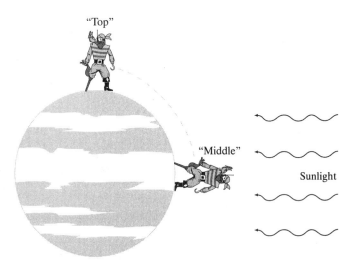

FIGURE 10–7

20. (III) You are standing on the surface of your new home—a motionless asteroid that is a solid sphere of radius $R = 100$ m and mass $M = 1.0 \times 10^{10}$ kg—holding a 7.0-kg bowling ball.
(a) With what speed v should you launch the bowling ball away from yourself so that it follows a circular orbit slightly above the asteroid's surface (i.e., the orbit's radius can be taken to be R)?
(b) As you launch the bowling ball, the asteroid will acquire a recoil angular velocity ω. Assuming your body's mass is very small in comparison with the asteroid's mass (so that you may neglect your contribution to the asteroid's moment of inertia), determine ω.
(c) Determine the time t it takes the bowling ball to return to you and the angle through which the asteroid has recoiled during this time.
[*Hint*: The ball's orbital period can be taken as a very good approximation for t.]

(d) Assume you catch the ball when it returns to you a time t after a launch. If you wanted to rotate the asteroid through a total angle of 90° by repeated launches of the bowling ball, how many launches would be required? If the launching operation required insignificant time in comparison with the orbit time, how many years would it take you to rotate the asteroid through 90° using this method?

21. (III) Competitive ice skaters commonly perform single, double, and triple axle jumps in which they rotate 1.5, 2.5, and 3.5 revolutions, respectively, about a vertical axis while airborne. For all of these jumps, a typical skater remains airborne for about 0.70 second. Consider the following model for an axle jump in which the skater's flight is divided into a 0.10 s initial section, a 0.50 s middle section, and a 0.10 s final section. In this model we assume the skater leaves the ground in an "open" position (e.g., arms outstretched) with moment of inertia I_o and rotational frequency $f_o = 1.2$ rev/s, maintaining this position for 0.10 s. The skater then immediately assumes a "closed" position (e.g., arms brought closer to body's center) with moment of inertia I, acquiring a rotational frequency f, which is maintained for 0.50 s. Finally, the skater immediately returns to the "open" position with moment of inertia I_o and rotational frequency $f_o = 1.2$ rev/s, maintaining this position for 0.10 s until landing (see Fig. 10–8).
(a) Assuming air resistance can be neglected, explain why angular momentum is conserved during the skater's jump.
(b) In this model, determine the required rotational frequency f during the flight's middle section in order for the skater to successfully complete a single and a triple axle.
(c) Show that, according to this model, a skater must be able to reduce his or her moment of inertia in mid-flight by factors of about 2 and 5 in order to complete a single and triple axle, respectively.

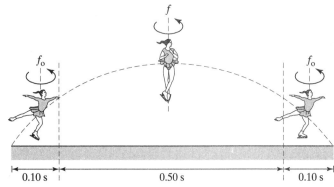

FIGURE 10–8

Section 10–10

22. (II) Earth rotates about its axis once per day ($= 86{,}400$ s) and hence possesses rotational kinetic energy. To get a feel for the magnitude of this energy, imagine a device has been invented that can siphon off a portion of Earth's rotational energy (and thereby reduce our planet's angular velocity) to provide power for the world's population at our present rate of consumption, which is about 15×10^{12} W. If, due to environmental concerns, the length of the day is not to be increased by more than 1.0 min, how long can this device be operated?

Section 10–11

23. (II) Consider again Example 10–23 in the text.
(a) By analysis of the forces and torques that act on the yo-yo-like cylinder, it was shown that the downward acceleration of this object's center of mass (CM) is the constant value of $a = \frac{2}{3}g$. If it starts from rest, what will the CM velocity be after it has fallen a distance h?
(b) Re-analyze this situation using conservation of energy to determine the cylinder's CM velocity after it has fallen a distance h, starting from rest. You should, of course, get the same answer as in part (a).

24. (III) At time $t = 0$, a bowler gives a spherical bowling ball of radius $R = 11$ cm an initial center-of-mass (CM) velocity $v_o = 4.0$ m/s and releases it with a "spin" of $f_o = 2.0$ rev/s about a horizontal axis through its CM as shown in Fig. 10–9, sending the ball sliding straight down the alley toward the ten pins. If the coefficient of kinetic friction between the ball and the alley is $\mu_k = 0.20$, the ball will slide for a total time T and over a distance x before it begins to roll without slipping.
(a) Taking both the CM velocity and spin into account, determine the direction of the initial velocity of the bottom of the ball relative to the alley. Based on this determination, in which direction does the kinetic friction force on the ball act?
(b) Over the time interval $t < T$ during which the ball is sliding, find an expression (in terms of $v_o, f_o, \mu_k,$ and/or R) for the ball's CM velocity v as a function of time t.
(c) During the time interval $t < T$, find an expression (in terms of $v_o, f_o, \mu_k,$ and/or R) for the ball's angular velocity ω as a function of time t.
(d) Use your results from parts (a) and (b) to find the numerical value of T (s), the time at which the ball starts to roll without slipping.
(e) The length of a bowling alley is 19 m. Does this bowling ball start to roll without slipping before reaching the ten pins?

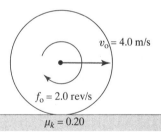

FIGURE 10–9

25. (III) For fun, bicycle and motorcycle riders "pop a wheelie," a trick in which a large acceleration causes the bike's front wheel to leave the ground. Let's analyze the physics behind this trick. Define M to be the total mass of the bike-plus-rider system. Let x and y be the horizontal and vertical distances, respectively, of the center of mass of this system from the rear wheel's point of contact with the ground (see Fig. 10–10).
(a) Determine the acceleration a required to barely lift the bike's front wheel off of the ground.
(b) To minimize the acceleration necessary to pop a wheelie, should x be made as small or as large as possible? What

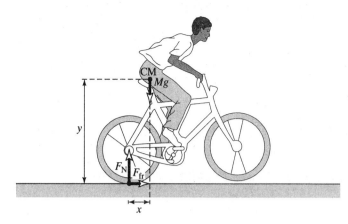

FIGURE 10–10

about y? How should a rider position his or her body on the bike in order to achieve these optimal values for x and y?
(c) If, for a bicycle, $x = 30$ cm and $y = 100$ cm, find a.

General Problems

26. (II) Two masses are connected by a string that hangs over a pulley of radius $R = 0.10$ m, which can rotate without friction about an axis at its center. The pulley's moment of inertia I about this axis is 0.010 kg·m². As shown in Fig. 10–11, mass $m_1 = 4.0$ kg rests on a frictionless table, while mass $m_2 = 5.0$ kg is initially held at a height $h = 2.0$ m above the ground. Mass m_2 is then released from rest and allowed to fall to the ground, causing the pulley to spin in the process. Let's analyze this situation from two points of view: first using forces and torques and then using conservation principles.
(a) As m_2 falls, what will be the resulting acceleration of the masses?
(b) Use the kinematic equations to find the masses' speed just before m_2 hits the floor.
(c) Use conservation of energy to find the speed of the masses just before m_2 hits the floor. You should get the same answer as in part (b).

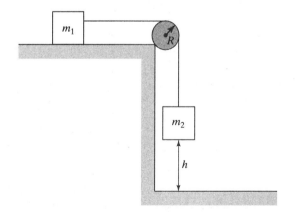

FIGURE 10–11

27. (II) An interesting result from quantum mechanics is that each electron in nature has an intrinsic angular momentum L of magnitude $\frac{h}{4\pi}$ about a chosen axis, where $h = 6.63 \times$

10^{-34} kg·m²/s is called Planck's constant. One possible model for explaining this angular momentum is to imagine an electron as a sphere spinning about its axis in the same way that Earth rotates about the North Pole. Let's explore this model quantitatively. Assume an electron to be a solid, uniform sphere of radius R. High-energy physics experiments have shown that the radius of an electron, if it has a nonzero value, can be no larger than about 1×10^{-18} m, so let's assume $R = 1 \times 10^{-18}$ m.
(a) Using this model, determine the angular velocity ω required to produce an angular momentum of $\dfrac{h}{4\pi}$.
(b) Again within this model, if ω has the value determined in part (a), calculate the linear velocity v of a point on the sphere's equator.
(c) Demonstrate that our "spinning sphere" model of the electron is invalid by comparing the linear velocity v found in part (b) with the speed of light $c = 3 \times 10^8$ m/s. Show that our model violates the theory of relativity, which prohibits matter from traveling faster than c. [Note that if we had assumed a smaller radius for the electron our model would be in even greater violation of relativity theory.]

28. (II) A motionless gate is open at 90° to a fence. A farmer then closes the gate by pushing on it with a perpendicular force (i.e., the force is perpendicular to the gate at all times) of 20 N at a point that is 3.0 m from the gate's hinge. Assume the gate to be a long uniform rod of mass $M = 35$ kg and length $L = 4.0$ m with an axis at one end.
(a) Determine the gate's angular acceleration; then use the rotational kinematic equations to find the gate's angular velocity ω (rad/s) at the moment it aligns with the fence (i.e., at the moment it becomes closed).
(b) By evaluating $\int \tau \, d\theta$, determine the work done by the farmer while shutting the gate.
(c) Using the value for the gate's final angular velocity found in part (a), show that the work done by the farmer is equal to the change in the gate's rotational kinetic energy.

29. (II) Every so often, Earth and Mars line up on the same side of the Sun and so Mars appears opposite the Sun in our sky. This event is called an "opposition of Mars." Assume that Earth and Mars move in circular orbits about the Sun with radii r_E and r_M, respectively, and that their orbital periods T_E and T_M are related by Kepler's third law (Eq. 6–7 in the text).
(a) During the time interval t between successive oppositions, Earth will travel one extra revolution about the Sun in comparison with Mars. For our model of circular orbits, show that

$$t = \frac{T_E}{\left[1 - (r_E/r_M)^{3/2}\right]}.$$

(b) Given that the orbital radius of Mars about the Sun is 1.52 times larger than Earth's orbital radius, show that an opposition of Mars occurs about every other year.

30. (III) If a billiard ball is hit in just the right way by a cue stick, the ball will roll without slipping immediately after losing contact with the stick. Consider the following situation: A billiard ball (radius R, mass M) is at rest on a horizontal pool table. A cue stick applies a constant horizontal force F on the ball for a time T at a point that is a height h above the table's surface (see Fig. 10–12). Assume that the coefficient of kinetic friction between the ball and table is μ. Determine the correct value for h (in terms of R, M, F, and μ) such that the ball will roll without slipping immediately after losing contact with the stick.

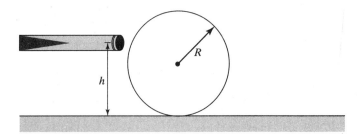

FIGURE 10–12

31. (III) How much easier is it to swing a baseball bat when gripping the "thick" end rather than the "handle" end? Assume the length of a baseball bat is $L = 0.84$ m and that the bat's mass is distributed according to the following relation

$$\lambda = 0.56 \frac{\text{kg}}{\text{m}} + \left(3.5 \frac{\text{kg}}{\text{m}^3}\right) x^2$$

where x is the distance (m) from the end of the handle. The linear mass density λ is defined as the amount of infinitesimal mass dM per infinitesimal distance dx, i.e., $\lambda = dM/dx$. λ gives the amount of mass contained in an extremely thin slice of the bat located at position x.
(a) Determine the bat's moment of inertia about an axis located at the handle end (Fig. 10–13a).
(b) Determine the bat's moment of inertia about an axis located at the thick end (Fig. 10–13b).
(c) How many times easier is it to provide the baseball bat with a given angular acceleration when swinging it about the thick end rather than the handle end?

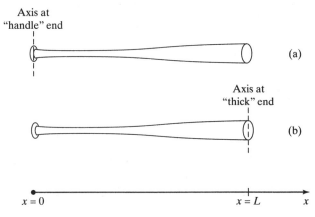

FIGURE 10–13

CHAPTER 11

General Rotation

Section 11–1

1. (II) Given three vectors, $\mathbf{A} = 1.0\mathbf{i} + 2.0\mathbf{j} + 3.0\mathbf{k}$, $\mathbf{B} = 1.0\mathbf{i} + 1.0\mathbf{j} + 0.0\mathbf{k}$, and $\mathbf{C} = 0.0\mathbf{i} + 2.0\mathbf{j} + 0.0\mathbf{k}$, demonstrate the distributive property of the cross product (Eq. 11–2c in the text). That is, show that $\mathbf{A} \times (\mathbf{B} + \mathbf{C}) = (\mathbf{A} \times \mathbf{B}) + (\mathbf{A} \times \mathbf{C})$ for these vectors.

2. (II) Prove $\mathbf{A} \times \mathbf{B} = -\mathbf{B} \times \mathbf{A}$.
[*Hint*: Write the vectors as $\mathbf{A} = A_x\mathbf{i} + A_y\mathbf{j} + A_z\mathbf{k}$ and $\mathbf{B} = B_x\mathbf{i} + B_y\mathbf{j} + B_z\mathbf{k}$.]

3. (II) Find the angle θ between two vectors $\mathbf{A}$ and $\mathbf{B}$, for the particular case when $|\mathbf{A} \times \mathbf{B}| = \mathbf{A} \cdot \mathbf{B}$.

4. (II) The time-dependent position of a point object that moves counterclockwise along the circumference of a circle (radius R) in the x-y plane with constant speed v is given by
$$\mathbf{r} = R\cos(\omega t)\mathbf{i} + R\sin(\omega t)\mathbf{j}$$
where the constant $\omega = v/R$. Determine the velocity $\mathbf{v}$ and angular velocity $\boldsymbol{\omega}$ of this object. Then show that these three vectors obey the relation $\mathbf{v} = \boldsymbol{\omega} \times \mathbf{r}$.

Section 11–3

5. (II) The time-dependent position of a particle with mass m traveling on a helical path (see Fig. 11–1) is given by
$$\mathbf{r} = R\cos\left(\frac{2\pi z}{p}\right)\mathbf{i} + R\sin\left(\frac{2\pi z}{p}\right)\mathbf{j} + z\mathbf{k}$$
where R and p are the radius and pitch of the helix, respectively. Determine the time-dependent angular momentum l of the particle about the origin, assuming that its velocity in the z-direction is constant (i.e., $dz/dt = v_o$, where v_o is a constant).

Sections 11–4 and 11–5

6. (II) A rigid object has a moment of inertia I about a fixed axis and rotates about that axis with constant angular acceleration α. If ω_o is the object's angular velocity at time $t = 0$, determine an expression for the object's time-dependent angular momentum $L(t)$. Then show that $\dfrac{dL}{dt} = \sum \tau$, where $\sum \tau$ is the net external torque on the object.

7. (II) Consider the "tetherball-like" situation viewed from above in Fig. 11–2. A point mass m on a frictionless horizontal surface is attached by a massless string to a cylindrical post of radius R. At time $t = 0$, the mass is given an initial velocity of magnitude v_o directed perpendicularly to the taut string. As the mass moves, the instantaneous velocity v of the mass will always be at a right angle to the string and the string will wrap itself around the post, causing the mass to spiral inward toward the post. At time t, define r to be the distance that the mass is from the center of the post and D to be the length of the portion of the string that is not yet in contact with the post. At time $t = 0$, $r = r_o$ and $D = D_o$.
(a) As the mass moves on the spiral path, is its angular momentum conserved about the center of the post?
(b) As it moves on the spiral path, how much work is done on the mass by the forces acting on it (i.e., tension, weight, and normal force)?
(c) Use your results from part (a) and/or (b) to determine the speed v of the mass when the length of the string is D.

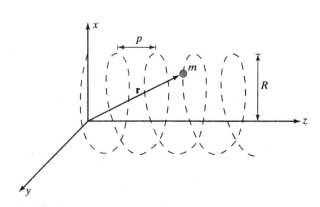

FIGURE 11–1

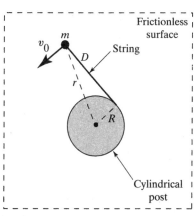

FIGURE 11–2

8. (III) Consider the "tetherball-like" situation viewed from above in Fig. 11–2. A point mass m on a frictionless horizontal surface is attached by a massless string to a cylindrical post of radius R. At time $t = 0$, the mass is given an initial velocity of magnitude v_o directed perpendicularly to the taut string. As the mass moves, its speed will remain v_o because its instantaneous velocity $\mathbf{v}$ is always at a right angle to the string and the string will wrap itself around the post, causing the mass to spiral inward toward the post. At time t, define r to be the distance that the mass is from the center of the post and D to be the length of the portion of the string that is not yet in contact with the post. At time $t = 0$, $r = r_o$ and $D = D_o$.

(a) As it moves along its spiral path, at time t, the mass is momentarily moving on a circle of radius D, which is centered at the point where the string barely fails to make contact with the post. Apply $\sum \boldsymbol{\tau} = d\mathbf{l}/dt$ about the center of the post and show that this principle gives

$$v_o R = -D \frac{dD}{dt}.$$

(b) Integrate both sides of the expression in part (a) to show that the length of the string at time t is given by

$$D = \sqrt{D_o^2 - 2Rv_o t}.$$

(c) Show that the elapsed time T until the mass hits the post is

$$T = \frac{D_o^2}{2Rv_o}.$$

9. (III) Let's explore why tall, narrow vehicles such as SUVs, vans, and buses are prone to "rollover" if driven too quickly around a curve. Consider a vehicle of mass M rounding a left-hand curve of radius R on a flat road as shown from a back view in Fig. 11–3. Assume that the vehicle's speed v_c (directed into the page) is such that it is just on the verge of rollover (i.e., clockwise rotation in the figure). At this critical speed, the tires on the "inside" of the curve are on the verge of leaving the ground and so the friction and normal forces on these two tires are zero. Assume the vehicle is not skidding and that a total static friction force F_{fr} (not necessarily the maximum static friction force available) and a total normal force F_N act on the two "outside" tires. Finally, define the distance between the centers of the vehicle's opposing tires to be the "track width" w and let h be the height of its center of mass (CM) above the ground.

(a) Analyze this situation using Newton's laws for both linear and rotational motion to show that the critical rollover speed v_c is given by

$$v_c = \sqrt{Rg\left(\frac{w}{2h}\right)} = \sqrt{Rg(SSF)}$$

where the Static Stability Factor $SSF = w/2h$ is a measure of how resistant a vehicle is to rollover based on the geometry of its construction. A low SSF value corresponds to a "top-heavy" vehicle.

[Hint: Since the vehicle is only on the verge of rollover at the critical speed, the angular momentum about its CM remains zero at all times.]

(b) SSF values across all vehicle types range from 1.00–1.50. Most passenger cars have values in the 1.25–1.50 range. Higher-riding SUVs, pick-up trucks, and vans usually have values in the 1.00–1.25 range. Assume a highway-driving situation where a vehicle is traveling at a speed of 25 m/s (= 90 km/h) and the driver suddenly is forced to perform a "high-speed collision avoidance maneuver," turning the vehicle onto a circular path of radius R. Show that a passenger car with $SSF = 1.40$ can turn on a much tighter (i.e., smaller radius) circle than an SUV with $SSF = 1.05$ without rolling over.

10. (III) Consider an extended object of total mass M that rotates with angular velocity $\boldsymbol{\omega}$ while its center of mass (CM) undergoes translational motion with velocity $\mathbf{v}_{CM}$. A general theorem is derived in Section 10–11 of the text that shows this object's kinetic energy is given by

$$K = \frac{1}{2} M v_{CM}^2 + \frac{1}{2} I_{CM} \omega^2$$

where I_{CM} is the object's moment of inertia about its CM.

(a) The total angular momentum $\mathbf{L}$ of this object about the origin O is defined as

$$\mathbf{L} = \sum \mathbf{r}_i \times \mathbf{p}_i = \sum \mathbf{r}_i \times (m_i \mathbf{v}_i).$$

Starting with this relation and using the derivation given in Section 10–11 for total kinetic energy as a guide, show that

$$\mathbf{L} = M \mathbf{r}_{CM} \times \mathbf{v}_{CM} + \sum m_i \mathbf{r}_i^* \times \mathbf{v}_i^*.$$

For each particle, $\mathbf{r}_i^*$ is the particle's position vector relative to the CM and $\mathbf{v}_i^*$ is its velocity about the CM.

[Hint: Because of the cross product's distributive property, $\sum (m_i \mathbf{r}_i^* \times \mathbf{A}) = \left(\sum m_i \mathbf{r}_i^*\right) \times \mathbf{A}$, where $\mathbf{A}$ is some vector.]

(b) Consider now a rigid (i.e., nondeformable) object, which rotates about its CM. Then, for each particle m_i within this object, the particle's velocity $\mathbf{v}_i^*$ will be perpendicular to its position $\mathbf{r}_i^*$. For this situation, show that

$$\mathbf{L} = M \mathbf{r}_{CM} \times \mathbf{v}_{CM} + I_{CM} \boldsymbol{\omega}.$$

The total angular momentum about the origin O is the sum of the angular momentum of the CM about O plus the object's "spin" about its own CM.

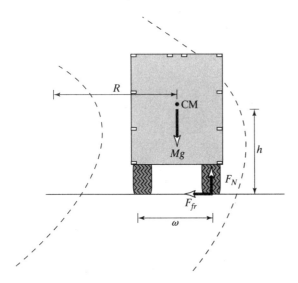

FIGURE 11–3

Section 11–6

11. (II) A point-like object of mass m rotates on the circumference of a circle (radius R) with angular velocity $\boldsymbol{\omega} = \omega \mathbf{k}$, where ω is a constant. The circle lies in the plane defined by $z = z_o$, which is parallel to the x-y plane (see Fig. 11–4). Assume that at time $t = 0$ the object is located at $\mathbf{r} = R\mathbf{i} + 0\mathbf{j} + z_o\mathbf{k}$.
(a) Determine the expression for the object's time-dependent angular momentum $\mathbf{L}$ about the origin.
(b) Find the time-independent angle θ between $\mathbf{L}$ and $\boldsymbol{\omega}$.
(c) For what value of z_o will $\mathbf{L}$ and $\boldsymbol{\omega}$ be parallel?

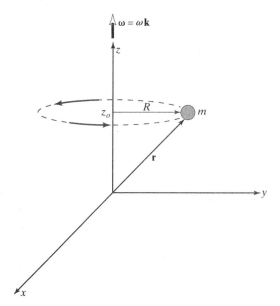

FIGURE 11–4

Section 11–7

12. (II) Reconsider the situation of Example 10–24 in the text, where a bowler releases a bowling ball of mass M and radius R with a linear speed v_o, but having no spin. The ball then slides down a bowling alley and, after a time t_1, begins to roll without slipping. Previously, this situation was analyzed using Newton's method (consideration of forces and torques, followed by use of the kinematic equations); let's now do the analysis via conservation laws.
(a) Defining the origin O to be the point on the alley at which the ball is released from the bowler's hand, briefly explain why no net torque acts on the ball about point O as it moves down the alley.
(b) Since no net torque acts on the ball about O, the ball's total angular momentum $\mathbf{L}$ is conserved about this point. Use conservation of angular momentum to show that, at the moment when the ball begins to roll without slipping, the linear speed of its CM is given by $v_{\text{CM}} = \dfrac{5}{7} v_o$.

[*Hint*: The ball possesses two types of angular momentum: the first due to the linear speed v_{CM} of its CM relative to point O and the second due to the spin at angular velocity ω about its own CM. The ball's total $\mathbf{L}$ about O is the sum of these two angular momenta.]

(c) During the time interval when the ball is sliding, $v_{\text{CM}} = v_o - \mu_k g t$. Use the result of part (b) to show $t_1 = \dfrac{2v_o}{7 \mu_k g}$.

13. (II) Tom, a child of mass $m = 30$ kg, stands on the edge of a stationary disk-shaped merry-go-round of mass $M = 100$ kg and radius $R = 2.0$ m. Tom's friend Mary (who is on the ground) throws a ball of mass $m = 1.0$ kg to Tom, who catches it. Just before the ball is caught, it has a horizontal velocity $v = 12$ m/s, directed along a line tangent to the outer edge of the merry-go-round, as shown in Fig. 11–5.

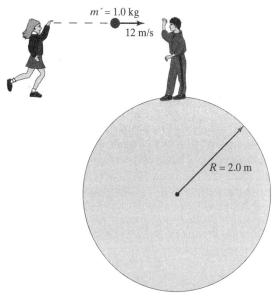

FIGURE 11–5

(a) What is the angular speed ω of the merry-go-round just after Tom catches the ball? Assume that the merry-go-round rotates without friction about a vertical axis at its center.
(b) What percentage of the ball's initial kinetic energy is transformed into rotational kinetic energy in this situation?

14. (II) On a level billiards table, a cue ball of radius R is struck so that it leaves the cue stick (at point O on the table) with a center-of-mass (CM) speed v_o and a "reverse" spin of angular speed ω_o (see Fig. 11–6). A kinetic friction force then acts on the ball as it initially skids across the table.

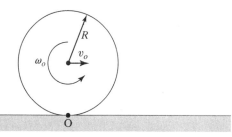

FIGURE 11–6

(a) Explain (briefly) why the ball's angular momentum is conserved about point O as the ball skids across the table.
(b) Using conservation of angular momentum, find the critical angular speed ω_c such that, if $\omega_o = \omega_c$, kinetic friction will bring the ball to a complete (as opposed to momentary) stop.
(c) If ω_o is 10% smaller than ω_c (i.e., $\omega_o = 0.90\omega_c$) determine the ball's CM velocity v_{CM} when it starts to roll without slipping.

(d) If ω_o is 10% larger than ω_c (i.e., $\omega_o = 1.10\omega_c$) determine the ball's CM velocity v_{CM} when it starts to roll without slipping.

[*Hint*: The ball possesses two types of angular momentum: the first due to the linear speed v_{CM} of its CM relative to point O and the second due to the spin at angular velocity ω about its own CM. The ball's total **L** about O is the sum of these two angular momenta.]

15. (III) To get the speed they need to do aerial tricks, skateboard riders use a technique called "pumping," the physical basis of which we will investigate in this problem. Consider a skateboarder riding on a halfpipe, a U-shaped ramp with a horizontal middle section sandwiched between two upwardly curved ends. Each of the two ends is circularly shaped with radius R (see Fig. 11–7a). To pump in a halfpipe, a rider drops into a crouch while crossing the flat middle section. Then, upon arrival at the beginning of a curved end section (called a "transition"), he quickly stands up, thereby raising his center of mass (CM). To demonstrate that the rider's mechanical (i.e., kinetic plus potential) energy increases via pumping, we take the following model of an "instantaneous pump." Fig. 11–7b shows the rider of mass m just upon entering the transition in a crouched position with CM at height x_1 above the ground and moving at speed v_1. Then, over a negligibly small time interval, the rider straightens his legs, so that his CM is now at height x_2 above the ground and moving at speed v_2 as in Fig. 11–7c.
    (a) Let O be the center of the circle defined by the halfpipe's curved end. Explain why, when at the transition, no net torque acts on the rider about point O.
    (b) Since no net torque acts on the rider at the transition, his angular momentum about point O will be conserved during the instantaneous pump. Applying angular momentum conservation between the initial and final situations pictured in Figs. 11–7b and 11–7c, respectively, show that
    $$v_2 = \frac{R - x_1}{R - x_2} v_1.$$
    (c) Show that the difference between the rider's mechanical energies E_1 and E_2 in Figs. 11–7b and 11–7c, respectively, which equals the work W performed during the pump, is given by
    $$W = E_2 - E_1 = \frac{1}{2} m v_1^2 \left[\left(\frac{R - x_1}{R - x_2} \right)^2 - 1 \right] + mg(x_2 - x_1).$$
    (d) Explain why the expression in part (c) implies that $W > 0$; i.e., the rider's mechanical energy increases via the work he does during the pump.

16. (III) A skateboard rider is riding on a halfpipe as described in Problem 15. Assume that the radius of the halfpipe's curved ends is $R = 2.0$ m and that, when standing erect on the skateboard, the rider's center of mass (CM) is $x_2 = 1.0$ m above the ground. Also assume that no friction forces act. With these assumptions, let's determine "how much air the rider can catch" (i.e., how far he can rise above the ground) using the pumping technique (see Fig. 11–8).
    (a) Assume that the rider starts at rest at the top of the halfpipe's left-hand curved end with his CM at height h_o above the ground. He then rides the skateboard down to the flat midsection, where, while standing erect on the board, his CM has a horizontal speed of v_1. Show that
    $$v_1 = \sqrt{2g(h_o - x_2)}.$$
    (b) While standing erect and traveling with horizontal speed v_1 on the midsection, the rider then crouches, lowering his CM from a height x_2 to x_1 above the ground. Using the fact that no net force acts on the rider while on the flat midsection, explain why the rider's horizontal speed will still be v_1 after crouching.
    (c) When the crouched rider with speed v_1 arrives at the right-hand transition, he performs an "instantaneous pump," increasing his mechanical energy E by the amount W given by (see Problem 15)
    $$W = E_2 - E_1 = \frac{1}{2} m v_1^2 \left[\left(\frac{R - x_1}{R - x_2} \right)^2 - 1 \right] + mg(x_2 - x_1).$$
    Using this expression, show that the maximum height h he will attain after being launched from the right-hand curved end is given by
    $$h = h_o + (x_2 - x_1) + (h_o - x_2) \left[\left(\frac{R - x_1}{R - x_2} \right)^2 - 1 \right].$$
    (d) Taking $h_o = R + x_2 = 3.0$ m and $x_1 = 0.80$ m, determine h. [*Note*: Pumping over a finite time interval and air resistance will diminish this height somewhat.]

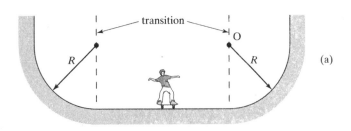

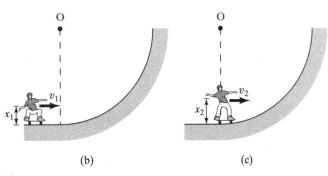

(b)   (c)

FIGURE 11–7

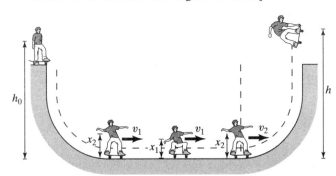

FIGURE 11–8

Section 11–8

17. (II) A spherical asteroid with radius $R = 100$ m and mass $M = 1.0 \times 10^{10}$ kg rotates about an axis at four revolutions per day. A "tug" spaceship attaches itself to the asteroid's south pole (as defined by the axis of rotation) and fires its engine, applying a force F tangentially to the asteroid's surface always directed as shown in Fig. 11–9. If $F = 2.5$ N, how long will it take the tug to rotate the asteroid's axis of rotation through an angle of 10° by this method? [Researchers have proposed this scenario as a viable method for manipulating a rotating asteroid.]

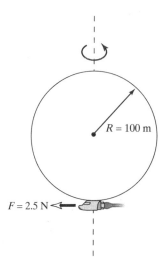

FIGURE 11–9

18. (II) A popular lecture demonstration illustrating precession uses a 65-cm diameter bicycle wheel (mass M) that can rotate nearly friction-free on its axle; 20-cm long wooden handles are attached straight outward from each end of the axle. The instructor then ties a rope to a small hook on the end of one of the handles, spins the bicycle wheel with a flick of the hand, and then releases the wheel while supporting the rope vertically as shown in Fig. 11–10. The wheel then (rather surprisingly) precesses about the vertical axis defined by the rope (instead of falling to the ground as most onlookers might expect).
(*a*) Model the wheel as a hoop. Assume that the distance from the wheel's CM to the rope is 20 cm, that the instructor causes the wheel to rotate at a rate of 2.0 rev/s, and that the wheel's axle is completely horizontal during the precession. Determine the wheel's rate of precession in rev/s.

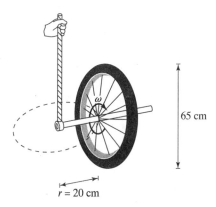

FIGURE 11–10

(*b*) If the wheel spins in the direction shown in Fig. 11–10, in which direction (clockwise or counterclockwise, when viewed from above) does it precess about the rope?

19. (II) In the Nuclear Magnetic Resonance (NMR) experimental technique and the medical-related Magnetic Resonance Imaging (MRI) method, short pulses of radio waves are used to "flip the spin" of protons within a sample. Consider the following mechanical analogy involving a spin flip. Imagine a sphere (mass M, radius R) with a very small hole drilled through it from its north to south pole. An axle threads through this hole and the sphere is initially rotating without friction about the axle with angular velocity $\boldsymbol{\omega} = \omega \mathbf{k}$. The axle is initially aligned with the z-axis, but it is free to move about in the x-y-z coordinate system. At time $t = 0$, a pair of equal-magnitude forces $\mathbf{F}$ is then applied to the axle at the north and south poles, each directed as shown in Fig. 11–11a. The magnitude F of each force, as well as its direction relative to the axle at its respective pole, remains constant until $t = T$, at which time the location of the north and south poles have reversed (compared with their initial orientation) as shown in Fig. 11–11b.
(*a*) During the time interval T, is the sphere's angular momentum vector rotating about the x-, y-, or z-axis?
(*b*) Determine T, the time interval required to "flip" the spin (i.e., angular momentum vector) of the sphere.

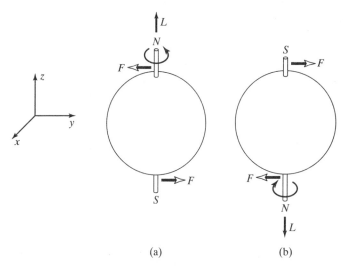

FIGURE 11–11

20. (III) A spinning top's angular momentum $\mathbf{L}$, of constant magnitude L_o and fixed angle ϕ with respect to the z-axis, precesses about the z-axis with constant angular velocity Ω so that at time t

$$\mathbf{L} = (L_o \sin \phi \cos \Omega t)\mathbf{i} + (L_o \sin \phi \sin \Omega t)\mathbf{j} + (L_o \cos \phi)\mathbf{k}.$$

(*a*) Use the time-dependent expression for $\mathbf{L}$ to show that the magnitude of this vector is indeed equal to the constant L_o at all times.
(*b*) Find an expression for $d\mathbf{L}$ and then prove $d\mathbf{L}$ is perpendicular to $\mathbf{L}$ at all times.

[Hint: $dL_x = \dfrac{dL_x}{dt} dt$]

(*c*) Determine dL, the magnitude of $d\mathbf{L}$. Then show that $\dfrac{dL}{L_o \sin \phi}$ equals Ωdt as suggested by Eq. 11–11a in the text.

Section 11-9

21. (II) Ice skaters commonly perform jumps in which they must move their outstretched hands toward their bodies' centers while rotating about a vertical axis. Skaters can produce such high angular speed at take-off, however, that they may not be physically strong enough to perform this maneuver. If a 50-kg skater leaves the ice with rotational frequency $f = 1.2$ rev/s with a hand outstretched 60 cm from her body's center, determine the magnitude of the pseudoforce F that the skater will experience (at $r = 60$ cm) as she begins to move her hand inward. Assume that 1.0% of the body's mass is contained within the skater's hand. Compare (as a ratio) F with the weight of the skater's hand.

22. (III) Consider the pumping technique used by skateboard riders to increase their mechanical energy on a halfpipe. The underlying physics of this technique can be explained via conservation of angular momentum (see Problem 15). In this problem, we will show that the concept of a pseudoforce can also be used to provide an alternate explanation. Consider the transition from a halfpipe's flat midsection to its circularly curved end of radius R (Fig. 11–7a). Assume that a crouched rider enters the transition with her center of mass (CM) a height x_1 above the ground and traveling at speed v_1. At the transition, she performs an "instantaneous pump"; i.e., she quickly stands erect, raising her CM to a height x_2 above the ground.

(a) At the transition, the rider is moving in a circular path and so is in an accelerated frame of reference. Thus, as she stands erect, her leg muscles generate a force F, which does work against her weight force as well as the centrifugal pseudoforce. Assuming that, during the pump, her CM always travels at close to the speed v_1 and follows a trajectory with an approximate radius R (an assumption that will be valid if the rider's height is small in comparison with the halfpipe's radius of curvature), show that the work the rider performs during the pump can be approximated as

$$W \approx \left[\frac{mv_1^2}{R} + mg\right](x_2 - x_1).$$

From this result, explain why pumping is most effective when performed at the halfpipe's transition (i.e., at the bottom of the curved end).

(b) In Problem 15, conservation laws were used to determine that the work done during an instantaneous pump is

$$W = \frac{1}{2}mv_1^2\left[\left(\frac{R-x_1}{R-x_2}\right)^2 - 1\right] + mg(x_2 - x_1).$$

Show that this expression reduces to that found in part (a) using pseudoforces when the rider's height is small in comparison with the halfpipe's radius of curvature; i.e., in the limit $x_1 \ll R$ and $x_2 \ll R$.

[*Hint*: the binomial expansion gives $(1 - z)^n \approx 1 - nz$, if $z \ll 1$.]

General Problems

23. (II) The time-dependent position of a point mass m, which is uniformly accelerated as it moves counterclockwise along the circumference of a circle (radius R), is given by

$$\mathbf{r} = R\cos(\theta)\mathbf{i} + R\sin(\theta)\mathbf{j}$$

with $\theta = \omega_o t + \frac{1}{2}\alpha t^2$, where the constants ω_o and α are the initial angular velocity and angular acceleration, respectively. Determine the object's tangential acceleration **a**. Then use $\boldsymbol{\tau} = \mathbf{r} \times \mathbf{F}$ to determine the torque acting on the object and show that your result is equivalent to the torque obtained by applying $\boldsymbol{\tau} = I\boldsymbol{\alpha}$ to this situation.

24. (II) A point projectile with mass m is launched from the ground level and follows the trajectory given by

$$\mathbf{r} = (v_{xo}t)\mathbf{i} + \left(v_{yo}t - \frac{1}{2}gt^2\right)\mathbf{j}$$

where v_{xo} and v_{yo} are the initial velocities in the x- and y-direction, respectively, and g is the acceleration due to gravity. The launch position is defined as the origin. Determine the torque acting on the projectile about the origin using
(a) $\boldsymbol{\tau} = \mathbf{r} \times \mathbf{F}$.
(b) $\boldsymbol{\tau} = \dfrac{d\mathbf{l}}{dt}$.

25. (II) Consider a planet (mass m) orbiting due to the gravitational influence of a stationary central star at the origin, where **r** and **v** are the planet's time-dependent position and velocity, respectively (Fig. 11–12). Regarding the gravitational force on the planet, assume only that $\mathbf{F} = -f(r)\hat{\mathbf{r}}$, where f can be any function of r (the magnitude of **r**) and $\hat{\mathbf{r}}$ is the unit vector pointing outward from the origin along the direction of **r**. Such a force is called a "central force." Prove that the angular momentum of the planet will be conserved in this situation
(a) by starting with $\mathbf{L} = \mathbf{r} \times \mathbf{p}$ and explicitly calculating $d\mathbf{L}/dt$.
(b) by calculating the torque on the planet and applying $d\mathbf{L}/dt = \sum \boldsymbol{\tau}$ (Eq. 11–7 in the text).

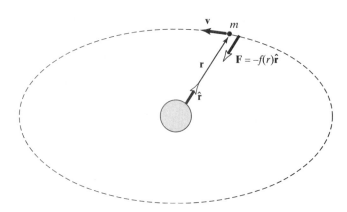

FIGURE 11–12

26. (II) While most of our solar system's mass is contained in the Sun, the planets possess almost all of the solar system's angular momentum. This observation plays a key role in theories attempting to explain the formation of our solar system. Estimate the fraction of the solar system's total angular momentum that is possessed by planets using the following simplified model, which includes only the large outer planets with the most significant angular momentum: The central Sun (mass 1.99×10^{30} kg, radius 6.96×10^8 m) spins about its axis once every 25 days and the planets Jupiter, Saturn, Uranus, and Neptune move in circular orbits around the Sun with orbital data given in Table 11-1. You may assume each planet does not spin about its own axis, as this form of angular momentum is negligible compared with the planet's orbital angular momentum.

TABLE 11–1 Large Planet Orbital Data

Planet	Mean Distance from Sun ($\times 10^6$ km)	Orbital Period (Earth Years)	Mass ($\times 10^{25}$ kg)
Jupiter	778	11.9	190
Saturn	1427	29.5	56.8
Uranus	2870	84.0	8.68
Neptune	4500	165	10.2

27. (II) Let's determine the relative sizes of the various angular momenta present in the Earth-Moon system.
   (a) Earth and the Moon both orbit about the Earth-Moon system's center of mass (CM) once every 27.3 d. Define an axis passing through the center of Earth and the Moon with the system's CM at the origin and determine the locations of Earth and the Moon (each considered as point masses at their spherical centers) on this axis. Then, assuming circular orbits, calculate the "orbital" angular momentum L_o about the system's CM for Earth and also for the Moon.
   (b) Earth spins about its axis once every 24 h, while the Moon spins about its axis once every 27.3 d. Assuming each object is a uniform sphere, determine the "spin" angular momentum L_s for Earth and also for the Moon.
   (c) Determine which object in the Earth-Moon system possesses the most angular momentum by calculating the ratio $\dfrac{(L_o + L_s)_{Earth}}{(L_o + L_s)_{Moon}}$.
   (d) Determine the relative amount of angular vs. orbital angular momentum possessed by each object by calculating the ratios $\left(\dfrac{L_o}{L_s}\right)_{Earth}$ and $\left(\dfrac{L_o}{L_s}\right)_{Moon}$.

28. (II) (a) In one of the seminal ideas of quantum theory, Max Planck postulated that the energy E of a mass vibrating on the end of a spring is equal to an integer multiple of a basic amount of energy given by the product hf, where f is the vibrational frequency and h is a fundamental constant of nature now called Planck's constant (see Fundamental Constants table at front of book). That is, $E = n(hf)$, where n is an integer. From the form of Planck's equation for E, determine the units for h and then show that these units are equivalent to the units of angular momentum. (b) Building on Planck's idea, Niels Bohr postulated that, in the hydrogen atom's lowest-energy ("ground") state, an electron of mass m revolves at constant speed v in a circular orbit of radius R about a stationary proton such that the electron's angular momentum equals $h/2\pi$ (i.e., Planck's constant divided by 2π). From experiments at that time, it was known that the diameter of a hydrogen atom in its ground state is about 1×10^{-10} m and that the mass of the electron is 9×10^{-31} kg. Given this experimental information, how fast does the electron move in its ground-state orbit in Bohr's model of the hydrogen atom?

29. (II) In the classic example of angular momentum conservation, a slowly revolving ice skater with outstretched arms, pulls her arms in close to her body, resulting in a fast rotation. In real life, however, one might be concerned that the torque caused by friction as the skate blades move on the ice should negate the use of angular momentum conservation in this situation. Let's explore this point further.
   (a) Assume that a skater initially has her arms outstretched so that her moment of inertia about a vertical axis is 2.8 kg·m^2 and that she is revolving about this axis at a rate of 1.0 rev every 2.0 s. Determine the magnitude L_o of the skater's initial angular momentum.
   (b) The 50-kg skater then pulls her arms inward over a time interval of 1.0 s during which the blades of her skates travel in a circular path of radius $R = 8.0$ cm about the vertical axis. Assuming the coefficient of kinetic friction between the blades and ice is $\mu_k = 0.010$, determine the change in the skater's angular momentum ΔL during the 1.0-s time interval due to the frictional torque.
   (c) Calculate the percent ratio $(\Delta L/L_o) \times 100\%$ as a measure of the loss of angular momentum during the 1.0-s time interval. Based on your result for this ratio, is it valid to apply conservation of angular momentum to the revolving skater?

30. (III) Consider the following model for the Earth-Moon system: Earth (mass M_E, radius R_E), rotates about its own axis with rotational period $T_E = 24$ h and is centrally located, while the Moon (mass M_M) orbits at orbital radius r_M with orbital period $T_M = 27.3$ d (sidereal month). The motions associated with the spin of the Moon about its axis and the orbital motion of Earth about the Earth-Moon system's CM are ignored in this model because these produce negligible angular momenta.
   (a) A tidal torque on Earth due to the Moon's gravitational attraction is slowing down Earth's rotation about its axis, changing the rotational period at a rate of dT_E/dt. Given this "day-lengthening" rate, show that Earth's angular momentum L_E changes according to

$$\frac{dL_E}{dt} = -\frac{2}{5}\frac{2\pi M_E R_E^2}{T_E^2}\frac{dT_E}{dt}.$$

Thus, as its day lengthens (i.e., $dT_E/dt > 0$), Earth loses angular momentum.
   (b) Prove that the Moon can gain angular momentum by increasing its orbital radius. In particular, show

$$\frac{dL_M}{dt} = \frac{1}{2}\frac{2\pi M_M r_M}{T_M}\frac{dr_M}{dt}.$$

[Hint: r_M and T_M are related by Kepler's third law (see Eq. 6–6 in the text).]
   (c) Assuming no external torques act, angular momentum must be conserved by the Earth-Moon system. Show that this principle implies

$$\frac{dr_M}{dt} = \frac{4}{5}\left(\frac{M_E}{M_M}\right)\left(\frac{R_E^2}{r_M}\right)\left(\frac{T_M}{T_E^2}\right)\frac{dT_E}{dt}.$$

   (d) By how much does the Earth-Moon distance increase per year, if Earth's rotational period lengthens by $dT_E/dt = 20\ \mu s/y$?

CHAPTER 12

Static Equilibrium; Elasticity and Fracture

Sections 12–1 to 12–3

1. (II) The force required to pull the cork out of the top of a wine bottle is in the range of 200 N to 400 N. A common bottle opener is shown Fig. 12–1. What range of forces F is required of the operator when opening a wine bottle with this device?

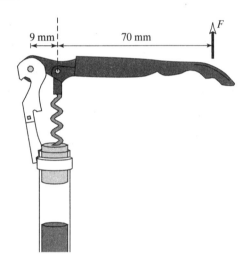

FIGURE 12–1

2. (II) A man holds a 4.0-m uniform wooden board (weight w) so that it is horizontally oriented. The man grasps the board such that the magnitude of forces F_R and F_L due to his right and left hands are a distance x and $(x + 60\ \text{cm})$ from one end of the board, respectively (see Fig. 12–2).
   (a) If $x = 20\ \text{cm}$, determine F_R and F_L (as multiples of w), as well as the direction of the right- and left-hand forces.
   (b) For what x is $F_R = 0\ \text{N}$?
   (c) For what x is $F_R = F_L$? What are the magnitudes and directions of these forces in this case?

3. (II) When placed on a horizontal floor, the center of gravity (CG) of a couch with weight w is a height $h = 40\ \text{cm}$ above the midpoint between its legs, which are separated by a distance $L = 2.0\ \text{m}$ (see Fig. 12–3a). It's moving day and two friends are carrying this couch up a flight of stairs as shown in Fig. 12–3b such that the couch is inclined at an angle $\theta = 35°$ relative to the horizontal and the forces F_B and F_T applied to the couch by the bottom-side and top-side friend, respectively, are both vertically oriented. Determine F_B and F_T (as fractions of w). Which friend must supply the greatest force?

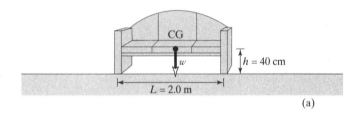

(a)

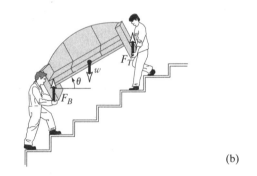

(b)

FIGURE 12–3

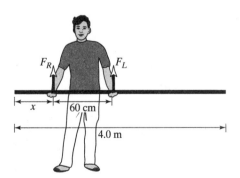

FIGURE 12–2

4. (II) During a break from her physics homework in the library, Katie finds she can balance her pencil horizontally, if her finger is placed under the pencil at a distance of 11 cm from its tip. Amazingly, she then stands the pencil up vertically on her library desk, perfectly balancing it on its blunted tip. If the pencil's "tip" is a flattened surface of diameter 0.50 mm and the CG of the pencil is on the axis defined by the center of this tip, at what angle with respect to the vertical would a breeze in the library have to displace the pencil in order to topple it?

5. (II) Erik's dog, Buddy, refuses to get on the bathroom scale to get weighed, but he loves to sit next to Erik on the living room couch. So, Erik places a bathroom scale under each end of the couch, and when he and Buddy are sitting as shown in Fig. 12–4, Erik notes that the scale nearest him reads 1060 N and that the other scale reads 560 N. Erik weighs 820 N. If he assumes the couch to be of uniform construction, what does Erik calculate Buddy's weight to be?

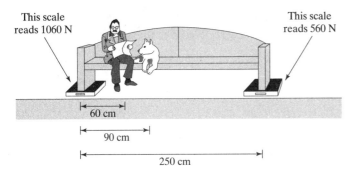

FIGURE 12–4

6. (II) A 500-N person walks up a 120-N uniform ladder of length L. The ladder's upper and lower ends rest on frictionless surfaces (i.e., surfaces that can only exert a normal force) as shown in Fig. 12–5. The lower end of the ladder is fastened to the wall by a horizontal rope that can support a tension up to 300 N before breaking. Find the maximum distance x that the person can walk up the ladder before the rope breaks, expressing your answer as a fraction of the length L.

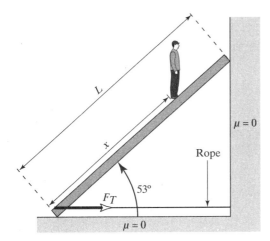

FIGURE 12–5

7. (II) A uniform cylinder (weight w, radius R) is placed against a step of height $R/2$ and a force is applied to the cylinder's center at an angle θ above the horizontal as shown in Fig. 12–6.
   (a) For a given θ, determine the force magnitude F required to barely lift the cylinder off the ground.
   (b) To lift the cylinder off the ground with the least value of F, what angle θ should be used?

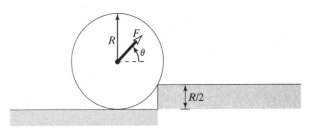

FIGURE 12–6

(c) How does F at the optimal angle that you found in part (b) compare with the required F at $\theta = 90°$ (i.e., when simply lifting the cylinder straight upward)?

8. (II) In a mountain-climbing technique called the "Tyrolean traverse," a rope is anchored on both ends (to rocks, strong trees, etc.) across a deep chasm and then a climber traverses the rope while attached by a sling as shown in Fig. 12–7. Because this technique can generate tremendous forces that stress the rope and anchors, a basic understanding of physics can be helpful in implementing it safely. A typical climbing rope can provide a tension force of up to 29 kN before breaking and, for safe use, a "safety factor" of 10 is usually recommended (i.e., the rope should only be required to supply a tension force of up to 2.9 kN). As we will demonstrate, in setting up the Tyrolean traverse, the length of rope used must allow for some "sag" in order for it to remain in its recommended safety range. Consider a 90-kg climber at the center of a Tyrolean traverse, which spans a 30-m chasm.
   (a) In order for the rope to be within its recommended safety range, find the distance x that the rope sags in this situation.
   (b) If the Tyrolean traverse is incorrectly set up so that the rope sags by only one-fourth the distance found in part (a), determine the tension force in the rope. Will the rope break?

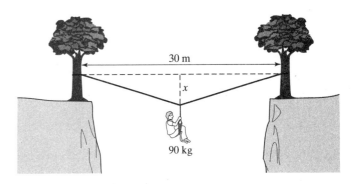

FIGURE 12–7

9. (III) The bottom of a vertically oriented post of length L rests on the horizontal ground. A taut cable with tension F_{T_1} at an angle of 30° with respect to the vertical is attached to the post's top and a horizontal cable with tension F_{T_2} is attached to the post at a height αL above the ground, where $0.0 \leq \alpha \leq 1.0$ (see Fig. 12–8). If the coefficient of static friction between the post and the ground is $\mu_s = 0.80$, determine the range of α such that the post is maintained in static equilibrium. Ignore the weight of the post.

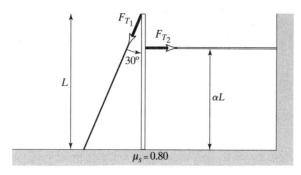

FIGURE 12–8

10. (II) Arnold and Sylvester are having a contest to see who can support the largest mass m in his hand when his forearm is horizontal as shown in Fig. 12–9. Assume that the maximum force F_M each man's bicep muscle can exert is identical and also that the dimensions of both men's arms are exactly the same except for the fact that the bicep muscle's point of insertion on the forearm is a distance (measured from the elbow) of $d_S = 5.0$ cm and $d_A = 5.5$ cm for Sylvester and Arnold, respectively. If Sylvester is able to support a mass $m_S = 50$ kg in this contest, what mass m_A will Arnold be able to support? Neglect the weight of the forearm and hand.

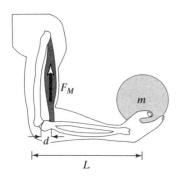

FIGURE 12–9

11. (II) Consider a person of weight w_P lifting an object of weight w (such as a small child), while bent forward at the waist as shown in Fig. 12–10a. A simple model of the forces acting on the person's back in this situation is given in Fig. 12–10b. Here the spine of length L is assumed to pivot about the fifth lumbar vertebra (point A). The weight of the person's upper body is taken to be two-thirds of the body's total weight centered a distance $0.60L$ from A; the back is supported by force F_M due to the erector spinalis muscle, which is assumed to be attached two-thirds of the way up the spine acting at a 12° angle relative to it. F_x and F_y are the compressional and shear forces on the lower back due to the sacrum (attached to the pelvis). The force w is assumed to act at the top of the spine. Taking typical values of $w_P = 700$ N and $w = 200$ N, determine F_M and F_x. Compare F_M and F_x with the person's weight (i.e., determine F_M/w_P and F_x/w_P) to demonstrate why lifting in this manner can cause a lower back injury.

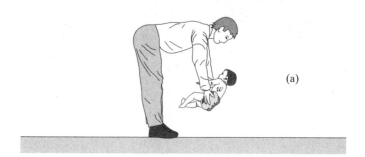

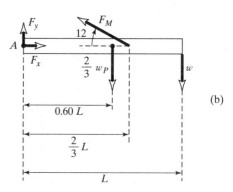

FIGURE 12–10

12. (II) Consider the following method to determine the location of the center of gravity (CG) in a person's lower leg. A horizontal board is supported by a fulcrum and a scale (calibrated in kg) separated by a distance L (Fig. 12–11). A person lies prone on the board so that the top of her head is aligned with the fulcrum. In this position, the scale reading S is taken. The person then bends her knee so that the lower leg is straight upward and, while in this position, the scale reading S' is taken. Assume that the lower leg has a mass m and that, in the prone position, the lower leg's CG and the knee are horizontal

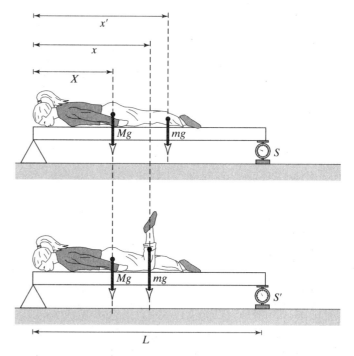

FIGURE 12–11

distances x and x' from the top of the head, respectively. Also define M to be the total mass of the board and the person's body, excluding the lower leg, and assume the CG of this distributed mass is a horizontal distance X from the fulcrum.
(a) Show that the lower leg's CG is located a distance d below the knee where $d = L(S - S')/m$.
(b) Assume, based on volume measurements, it is known that m is equivalent to 5.7% of the total body's mass. If $L = 250$ cm and, for a 55-kg person, S and S' are found to differ by 0.23 kg, determine d.

13. (II) Consider the possible lever systems A and B shown in Fig. 12–12. In both of these systems, a force F_P is applied to the (massless) lever a distance R from a fulcrum in order to produce a force F_L on a "load," which is a distance r from the fulcrum. Assume $r < R$ for system A.
(a) The amplification of force by a lever system is described by its mechanical advantage defined as $M = F_L/F_P$. Show that $M_A > 1$, while $M_B < 1$. Which system is optimized to amplify the applied force F_P?
(b) Let v_P and v_L be the linear velocities of the lever (directed perpendicular to itself) at the location of the forces F_P and F_L, respectively, when the lever rotates about the fulcrum. Defining velocity amplification as $V = v_L/v_P$, show that $V_A < 1$, while $V_B > 1$. Which system is optimized to amplify the linear velocity v_P?
(c) The human forearm can be modeled as lever system B, where the bicep muscle applies force F_P a distance $r = 5$ cm from the elbow (fulcrum) in order to exert a force F_L at the location of the hand ($R = 35$ cm). Determine M and V for this system. Given that this system is optimized to amplify velocity, if the bicep muscle's length can contract at a rate up to 0.25 m/s, how fast can the hand be made to move?

14. (III) Consider a person of weight w_P lifting an object of weight w (such as a small child), while bent forward at the waist as shown in Fig. 12–10a. A simple model of the forces acting on the person's back in this situation is given in Fig. 12–10b. Here the spine of length L is assumed to pivot about the fifth lumbar vertebra (point A). The weight of the person's upper body is taken to be two-thirds of the body's total weight centered a distance 0.60L from A; the back is supported by force F_M due to the erector spinalis muscle, which is assumed to be attached two-thirds of the way up the spine acting at a 12° angle relative to it. F_x and F_y are the compressional and shear forces on the lower back due to the sacrum (attached to the pelvis). The force w is assumed to act at the top of the spine.
(a) Show that the compressional force on the spine when lifting in this manner is a linear function of the weight being lifted.
(b) Rewrite your expression for F_x as a function of w in dimensionless form by normalizing these quantities to the person's weight w_P; i.e., rewrite the expression in the form of (F_x/w_P) as a function of (w/w_P).
(c) If the lifted object in Fig. 12–10a has a weight equal to one-half the person's weight, determine F_x/w_P. Your answer should demonstrate why lifting in this manner can easily cause a lower back injury.

15. (III) In Example 12–8 of the text, a uniform ladder leans against a wall, which for simplicity is assumed to be frictionless. Under this assumption, the ladder's stability is maintained solely by a frictional force due to the ground and it is shown that the ladder will slip when a 58-kg painter's vertical height exceeds 2.2 m. More realistically, one can assume that, in addition to the ground, the wall is frictional, where the coefficients of static friction between the ladder-ground and ladder-wall are μ_g and μ_w, respectively. With this assumption, the ladder will be on the verge of slipping when both of these frictional forces take on their maximum values. Re-analyze the situation of Example 12–8 under the assumption of a "frictional wall" and determine the maximum vertical height to which a 58-kg painter can climb without the ladder slipping. Take $\mu_g = \mu_w = 0.40$.

16. (III) A uniform ladder of mass m and length L leans against a wall at an angle θ with respect to the ground (Fig. 12–13). The coefficients of static friction between ladder-ground and ladder-wall are μ_g and μ_w, respectively. The ladder will be on the verge of slipping when both the static friction forces due to the ground and due to the wall take on their maximum values.
(a) Show that the ladder will be stable if $\theta \geq \theta_m$, where the minimum angle θ_m is given by $\tan \theta_m = \dfrac{1}{2\mu_g}(1 - \mu_g \mu_w)$.
(b) Commonly, "leaning ladder problems" are analyzed under the seemingly unrealistic assumption that the wall is frictionless (see Example 12–8 in the text). Through the following procedure, investigate the magnitude of error introduced by modeling the wall as frictionless, if in reality it is frictional: Using the relation found in part (a), calculate the true value of θ_m for a frictional wall, taking $\mu_g = \mu_w = 0.40$. Then, determine the approximate value of θ_m for the "frictionless wall" model by taking $\mu_g = 0.40$ and $\mu_w = 0$. Finally, determine the percent deviation of the approximate value of θ_m from its true value.

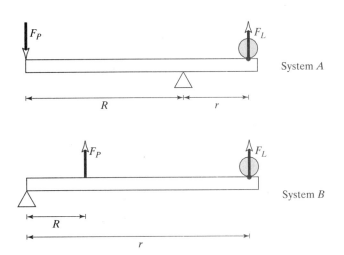

FIGURE 12–12

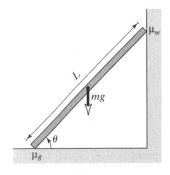

FIGURE 12–13

17. (III) If 35 kg is the maximum mass m that a person can hold in a hand when the arm is positioned with a 100° angle at the elbow as shown in Fig. 12–14 (i.e., 100° between the horizontal forearm and the upper arm), what is the maximum force F_M that the bicep muscle can exert on the forearm? Assume the forearm and hand have a total mass of 2.0 kg with a CG that is 15 cm from the elbow. Also assume that the bicep muscle attaches in a straight line from the top of the 30-cm long upper arm to the forearm at a location 5.0 cm from the elbow.
[*Hint*: The law of cosines and the law of sines can be used to find the angle θ.]

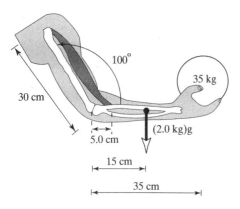

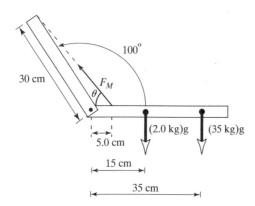

FIGURE 12–14

18. (III) A uniform cylinder (radius R, weight w) rests in the corner between a horizontal floor and vertical wall. The coefficient of static friction between the cylinder and each of these surfaces is μ. Determine the maximum vertical force F that can be applied tangentially to the cylinder as shown in Fig. 12–15 without causing the cylinder to rotate.

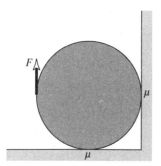

FIGURE 12–15

19. (III) A weightlifter constructs a homemade training device that consists of two uniform rods, each of length L and weight W, hinged together and hung vertically from a hinged ceiling support as shown in Fig. 12–16a. All hinges are well oiled and can be assumed to be frictionless. When the weightlifter pushes on its bottom end, what force (magnitude and direction) is required in order to hold the device at rest with its rods perpendicular to each other and its bottom end directly below the ceiling support (see Fig. 12–16b)?
[*Hint*: It may be easiest to first represent the weightlifter's force by its horizontal and vertical components.]

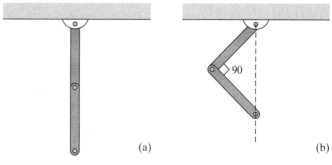

FIGURE 12–16

20. (III) When placed on a horizontal floor, the center of gravity (CG) of a couch with weight w is a height $h = 40$ cm above the midpoint between its legs, which are separated by a distance $L = 2.0$ m (see Fig. 12–3a). It's moving day and two friends are carrying this couch up a flight of stairs as shown in Fig. 12–3b such that the couch is inclined at an angle θ relative to the horizontal and the forces F_B and F_T applied to the couch by the bottom-side and top-side friend, respectively, are both vertically oriented.
(a) Show that the ratio of the forces $R = F_B/F_T$ supplied by the two friends is given by
$$R = \left(1 + \frac{2h}{L}\tan\theta\right) \bigg/ \left(1 - \frac{2h}{L}\tan\theta\right).$$
(b) Sketch R vs. θ over the range $0 \leq \theta \leq 60°$. At what angle is the force due to the bottom-side friend equal to twice the force due to the top-side friend?

21. (III) A uniform sphere (weight w, radius R) is tethered to a wall by a rope (length L) as shown in Fig. 12–17. When the sphere is in static equilibrium, the rope makes an angle θ with respect to the wall and the distance between the point where the rope is tied to the wall and the contact point of the sphere with the wall is h. The coefficient of static friction between the wall and the sphere is μ.
(a) Determine the value of the frictional force on the sphere due to the wall (in terms of any of the relevant quantities L, w, R, θ, and h).
[*Hint*: A wise choice of axis will make this calculation easy.]
(b) Imagine an experiment in which h and θ are varied until the value $h = h_S$ along with an associated $\theta = \theta_S$ are found for which the sphere is on the verge of slipping (at its point of contact with the wall). Derive an expression that could then be used to determine μ using the experimentally determined values h_S and θ_S.

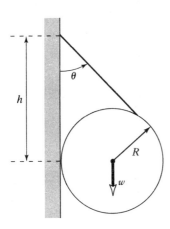

FIGURE 12-17

Section 12-5

22. (II) Over a certain range of applied pressures, a given material may be considered "incompressible." Assume a sample with volume V_o is subjected initially to pressure P_o and then the pressure is increased to $(P_o + \Delta P)$. One might reasonably define the sample as "incompressible" if the resulting change of its volume is less than 0.5%.
    (a) Over what range of pressure change ΔP_W is water (initially at normal atmospheric pressure) incompressible according to this criterion?
    (b) Over what range of pressure change ΔP_A is air (initially at normal atmospheric pressure) incompressible according to this criterion?
    (c) Demonstrate that water is much more incompressible than air by calculating the ratio $\Delta P_W / \Delta P_A$.

23. (II) Let's investigate how a solid generates a normal force. Imagine a rod-shaped piece of steel with radius $R = 15$ cm and length L_o placed upright on a firm surface. A person of mass $m = 70$ kg then stands motionless atop this rod, supported by the normal force it provides.
    (a) The normal force is caused by microscopic distortions of the rod. Determine the percent decrease in the rod's length due to the person.
    (b) X-ray diffraction studies have shown that when a metal is compressed, each atom throughout its bulk moves closer to its neighboring atom in exactly the same fractional amount by which the metal as a whole is compressed. If iron atoms in steel are normally 2.0×10^{-10} m apart, by what distance did this interatomic spacing have to change in order to produce the normal force required to support the 70-kg person? [Neighboring atoms repel each other when pressed closer together; this repulsion is the atomic source of the macroscopic normal force.]

24. (II) A long rigid pipe of constant diameter d, which transports water throughout a house, develops an obstruction at an unknown location, plugging the pipe so that no water can flow. A piston is placed in one end of the pipe and, when used to compress the trapped water, it is found that the water pressure increases by $\Delta P = 3 \times 10^5$ N/m² when the piston is moved a distance 0.5 cm into the pipe. How far from the end of the pipe do you estimate the obstruction to be?

25. (II) Let d be the typical equilibrium distance between atoms in a solid material of thickness L. A simple atomic model for a rectangular slice through this solid of cross-sectional area $A = d^2$ is as follows: $(N + 1)$ atoms in a line are connected by N springs, each with spring constant k. In equilibrium, neighboring atoms are a distance d apart with their associated springs unstretched so that $L = Nd$ (Fig. 12–18). If this slice is then subjected to a tensile stress by applying opposing forces F on its two ends, assume that its total increase in length ΔL is equally divided among all springs so that each spring is stretched by an amount $x = \Delta L/N$. Assume the stress acts only in the solid's elastic region so that $\Delta L \ll L$, allowing L to be approximated as Nd. Use this model to derive an expression for Young's modulus of a solid in terms of its atomic parameters k and d. Using typical values of $d = 3 \times 10^{-10}$ m and $Y = 7 \times 10^{10}$ N/m², estimate the size of an atomic spring constant k.

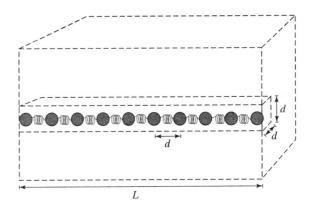

FIGURE 12-18

Section 12-6

26. (II) A nearly complete skeleton of a Brachiosaurus, one of the largest known dinosaurs, is on display in the Humboldt University Natural History Museum in East Berlin. This skeleton suggests that the midsection of a Brachiosaurus's body was about 4 m in diameter and 6 m long, while its neck and tail were both about 1 m in diameter with lengths of 10 m and 7 m, respectively.
    (a) Model the body of a Brachiosaurus as consisting of three cylinders with the given dimensions to estimate its volume. Then, assuming, as is true for modern-day animals, that the density of a dinosaur was about equal to the density of water (1000 kg/m³), estimate the mass of a Brachiosaurus.
    (b) From the skeleton it is found that the circumferences of the Brachiosaurus's humerus (front leg bone) and femur (rear leg bone) are both about 70 cm. Estimate the compressive stress these bones would experience when this animal stood squarely on its four feet. If these bones had comparable strength to the bones in modern-day animals, would they be able to support the Brachiosaurus's weight without breaking? [Researchers, using a more complicated analysis involving shear stress, have also shown that the Brachiosaurus's leg bones were likely strong enough to support walking at speeds up to 1 m/s.]

27. (II) The tallest trees in the world attain a height on the order of 100 meters. Model a tree as a wooden cylinder of cross-sectional area A; then determine the maximum height h for this cylinder such that its weight would cause the compressive stress at its base to exceed the ultimate strength of wood. Assume the density of wood is 700 kg/m³. Based on your result, is the upper limit for tree height determined by the ultimate compressive strength of wood?

General Problems

28. (II) A horizontal rod of length L and negligible weight is rigidly attached to two completely rigid walls a distance L apart. A horizontal force F is then applied at a distance $l = \alpha L$ from the left end of the rod, where $0 < \alpha < 1$, as shown in Fig. 12–19. After a slight deformation of the rod due to the application of F, the rod is maintained in static equilibrium. Determine the equilibrium values of forces F_L and F_R on the rod (in terms of the applied force F and α) due to the left and right walls, respectively. Assume the rod has a cross-sectional area A and Young's modulus Y.

[Hint: Imagine dividing the rod into a left and right portion. When F is applied to the rod, these left and right portions are compressed and expanded by an equal amount Δl, respectively.]

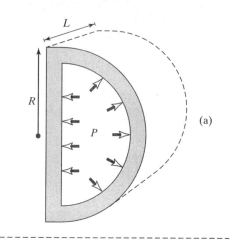

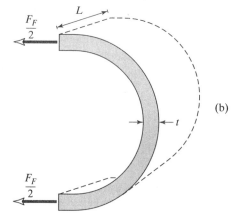

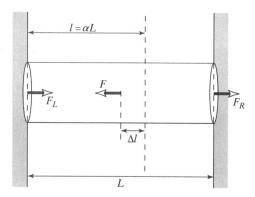

FIGURE 12–19

FIGURE 12–20

29. (III) Let's investigate the stress experienced by a cylindrically shaped vessel containing a fluid that exerts a pressure P (= force per area) on its interior walls.
(a) Consider first a fluid-containing vessel of length L constructed with one flat wall and one cylindrically shaped wall of radius R (Fig. 12–20a). Given that the fluid pressure exerts a force at right angles to the vessel walls at all locations, prove that in static equilibrium the fluid exerts a force of magnitude $F_F = P(2RL)$ on the cylindrical wall, in the direction perpendicular to (and way from) the flat wall.
[Hint: Find the force on the flat wall due to the fluid and then apply the condition for equilibrium.]
(b) In order to counteract the fluid force so that the vessel does not rupture, the flat wall must exert a force of magnitude F_F on the curved wall. Assume that the cylindrical wall has a thickness t and that the force F_F is equally distributed over the top and bottom regions of contact between the flat and the curved wall (i.e., $F = F_F/2$ acts on each region) as shown in Fig. 12–20b. Show that the cylindrical wall then experiences a tensile stress of magnitude PR/t directed along the circumference of the cylindrical wall.
(c) For a completely cylindrical vessel in static equilibrium, the contained fluid exerts the same force F_F found above on each cylindrical half of the vessel. Thus, our result in part (b) remains valid in this case. An aluminum soft-drink can of radius 3.4 cm is designed not to rupture when holding fluids with a pressure of up to $6.2 \times 10^5 \text{ N/m}^2$. Determine the required aluminum thickness of a soft-drink can.

30. (III) In a mountain-climbing technique called the "Tyrolean traverse," a rope is anchored on both ends (to rocks, strong trees, etc.) across a deep chasm and then a climber traverses the rope while attached by a sling as shown in Fig. 12–7. When setting up the Tyrolean traverse, the rope must be properly "pre-tensioned" so that it will always remain within its safe operating range, and yet not sag too much, when the climber is traversing it. Take the rope's diameter and Young's modulus to be 11 mm and $7.5 \times 10^8 \text{ N/m}^2$, respectively. A typical climbing rope can provide a tension force of up to 29 kN before breaking and, for safe use, a "safety factor" of 10 is usually recommended (i.e., the rope should only be required to supply a tension force of up to 2.9 kN).
(a) Consider a 90-kg climber at the center of a Tyrolean traverse, which spans a 30-m chasm, and assume that in this situation the rope has a tension of 2.9 kN (i.e., it is at the maximum of its recommended safety range). Find the total length L of (stretched) rope between the anchors; then determine the unstretched length L_o of this portion of rope (i.e., the correct length of unstretched rope to be used in setting up the Tyrolean traverse).
(b) When the Tyrolean traverse is set up, (i.e., before a climber is attached to it), the rope is tautly fixed between the two anchors and thus has length of 30 m. Using your result from part (a), under what tension should the rope be placed (called "pre-tensioning") before a climber is attached?

CHAPTER 13

Fluids

Section 13-1

1. (I) Show that the average density of the Sun is about the same as the density of common liquids on Earth.

2. (II) Here is one way to visualize how much air is in the average-sized room: If air is cooled to 77 K, all of its major components (N_2, O_2, and Ar) will condense into a liquid mixture with density of 875 kg/m³. Assume a room with typical dimensions of 4.0m × 4.0m × 3.0m, which is initially filled with air at 20°C and 1 atm. If all of the air in this room is condensed into 77 K liquid air, how many 2.0-L soda bottles could this liquid fill?

3. (II) The latest findings in astrophysics indicate that the observable universe can be modeled as a sphere of radius $R = 14$ billion light-years containing mass with an average density of 9.5×10^{-27} kg/m³, where 4% of the universe's total mass is due to protons and neutrons in the nuclei of atoms. Use this information to estimate the total number of protons and neutrons in the observable universe. 1 light-year = 9.5×10^{15} m.

4. (III) Earth is not a sphere of uniform construction, but rather, has regions of varying density. Consider a simple model of Earth's composition in which our planet is divided into three regions—inner core, outer core, and mantle—where each region is taken to have a unique constant density (= the average density of that region in the real Earth). Earth's crust is ignored in this model because it is only about 10 km thick. The data for this model are given in Table 13-1.

TABLE 13-1

Region	Radius (km)	Density (kg/m³)
Inner Core	0 to 1220	13,000
Outer Core	1220 to 3480	11,100
Mantle	3480 to 6371	4,400

(a) Use this model to predict the average density of the entire Earth.
(b) It is known from various experiments that the radius of Earth is 6371 km and its mass is 5.98×10^{24} kg. Use these data to determine the actual average density of Earth and compare this value (as a percent difference) with the one you determined in part (a).

5. (III) A two-component model for the human body is widely used to determine a person's percent body fat. In this model, a fraction $f < 1$ of the body's total mass m is assumed to be composed of fat with a density $\rho_f = 0.90$ g/cm³ and the remaining mass of the body is composed of fat-free tissue with a density $\rho_t = 1.10$ g/cm³. If SG is the specific gravity of the entire body, show that the percent of the body's total mass due to fat ($= f \times 100$) is given by

$$\%\text{Body Fat} = \frac{495}{\text{SG}} - 450.$$

Sections 13-2 to 13-5

6. (II) Who is more "dangerous" to the newly refinished hardwood floors in your house—a 50-kg woman in high heels or a 5000-kg elephant? For a high-heel shoe, assume the contact areas with the floor for the heel and the toe-section are 1 cm² and 3 cm², respectively, so that, when standing "flat," its total contact area is 4 cm². Take the diameter of the elephant's foot to be 30 cm. Assuming people walk with one foot on the ground at a time, while elephants walk with two, determine the resulting pressure
(a) each exerts on the floor while walking with maximum available contact area. How many times larger is the pressure due to the woman than due to the elephant?
(b) if the woman walks with all of her weight placed uniformly over only one heel.
(c) if the woman, while walking, momentarily places all of her weight on only a small portion (say, one-fourth) of the area of one heel. [An applied pressure above 7×10^6 N/m² can damage a hardwood floor.]

7. (II) Commercial airliners cruise at an altitude of about 10 km, where the outside air pressure is about 0.25 atm. The air in the airliner's passenger cabin is pressurized at about 0.75 atm. For the purposes of this problem, assume the air inside and outside of the airliner are at rest with respect to each other.
(a) Assume the door through which passengers enter the aircraft is a rectangle 1.0 m wide and 2.0 m high. At cruising altitude, what is the net force (magnitude and direction) on this door due to air pressure?
(b) Assume the air in airliner cabins was pressurized at 1.00 atm (rather than 0.75 atm). By what factor would the net force on the door in part (b) increase? [The air in passenger cabins is pressurized below the sea-level value of 1 atm so that airliners can be built with less strength, and thus, less weight.]

8. (II) To avoid adverse health effects among passengers due to oxygen deprivation, the Federal Aviation Regulations stipulate that, at a minimum, air pressure within a commercial airliner's passenger cabin must be equivalent to that at an altitude of 8000 ft (2438 m) in Earth's atmosphere. Estimate this minimum pressure (atm).

9. (II) During ascent and especially during descent in elevation, volume changes of trapped air in the middle ear can cause ear discomfort. As exterior pressure decreases during ascent, the middle ear's trapped air expands until the Eustachian tube is pushed open, letting some of the trapped air escape into nasal passages until the middle-ear pressure and exterior pressure are equalized. This process usually happens automatically and without pain in most people. As exterior pressure increases during descent, however, a person must consciously open the Eustachian tube (by swallowing, yawning, etc.), allowing air to enter the contracting middle ear and equalize pressure there with the exterior pressure. From experience with airplanes and high-rise elevators, it has been found that rapid descent at a rate of 7.0 m/s or faster causes ear discomfort for most people. Based on this fact, what is the maximum rate of decrease in atmospheric pressure (i.e., determine dP/dt) tolerable to most people? In 1956, architect Frank Lloyd Wright designed the 528-story Illinois Office Tower, which was to be one mile (1600 m) in height. If this building is ever built (so far it has not been), what will be the fastest possible descent time (in minutes) for an elevator traveling from the top floor to the ground floor, if the elevator is properly designed to account for human physiology?

10. (II) The health of a person's heart can be judged from the pressure with which it expels blood. In a blood pressure measurement, a doctor uses a pressurized cuff attached around a patient's upper arm to find blood pressure at that point in the body. Because the upper arm is level with the heart, the measured pressure can then be equated to blood pressure at the heart. If a doctor mistakenly attached the cuff around a standing patient's calf (about 1 m below the heart) and performed a measurement there, what error (in Pa) would be introduced if that measured blood pressure was equated to the pressure at the heart? Given that this technique yields a gauge pressure, would this error be significant? Average blood (gauge) pressure for a typical person is about 100 mmHg.

11. (II) At first glance, Pascal's principle may appear to suggest that an incompressible fluid can be use to magnify a force at no cost. In reality, another quantity is being sacrificed in order to increase force. Let's investigate. Consider a hydraulic lift with input and output pistons within the lift's input and output cylinders of radii R_1 and R_2, respectively, where $R_2 \gg R_1$. When a small downward force F_1 is applied to the input piston, a large upward force F_2 acts on the output piston. See Fig. 13–1.

(a) Apply Pascal's principle to show $F_2 = F_1 \left(\dfrac{R_2}{R_1}\right)^2 \gg F_1$.

(b) Assume that, as a result of the force F_1 being applied to the input piston, this piston moves downward a distance of x_1. Then, since it is incompressible, the volume of fluid that occupied the column height x_1 in the input cylinder is transferred over to the output cylinder where it occupies a column height of x_2. Determine x_2 and show $x_2 \ll x_1$.

(c) From parts (a) and (b) we found that force is magnified at the output piston at the expense of distance moved. Show, however, that energy is conserved in this device. That is, demonstrate that the work done by force F_1 equals the work done by F_2.

12. (III) What is the pressure at the center of Earth? To estimate the answer to this question, consider the simplest possible model for Earth: a sphere of radius R with a uniform density $\rho = 5520$ kg/m^3 (= average density of the real Earth).

(a) Inside Earth, when located a radius r away from the center, the acceleration g due to gravity is less than at Earth's surface. As described in Appendix C in the text, this reduction of g results from the fact that only the portion of Earth's mass within the sphere of radius r contributes to the gravitational force at that location. Applying Eq. 6–4 in the text to the interior of Earth, the acceleration due to gravity at a distance r from Earth's center is given by

$$g(r) = G\frac{M_{within}}{r^2} = \frac{G}{r^2}\left(\rho\frac{4}{3}\pi r^3\right) = \frac{4\pi G\rho r}{3}$$

where the mass within a sphere of radius r is $M_{within} = \rho\frac{4}{3}\pi r^3$, assuming the sphere has uniform density ρ (Fig. 13–2). Plugging the above expression into Eq. 13–4 from the text written as

$$\frac{dP}{dr} = -\rho g$$

show that the pressure P_o at the center of the sphere (i.e., at $r = 0$) is given by

$$P_o = \frac{2\pi\rho^2 GR^2}{3} + P_R$$

where P_R is the pressure at the surface of the sphere (i.e., at $r = R$).

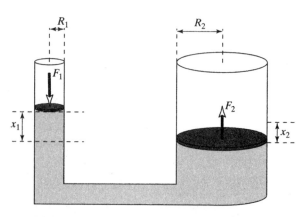

FIGURE 13–1

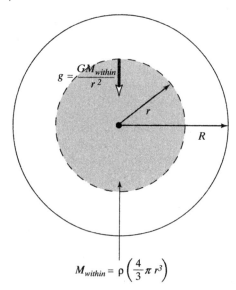

FIGURE 13–2

(b) Taking $P_R = 1.0$ atm, what value does this model predict for the pressure at the center of Earth? Sophisticated models that account for the variation of Earth's density with depth predict the pressure at Earth's center to be 360 GPa. By what factor does P_o from our simple model differ from this value?

Section 13–6

13. (II) Your fiancée presents you with a 24-carat pure gold ring that feels "a little light." With the encouragement of some close friends and your mother, you find that the ring "weighs" 34.00 g in air and has an apparent "weight" of 32.24 g when submerged under water. Is the ring made of gold? The densities for gold, lead, and plastic are 19.3, 11.3, and 1.2 g/cm³, respectively.

14. (II) A 3.0m × 3.0m × 20cm raft made of wood with density $\rho = 600$ kg/m³ floats on a freshwater lake. If at least 5 cm of the raft's 20-cm thickness is to remain above the lake surface (Fig. 13–3), how many 70-kg people can stand on it?

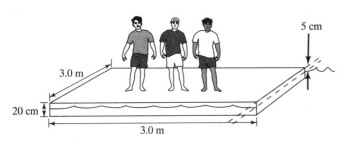

FIGURE 13–3

15. (II) As an object near Earth's surface falls, it is acted upon by a weight force, an air resistance force, and a buoyant force due to air. When does the buoyant force significantly influence the object's motion? One could plausibly say that when the magnitude of the buoyant force on an object exceeds 1% of an object's weight, then the buoyant force is a significant influence. Show that only substances of density ρ within a certain range fulfill this criterion. From Table 13–1 in the text, for what type of substances will the buoyant force be significant during free fall?

16. (II) A scuba tank, when fully submerged, displaces 15.7 L of seawater. The tank itself has a mass of 14 kg and, when "full," contains 3.0 kg of air. Assuming only a weight and a buoyant force act, determine the net force (magnitude and direction) on the fully submerged tank at the beginning of a dive (when it is full of air) and at the end of a dive (when it no longer contains any air).

17. (II) How many helium-filled balloons would it take to lift a person? Assume the person has a mass of 75 kg and that each helium-filled balloon is spherical with a diameter of 30 cm.

18. (II) The diameter of a spherically shaped helium-filled balloon must exceed the minimum value d_{min} or else the balloon will sink to the ground, rather than "float." Determine d_{min}, if the deflated latex balloon has a mass of 2.8 g.

19. (II) Show that the helium in a helium-filled balloon can lift up to 6.2 times its own "weight." That is, if a balloon filled with a mass m_{He} of helium is floating in air while supporting an attached load of mass m_L, including the mass of the (deflated) balloon itself, then show that $\dfrac{m_L}{m_{He}} \leq 6.2$.

20. (II) In the hydrostatic method for determining percent body fat, the volume of a person's body is found with the help of Archimedes' principle. The person's actual weight $w\,(= mg)$ is first determined on dry land. Then, in a large water tank or swimming pool, the person deeply exhales to minimize the residual volume V_R of air within the body (mainly in the lungs), and then his or her apparent weight $w'\,(= m'g)$ is found while entirely submerged under water. For the underwater weighing, a scale is suspended above water from a support (e.g., diving board) and the submerged person sits in a sling, which is attached to the scale via a rope or a chain.
(a) Typically scales are calibrated to give "weight" $m\,(= w/g)$ in the unit of kilograms. Show that if such a scale is used, the "mass-containing" volume V of the person's body (i.e., the volume excluding residual air pockets with in the body) is given by

$$V = \frac{m - m'}{\rho_W} - V_R$$

where ρ_W is the density of water.
(b) On dry land, an athlete is found to "weigh" $m = 70.2$ kg. Hanging from a scale, while submerged in a heated swimming pool (at 25°C, so $\rho_W = 997$ kg/m³), his "apparent weight" is determined to be $m' = 3.4$ kg, and V_R is estimated to be 1.3 L. Determine the specific gravity SG of the "mass-containing" volume of the athlete's body.
(c) Consider the human body as composed of both fat and fat-free components. Based on this simple two-component model, a person's percent body fat is given by

$$\%\text{Body Fat} = \frac{495}{\text{SG}} - 450.$$

Use this relation to find the above-mentioned athlete's percent body fat.

21. (II) The hydrostatic method for determining percent body fat (see Problem 20) is limited in that the technique fails for people whose percent body fat exceeds a certain maximum value.
(a) Noting that V_R and m typically fall in the ranges of 0.6–1.7 L and 50–100 kg, respectively, show that the maximum value measurable by this method is approximately

$$\%\text{Body Fat(Max Measurable)} \approx \frac{495}{\text{SG}_W} - 450$$

where SG_W is the specific gravity of water.
(b) In the swimming-pool temperature range of 20°C to 30°C, the specific gravity of water is given by

$$\text{SG}_W = 1.0034 - 0.0002559 \times T$$

where T is the water temperature in °C. Find the maximum value measurable for percent body fat when the pool's temperature is 20°C and when it is 30°C. Is a "warm" pool significantly better to use than a "cool" pool when implementing the hydrostatic method?

22. (II) A Japanese fishing float is a hollow glass sphere with outer diameter $d_O = 30$ cm (Fig. 13–4). If exactly half of the float is submerged when it floats in seawater, what is its inner diameter d_I? Include the weight of the enclosed air (with assumed room-temperature density of 1.2 kg/m³) in your calculation. For glass, $\rho = 2500$ kg/m³.

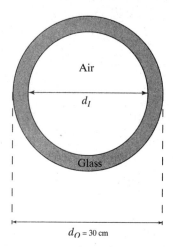

FIGURE 13–4

Sections 13–7 to 13–9

23. (III) Consider the siphon shown in Fig. 13–5, in which a filled tube, with a uniform cross-sectional area along its entire length, transfers liquid of density ρ from a liquid reservoir up through an intermediate point A that is a height h above the top surface of the reservoir, and then discharges liquid at a distance H below the reservoir surface. Assume the surface of the reservoir is at pressure P_o and is motionless. Also assume the "discharge side" of the siphon is at pressure P_o.

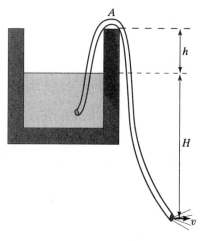

FIGURE 13–5

(a) Show that the speed that fluid flows through the siphon is given by $v = \sqrt{2gH}$.
(b) If the pressure of a liquid is reduced to its "vapor pressure" P_v, the liquid will boil. This situation must be avoided in order for the siphon to function. Show then that the maximum possible height h_{max} for the point A is given by

$$h_{max} = \frac{P_o - P_v}{\rho g} - H.$$

(c) If the reservoir contains water with $P_o = 1.0$ atm and the siphon is configured so that $H = 1.0$ m, what is the value of h_{max}? Assume the siphon contains room-temperature water for which $\rho = 998$ kg/m^3 and $P_v = 2330$ Pa.

24. (III) In a water clock, the passage of time is measured by the flow of water out of a hollow storage tank. Consider a water-clock tank constructed so that it is open to the atmosphere at its top and is closed at its bottom except for a centrally placed hole of radius $r = 1.5$ mm that is open to the atmosphere. Assume that the tank has an inner radius R, which varies with its height y above the hole according to the following relation:

$$R = Ay^{1/4}$$

where the constant $A = 0.50 m^{3/4}$. See Fig. 13–6.

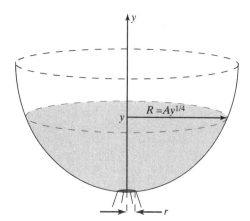

FIGURE 13–6

(a) Assume that this water-clock tank is filled with water and that at time t the water surface is a height y above the hole. Also, assume that the kinetic energy of water at height y can be neglected in comparison with its potential energy. Show that the change in height of the water surface per unit time is a constant for this clock, independent of the height y.
(b) How far will the water surface in the clock's tank descend each hour?
(c) To calibrate the clock, small "hour marks" could be placed at various heights y indicating the position of the water surface after elapsed times of 1 hour, 2 hour, etc. If you wanted the clock to run for 6 hours, to what initial height y_o would you fill it with water? At what heights would you place marks to indicate 1, 2, 3, 4, 5, and 6 hours?

25. (III) In a water clock, the passage of time is measured by the flow of water out of a hollow storage tank. Consider a water clock constructed so that its tank is a hollow cylinder with constant inner radius $R = 0.50$ m and is open to the atmosphere at its top. At the bottom it is closed except for a centrally placed hole of radius $r = 1.5$ mm, which is open to the atmosphere. Assume that this water-clock tank is filled with water and that at time t the water surface is a height y above the hole (Fig. 13–7). Also, assume that the kinetic energy of water at height y can be neglected in comparison with its potential energy.
(a) Show that, as water flows out the hole, the height y and time t obey the following relation:

$$\frac{dy}{\sqrt{y}} = -\frac{r^2\sqrt{2g}}{R^2}dt.$$

(b) Assume the height of the water surface is $y_o = 1.0$ m, at $t = 0$. Integrate the relation in part (a) to show

$$y = \left[1 - \frac{r^2}{R^2}\sqrt{\frac{g}{2}}t\right]^2.$$

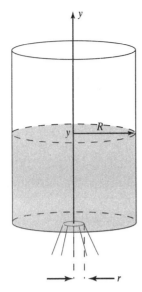

FIGURE 13–7

(c) Show that this clock will run for almost 14 hours.
(d) By what distance does the clock's water level descend during its first hour of operation? How about during its second hour? What about during its thirteenth hour?

26. (III) In deriving Bernoulli's equation (Eq. 13–8 in the text), the flowing fluid is assumed to be "incompressible." When is this assumption valid? Suppose that, within a horizontal pipe, a fluid of density ρ is at rest (i.e., $v_1 = 0$) at a location where its pressure is P_1 and at another location it has a pressure P_2 and speed v_2. For the sake of argument, let's define the fluid as incompressible if the volume V of a given amount of its mass changes by no more than 1% when moving from the first location to the second, i.e., $\frac{\Delta V}{V} \leq 0.01$.

(a) Show that this criterion implies that Bernoulli's equation will remain valid only if

$$v_2 \leq v_{\max} \equiv \sqrt{\frac{B}{50\rho}}$$

where B is the bulk modulus of the fluid (see Eq. 12–7a and Table 12–1 in the text).
(b) For air at atmospheric pressure and 0°C, determine $v_{\max}$.
(c) For water at atmospheric pressure and 4°C, determine $v_{\max}$.

Section 13–11

27. (II) The large quantity of heat produced by high-power lasers is commonly exhausted using a "flow-through" water-cooling system. In such a system, tap water at pressure P_1 flows into the cooling system's pipe, which passes inside the laser system where the flowing water absorbs the laser's waste heat, after which the heated water flows out of the system and is discharged at pressure P_2 into a drain. Assume the cooling system's pipe is a cylindrical tube with 3.0-mm diameter and 4.0-m length and, for proper cooling, 20 L of water must flow through this pipe every minute. If the pressure at the drain is $P_2 = 1.0$ atm, what is the required gauge pressure $(P_1 - P_2)$ for the tap water?

28. (III) Does the stretching of latex cause the pressure inside of an inflated balloon to differ significantly from atmospheric pressure? Consider the following method for answering this question (Fig. 13–8). A person inflates a spherically shaped latex balloon with air to its recommended diameter $d_o = 31$ cm and pinches off the neck with her fingers. She then releases her grasp on the neck and allows air to flow out of the balloon for 1.5 s, then pinches off the neck again. She measures the balloon's new diameter to be $d = 28$ cm. Finally, she estimates that when air was flowing out through the balloon's neck, the neck was cylindrically shaped with a length $L = 4.0$ cm and radius $R = 5.0$ mm. Determine the difference in pressure ΔP between inside the inflated balloon and outside the balloon. Show that your result implies that the balloon's internal pressure exceeds atmospheric pressure by less than 0.1%.

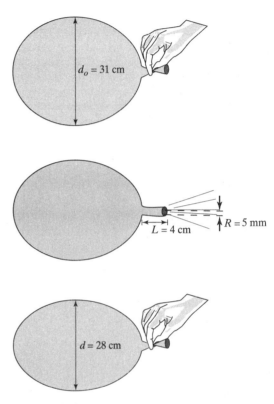

FIGURE 13–8

General Problems

29. (III) When can a fluid be reasonably approximated as "incompressible"? Assume a fluid, initially at volume V_o and pressure $P_o = 1.0$ atm, is subjected to a change in pressure ΔP so that its volume decreases to V. As a result of this compression process, which we assume occurs at constant temperature, the fluid's density changes from ρ_o to ρ.

(a) Show that $\Delta P = B\left[1 - \dfrac{\rho_o}{\rho}\right]$, where B is the bulk modulus of the fluid (see Eq. 12–7a and Table 12–1 in the text). Assume B remains constant throughout the compression.
(b) One might reasonably define a fluid as incompressible if, when subjected to a pressure change, its density changes by less than 1.0%. According to this criterion, what is the maximum ΔP (in atm) possible for water to be considered incompressible?
(c) According to the criterion given in part (b), what is the maximum ΔP possible for a gas such as air, initially at standard sea-level conditions, to be considered incompressible?

30. (III) Because it has a small mass, yet a large size, three forces act significantly on a freely floating helium-filled balloon— weight, air resistance (termed "drag" force), and buoyant force. Consider a spherical helium-filled balloon of radius $R = 15$ cm rising upward through room-temperature air of density $\rho_A = 1.2$ kg/m^3. Let $\rho_{He} = 0.166$ kg/m^3 be the density of room-temperature helium inside the balloon and $m = 2.8$ g be the mass of the (deflated) balloon itself. For all speeds v, except the very slowest ones, the flow of air past a rising balloon is turbulent (see Section 5–5 in the text), and the drag force F_D is given by the following relation

$$F_D = \frac{1}{2} C_D \rho_A \pi R^2 v^2$$

where the constant $C_D = 0.47$ is the "drag coefficient" for a smooth sphere of radius R. Assume a helium-filled balloon is held at rest slightly above the ground. If this balloon is released, it will accelerate very quickly (few tenths of a second) to its terminal velocity v_T, where the buoyant force is cancelled by the drag force and the balloon's total weight force. Assuming the balloon's acceleration takes place over a negligible time and distance, how long does it take the released balloon to rise a distance $h = 10$ m?

CHAPTER 14

Oscillations

Sections 14–1 and 14–2

1. (II) Tall buildings are designed to sway in the wind. In a 100-km/h wind, for example, the top of the 110-story Sears Tower oscillates horizontally with an amplitude of 15 cm. The building oscillates at its natural frequency, which has a period of 7.0 s. Assuming SHM, find the maximum horizontal velocity and acceleration experienced by a Sears employee as she works at her desk located on one of the top floors. Compare the maximum acceleration (as a percentage) with the acceleration due to gravity.

2. (II) An object of unknown mass m is hung from a vertical spring of unknown spring constant k, and the object is observed to be at rest when the spring has extended by a length of 10 cm. The object is then given a slight push and executes SHM. Determine the period T (in seconds) of this oscillation.

3. (II) An oxygen atom at a particular site within a DNA molecule can be made to execute simple harmonic motion, if illuminated by infrared light. The oxygen atom is bound with a spring-like chemical bond to a phosphorus atom, which is rigidly attached to the DNA backbone. The oscillation of the oxygen atom, called the "antisymmetric phosphate stretch," occurs with frequency $f = 3.7 \times 10^{13}$ Hz. If the oxygen atom at this site is chemically replaced with a sulfur atom, the spring constant of the bond is unchanged (because sulfur is just below oxygen in the periodic table). Predict the antisymmetric phosphate stretch frequency for a DNA molecule after the sulfur substitution.

4. (II) A mass is connected to the end of a spring and undergoes simple harmonic motion so that its position x (m) as a function of time t (s) is given by

$$x = (3.0\,\text{m}) \cos\left[(5.0\,\text{rad/s})t + \frac{\pi}{6} \right].$$

During portions of this cyclic motion, the mass travels exactly 2.0 m as it moves directly from $x = 0.0$ m to $x = 2.0$ m. How long does this traverse of 2.0 m take?

[*Hint*: It might be helpful to graph x vs. t.]

5. (II) Mass $m_1 = 1.0$ kg is suspended by a string from mass $m_2 = 3.0$ kg, which in turn is suspended from a vertical spring (Fig. 14–1). The spring with spring constant $k = 100$ N/m is rigidly supported from above. Below the threshold amplitude A_o, these masses will oscillate together as a single unit. Determine A_o.

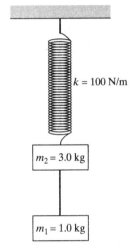

FIGURE 14–1

6. (II) On a miniature golf course, you are standing 2.0 m from a small tunnel into which you must putt your ball. Your putt has to be perfectly timed because the tunnel's entrance is alternately blocked and unblocked by a metal plate that slides back and forth in front of it with a frequency $f = \dfrac{\omega}{2\pi} = 1.0$ Hz. Assume the position of the plate is given by $x = A \cos(\omega t)$, and that the entrance is blocked and unblocked when $x = +A$ and $x = -A$, respectively (Fig. 14–2). Find the first three positive times t at which you can putt your ball and have it pass unimpeded into the unblocked tunnel if the ball rolls at constant speed $v = 1.5$ m/s after you putt it.

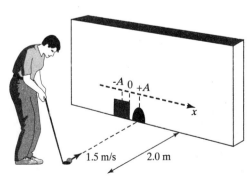

FIGURE 14–2

7. (II) Consider the one-dimensional motion of a mass on a spring over the time interval Δt from the time $t_o = 0$ to $t_o + \Delta t = T/2$, assuming the time-dependent position of the mass is given by $x(t) = A\cos(\omega t + \pi) = -A\cos(\omega t)$. The quantities T, A, and ω are the period, amplitude, and angular frequency of the motion, respectively.
(a) In Chapter 2 of the text, average velocity was defined as displacement divided by elapsed time (Eq. 2–2). Use this definition to determine $\bar{v}$ over the time interval Δt.
(b) More generally, the average value $\bar{f}$ of a time-dependent quantity $f(t)$ over the time interval from time $t_o = 0$ to $t_o + \Delta t$ is given by

$$\bar{f} = \frac{\int_{t_o}^{t_o+\Delta t} f(t)\,dt}{\Delta t}.$$

Find the time-dependent velocity $v(t)$ for the mass; then apply this general definition to determine $\bar{v}$ over the time interval Δt and show that you obtain the same answer as in part (a).

8. (II) Consider two objects A and B, both undergoing SHM, but with different frequencies, as described by the following equations:

$$x_A = (2.0\,\text{m})\sin[(2.0\,\text{rad/s})t]$$
$$x_B = (5.0\,\text{m})\sin[(3.0\,\text{rad/s})t].$$

After $t = 0$, find the next three times t at which both objects simultaneously pass through the origin.

9. (II) Many real springs at their natural length cannot be compressed. The tightly wound coils touch, preventing any significant compression of the spring's length (Fig. 14–3a). However, an opposing pair of two such springs, both stretched, can produce the Hooke's law force of an idealized spring. Consider two identical springs with spring constant k and natural length L anchored distance $D > 2L$ apart and attached to a very slender mass m, which is on a frictionless horizontal surface (Fig. 14–3b). Define an x-axis with the origin at the equilibrium position for this system (which is midway between the anchor points) and positive direction to the right.
(a) As long as the excursions of the mass from $x = 0$ never exceed $|x| = x_{\max}$, both springs will always be stretched. Determine $x_{\max}$.
(b) If $|x| < x_{\max}$, show that the mass experiences a Hooke's law force and find its period of oscillation, if displaced from equilibrium and released.

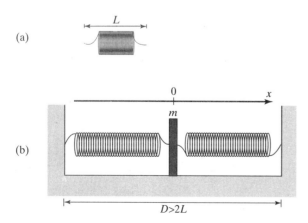

FIGURE 14–3

10. (III) What difference does a minus sign make? Assume that an object of mass m is constrained to move only in one dimension, and that an equilibrium position (a point at which no net force acts) exists for this object. Define the line along which the object moves to be an x-axis with its origin at the equilibrium position and the positive direction to the right.
(a) Assume that a net (Hooke's law) force given by $-kx$ acts on the object, where k is a positive constant. Explain why the minus sign in this expression means that, whichever way the object moves, the force will always "attract" the object to the equilibrium position. Then apply Newton's second law to this situation and obtain the differential equation for the object's time-dependent position $x(t)$. Finally, by direct substitution, show that the following function involving the cosine and sine functions is the solution to this differential equation:

$$x(t) = a\cos(\omega t) + b\sin(\omega t)$$

where a and b are constants and $\omega = \sqrt{k/m}$. [As mentioned in the text, this solution is an alternate, but equivalent, form of $x(t) = A\cos(\omega t + \phi)$.]
(b) Now, assume that a net force given by $+kx$ acts on the object, where k is a positive constant. Explain why the plus sign in this expression means that, whichever way the object moves, the force will always "repel" the object from the equilibrium position. Then apply Newton's second law to this situation and obtain the differential equation for the object's time-dependent position $x(t)$. Finally, by direct substitution, show that the following function involving the exponential function is the solution to this differential equation:

$$x(t) = a\exp(\omega t) + b\exp(-\omega t)$$

where a and b are constants and $\omega = \sqrt{k/m}$.
(c) For the situation in part (b), the object's initial conditions for position and velocity, in general, are $x(0) = x_o$ and $v(0) = v_o$, where x_o and v_o are constants. Show then that $x(t)$ is given by

$$x(t) = \frac{1}{2}\left(x_o + \frac{v_o}{\omega}\right)\exp(\omega t) + \frac{1}{2}\left(x_o - \frac{v_o}{\omega}\right)\exp(-\omega t).$$

If at $t = 0$, the object is not at the equilibrium position, or it is at $x = 0$ and moving, describe qualitatively where the object moves as time increases. What is the (unphysical) prediction for the object's velocity as $t \to \infty$?

11. (III) Consider the following model for the vibration of a diatomic molecule such as N_2. Two identical atoms, each of mass m, are chemically bound to each other with an equilibrium separation L (called the "bond length") as shown in Fig. 14–4a. The bond, however, is not completely rigid and can be modeled as a spring with spring constant k, connecting the two atoms. In the "stretch" vibrational mode of this molecule, the atoms each oscillate along the direction of the bond. To describe the stretch oscillation, define x_R and x_L to be the distances the right and left atoms, respectively, have moved from their equilibrium positions as shown in Fig. 14–4b, with positive defined to the right.
(a) In our model, at a given instant, the bond is stretched (or compressed) by a total amount $(x_R - x_L)$. Apply Newton's second law to each atom and obtain a pair of differential equations involving the time-dependent positions $x_R(t)$ and $x_L(t)$ of the right and left atoms, respectively.
(b) Show that the pair of equations $x_R = A\cos(\omega t)$ and $x_L = A\cos(\omega t + \pi)$, in which the atoms alternately move

toward and then away from each other with the same amplitude A and angular frequency ω, is a solution to the pair of differential equations in part (a), if $\omega = \sqrt{\dfrac{2k}{m}}$.

[Hint: $\cos(\omega t + \pi) = -\cos(\omega t)$]

(c) From experiments in which light is used to excite the stretch oscillation in nitrogen gas, it is found that the frequency f of this vibrational mode is 7.0×10^{13} Hz. Using our model, what is the spring constant for the N_2 bond? How many times greater is k for this bond than for a spring commonly used in an introductory physics lab (10 N/m)? The mass of the nitrogen atom is 14 u, where 1 u = 1.66×10^{-27} kg.

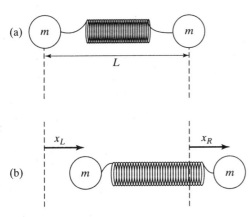

FIGURE 14–4

12. (III) Consider the following model for the vibration of a diatomic molecule. The two atoms, with masses m_1 and m_2, are chemically bound to each other with an equilibrium separation L (called the "bond length") as shown in Fig. 14–5a. The bond, however, is not completely rigid and can be modeled as a spring with spring constant k, connecting the two atoms. In the "stretch" vibrational mode of this molecule, each atom oscillates along the direction of the bond. To describe the stretch oscillation, define x_1 and x_2, the distances the atoms have moved from their equilibrium positions as shown in Fig. 14–5b, with positive defined to the right.

(a) In our model, at a given instant, the bond is stretched by a total amount $(x_2 - x_1)$. Apply Newton's second law to each atom and obtain a pair of differential equations involving the time-dependent positions $x_1(t)$ and $x_2(t)$ of both atoms.

(b) Using the fact that $\cos(\omega t + \pi) = -\cos(\omega t)$, show that the pair of equations $x_1 = A_1 \cos(\omega t)$ and $x_2 = A_2 \cos(\omega t + \pi)$, in which the atoms alternately move toward and then away from each other with different amplitudes but the same angular frequency ω, is a solution to the pair of differential equations in part (a), if the following two conditions are met:

$$(m_1 \omega^2 - k)A_1 = kA_2$$
$$(m_2 \omega^2 - k)A_2 = kA_1.$$

(c) Show that the two conditions in part (b) imply $\omega = \sqrt{\dfrac{k}{\mu}}$

and $\dfrac{A_2}{A_1} = \dfrac{m_1}{m_2}$, where the "reduced mass" is defined as

$\mu = \dfrac{m_1 m_2}{m_1 + m_2}$.

(d) From optical absorption experiments on hydrogen bromide it is found that the frequency f of the HBr stretch vibration is 7.9×10^{13} Hz. Using our model, what is the spring constant for the HBr bond and what is the ratio of hydrogen-to-bromine oscillatory amplitudes A_H/A_{Br}? The masses of the hydrogen and bromine atoms are 1.0 u and 80 u, respectively, where 1 u = 1.66×10^{-27} kg.

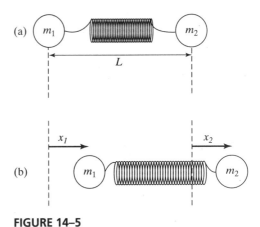

FIGURE 14–5

13. (III) Consider the following model for the vibration of a diatomic molecule. The two atoms, with masses m_1 and m_2, are chemically bound to each other with an equilibrium separation L (called the "bond length") as shown in Fig. 14–5a. The bond, however, is not completely rigid and can be modeled as a massless spring with spring constant k, connecting the two atoms. In the "stretch" vibrational mode of this molecule, each atom oscillates along the direction of the bond with the time-dependent position of each atom relative to its equilibrium position given by $x_1 = A_1 \cos(\omega t)$ and $x_2 = A_2 \cos(\omega t + \phi)$, where A_1 and A_2 are the (positive) amplitudes of each atom's oscillation, ω is the angular frequency, and ϕ is the phase difference of the second atom's oscillation relative to the first's (Fig. 14–5b). The positive x-axis is defined to the right.

(a) Apply conservation of momentum to this system. Assume that no outside forces act on this molecule and that we view it from the frame of reference in which its center of mass is at rest. Then, for any time t, the total momentum of this system must remain zero. Use this principle to show $\phi = \pi$ and $\dfrac{A_1}{A_2} = \dfrac{m_2}{m_1}$.

(b) Qualitatively explain the meaning of $\phi = \pi$ in terms of how m_2 moves relative to m_1.

(c) Using this model, what is the ratio of hydrogen-to-bromine oscillatory amplitudes A_H/A_{Br} when HBr vibrates in the stretch mode? The masses of the hydrogen and bromine atoms are 1.0 u and 80 u, respectively, where 1 u = 1.66×10^{-27} kg. What is the ratio of the atomic oscillatory amplitudes when a nitrogen (N_2) molecule vibrates in the stretch mode?

14. (III) Let the distance between a room's floor and ceiling be H. Two identical massless springs, one with an end connected to the ceiling, the other with an end connected to the floor, each have spring constant k and unstretched length L, where $L < \dfrac{H}{2}$ (Fig. 14–6a). Each spring is stretched and its free end is connected to a very slender mass m located near the midpoint between floor and ceiling. The equilibrium position of this spring-mass system is found to be a distance $d = 10$ cm

below the midpoint. Define an x-axis with the origin at this equilibrium position and positive downward (Fig. 14–6b). Assume the mass is displaced from $x = 0$ by a distance $x < \left(\dfrac{H}{2} - d - L\right)$, and is released. Apply Newton's second law to this situation and show that the resulting differential equation for the time-dependent position of the mass has the solution $x(t) = A\cos(\omega t + \phi)$. Determine the period (in seconds) of the oscillatory motion.

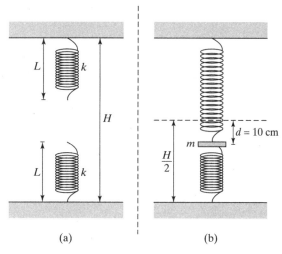

FIGURE 14–6

15. (III) A mass M on a frictionless surface is attached to a spring with spring constant k as shown in Fig. 14–7. This mass-spring system is then observed to execute simple harmonic motion with a period T. The mass M is changed several times and the associated period T is measured in each case, generating the data in Table 14–1.

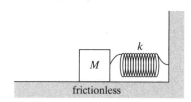

FIGURE 14–7

TABLE 14–1

Mass M (kg)	Period T (s)
0.50	0.445
1.00	0.520
2.00	0.630
3.00	0.723
4.50	0.844

(a) Starting with Eq. 14–7b in the text, show why graphing T^2 vs. M is expected to yield a straight-line plot. How can k be determined from the straight line's slope? What is the expected y-intercept?

(b) Using the data in Table 14–1, plot T^2 vs. M and show that this graph yields a straight line. Determine the slope and the (nonzero) y-intercept.

(c) Show that a nonzero y-intercept can be expected in our plot theoretically if, rather than simply using M for the mass in Eq. 14–7b, one uses $(M + M_o)$, where M_o is a constant. That is, repeat part (a) using $(M + M_o)$ for the mass in Eq. 14–7b. Then use the result of this analysis to determine k and M_o from your graph's slope and y-intercept.

(d) Offer a physical interpretation for M_o, a mass that appears to be oscillating in addition to the attached mass M.

16. (III) As a way of entertaining kids while families wait for pizza, a pizza shop owner designs the following kiddie ride. The child of mass m sits on top of a rectangular slab of mass $M = 35$ kg, which rests on the frictionless horizontal floor at the back of the pizza shop. The slab is attached to a horizontally oriented spring with spring constant $k = 400$ N/m, the other end of which is attached to an immovable wall (Fig. 14–8). The coefficient of static friction between the child and the top of the slab is $\mu = 0.40$. The shop owner's intention is that, when displaced from the equilibrium position by a parent and released, the slab and kid atop (with no slippage between the two) executes SHM with amplitude $A = 0.50$ m. Should there be a weight restriction for this ride (i.e. is there a range of m for which the child will not "stay stuck" to the slab during the operation of this ride)?

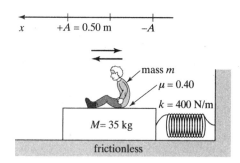

FIGURE 14–8

Section 14–3

17. (III) The average value $\overline{f}$ of a time-dependent quantity $f(t)$ over the time interval from $t = t_o$ to $t = t_o + \Delta t$ is given by

$$\overline{f} = \frac{\displaystyle\int_{t_o}^{t_o+\Delta t} f(t)\,dt}{\Delta t}.$$

Assume an object, under the influence of a Hooke's law force with spring constant k, is executing SHM with amplitude A. Given its position is $x(t) = A\cos(\omega t + \phi)$, show that the object's average kinetic energy $\overline{K}$ and average potential energy $\overline{U}$ over one cycle (take $t_o = 0$ and $\Delta t = T$, where T is the period) are both equal to $\dfrac{1}{4}kA^2$. Thus, on average, the object's energy spends half of its time in the form of kinetic energy and the other half in the form of potential energy.

$\left[\text{Hint: } \cos^2\theta = \dfrac{1 + \cos(2\theta)}{2} \text{ and } \sin^2\theta = \dfrac{1 - \cos(2\theta)}{2}\right]$

Section 14–5

18. (II) If you are ever marooned on a desert isle, here is a "trial-and-error" method for making a clock from a simple pendulum, where each of its swings corresponds to one second (i.e., one swing = half-period, so period T is two seconds). A major problem in constructing a pendulum clock is that the acceleration g due to gravity varies by as much as 0.5% over the surface of Earth, so the appropriate pendulum length L varies from place to place. We will assume only that we know the ballpark value for g is 9.8 m/s².
(a) Using $g = 9.8$ m/s², what is the approximate correct value to choose for L in constructing your clock?
(b) Suppose you now construct a simple pendulum using L found in part (a). Then you run your clock while watching the Sun's position in the sky and find out that instead of 86,400 swings between high-noon on two successive days, your clock executes N ($\neq$ 86,400) swings, which corresponds to an error of ΔT each pendulum period T. Show that this error ΔT can be eliminated by a change ΔL in the pendulum length given by

$$\frac{\Delta L}{L} \approx 2\frac{\Delta T}{T}.$$

This process could be iterated until the clock was very accurate.

$$\left[\text{Hint: }\Delta L \approx \frac{dL}{dT}\Delta T\right]$$

(c) If you constructed a clock with L given in part (a) and found that it executed $N = 86,300$ swings over a 24-hr time period, should its pendulum be lengthened or shortened? By how much?

19. (II) In the late 18th century, there were two competing proposals before the French Academy of Sciences for the definition of the meter. This length standard was proposed to be either the length L of a simple pendulum having a half-period of one second or one ten-millionth of the distance along a longitude line from Earth's equator to the North Pole.
(a) Although the longitude-line proposal was adopted by the Academy, show that the meter was defined consistent with the pendulum proposal. That is, show that L for a simple pendulum with a 1.0-s half-period is close (not coincidentally) to 1.00 m. Assume $g = 9.809$ m/s², the value appropriate for Paris. A half-period is the time for a pendulum to swing through just one of its arcs.
(b) While the pendulum approach to determining the length of a meter might seem much easier to implement than the longitude-line method, it was rejected because the acceleration due to gravity g varies slightly over the surface of Earth, affecting the period of the pendulum in a way that is hard to correct. Assume each of many scientists, spread around the world, built a simple pendulum and found the proper length L (at each location) so that the pendulum's half-period was 1.000 s. If each scientist then associated his or her pendulum length with the length of one meter, determine the resulting uncertainty ΔL in this length standard, if g varies by about 0.50% worldwide.

$$\left[\text{Hint: }\Delta L \approx \frac{dL}{dg}\Delta g\right]$$

20. (II) In the text, the approximation $\sin\theta \approx \theta$ is used to assist in solving the equation of motion for the simple pendulum. At small angles ($<15°$) where this approximation is valid, it is found that the bob executes SHM with period $T_o = 2\pi\sqrt{L/g}$, which is independent of the pendulum's angular amplitude θ_{max}. Using advanced calculus, it is possible to solve the pendulum's equation of motion for all angles, including large ones, resulting in the following general expression for the pendulum's period T, which does depend on θ_{max}:

$$T = T_o\left[1 + \left(\frac{1}{2}\right)^2\sin^2\left(\frac{\theta_{max}}{2}\right) + \left(\frac{1\cdot 3}{2\cdot 4}\right)^2\sin^4\left(\frac{\theta_{max}}{2}\right)\right.$$
$$\left. + \left(\frac{1\cdot 3\cdot 5}{2\cdot 4\cdot 6}\right)^2\sin^6\left(\frac{\theta_{max}}{2}\right) + \cdots\right]$$

where a larger number of terms within the square brackets must be retained for greater values of θ_{max}.
(a) Use the general expression for T (with the four terms within square brackets given above) to demonstrate that the period T, to a good approximation, is independent of the pendulum's angular amplitude at small angles of oscillation and is equal to T_o by showing that the deviation of T from T_o for both $\theta_{max} = 2°$ and $\theta_{max} = 4°$ is very small.
(b) Show that the percent difference between T_o and the actual period T is still less than 1% for $\theta_{max} = 15°$.
(c) What is the percent difference between T_o and the actual period T for the large amplitude oscillation with $\theta_{max} = 30°$?

Section 14–6

21. (III) Jeanne, a festival princess, wants some relief from waving to crowds during parades. As shown in Fig. 14–9, she constructs a mechanical arm, which pivots without friction about its "elbow-end" and is attached to two horizontal springs at a distance $d = 15$ cm above the pivot. Each spring has spring constant k and is of the perfect length so that when the arm is perfectly vertical, it is neither stretched nor compressed. During parades, Jeanne displaces the arm at a small angle θ from the vertical, releases it, and then relaxes while her device automatically waves to the crowd. Model the arm (= forearm plus gloved hand) as a uniform rod of length $L = 40$ cm and mass $M = 2.0$ kg.
(a) Prove that the mechanical arm oscillates with frequency

$$f = \frac{1}{2\pi}\sqrt{\frac{6kd^2}{ML^2} - \frac{3g}{2L}}.$$

(b) If Jeanne wants the mechanical arm to wave (back and forth) with a frequency of 1.0 Hz, what value of k should she choose to use when building her device?

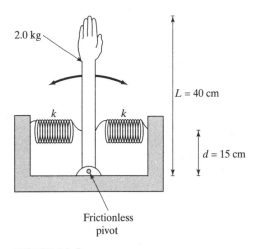

FIGURE 14–9

Section 14–7

22. (II) Designed by Thomas Jefferson, "The Great Clock" in the Entrance Hall of his home Monticello is a grandfather-type clock, which keeps time by counting the one-second swings of a pendulum (i.e., one swing = half-period, so oscillation period $T = 2.00$ s). To counteract frictional damping, energy is supplied to the clock via a cable attached to falling weights. The clock is wound every Sunday by raising the weights of mass M to the ceiling of the Hall; over the course of a seven-day period, they fall at constant velocity over a distance of about 5.0 m, until the clock is rewound the following Sunday. The lost potential energy of the weights is exactly sufficient to keep the clock's pendulum swinging with a time-independent angular amplitude $\theta_o = 12°$ from the vertical. Let's investigate how one might determine the correct value for M in order for Jefferson's clock to run for a full week.
(a) First, one needs to determine the level of damping within the clock. With the clock detached from the system of falling weights (i.e., without an outside energy source), assume that when the pendulum bob of mass $m = 1.3$ kg is displaced to angle $\theta_o = 12°$ and released, it is observed that the pendulum's angular amplitude decreases to $6°$ after 80 one-second swings. Model the clock as a damped simple pendulum of length L with the most significant damping due to the pendulum bob's motion through air so that the damping force per unit mass is $-b\omega$, where ω is the pendulum's angular velocity and b is the damping constant (see Example 14–10 in the text). Use the given data to find L (m) and b (in m/s).
(b) Next, one needs to determine the energy dissipated by friction per swing. Find the distance traveled by the bob during a single one-second swing with $\theta_o = 12°$; also find the bob's average angular speed $\overline{\omega}$ during the swing in radians per second. Then use the magnitude of the average frictional damping force $mb\overline{\omega}$ to estimate the work done by friction during one swing.
(c) Using your result from part (b), determine the necessary value for M if the falling weights are to power the clock for seven days. [At Monticello, the falling weights consist of six 18 lbs iron cannonballs with total mass of 49 kg.]

23. (II) An easy experimental method to determine the damping constant b for a mass m undergoing damped harmonic motion is as follows: measure the maximum displacement (i.e., amplitude) for some arbitrarily chosen cycle and for the cycle that immediately follows; also measure the time interval T between these maxima. Noting that the maxima occur at time t and $(t + T)$, show that b is given by

$$b = \frac{2m \ln(A_t / A_{t+T})}{T}$$

where A_t and A_{t+T} are the measured amplitudes at times t and $(t + T)$, respectively.

24. (III) Eden notices the marshmallow in her cup of hot chocolate is bobbing up and down at a well-defined frequency. She realizes that she might be able to perform a quick experiment to determine the acceleration due to gravity g and carries out the following steps: First, Eden places the cylindrical marshmallow in her hot chocolate so that it floats with its central axis vertically directed. While floating at rest, she notes that a depth $d \approx 1$ cm of the marshmallow is submerged. Eden then taps the top of the marshmallow, causing it to submerge a bit further, after which it oscillates up and down about twice each second, so the observed oscillation period $T \approx 0.5$ s. The oscillation amplitude appears to decrease to 37% of its initial value after about 2 s. Eden decides to try modeling this damping force as being proportional to the marshmallow's speed. Let A and ρ be the cross-sectional area of the marshmallow and the density of hot chocolate, respectively.
(a) Use this information to find the value Eden determines for g (m/s²).
(b) Eden is disappointed with the value she obtains for g. In broad qualitative terms, where do you think Eden erred in the theoretical model she chose for analyzing her experiment?

Section 14–8

25. (II) A 100-g mass, hung from a vertical spring with spring constant $k = 90$ N/m, is acted on by a velocity-dependent air resistance force $F_{air} = -bv$, where the damping constant $b = 0.10$ N·s/m. An external "off-resonance" (i.e., $\omega \neq \omega_o$) sinusoidal force $F_{ext} = (10 \text{ N}) \cos[(25 \text{ rad/s})t]$ is applied to the mass, and after a short time, its position oscillates according to $x = A_o \sin(\omega t + \phi)$.
(a) What is the oscillation frequency f (Hz) of the mass?
(b) Find the oscillation amplitude A_o.
(c) What is the phase difference (degrees) between the driving force and the displacement of the mass from the origin?
[Hint: Convert $x(t) = A_o \sin(\omega t + \phi_o)$ to a cosine function using $\sin(\theta) = \cos(\theta - 90°)$.]

General Problems

26. (II) While an Alaskan is out walking his pet ptarmigan (mass m), the bird accidentally slips over the edge of a shallow, icy (frictionless) spherically shaped basin of radius $R = 50.0$ m and depth $d = 1.5$ m as shown in Fig. 14–10. The ptarmigan's initial speed at the edge (Point A) is zero and its position while in the basin is described by the angle θ with respect to the vertical.
(a) Show that, as the ptarmigan moves within the basin, θ is always less than 15°.
(b) Apply Newton's second law to find the (shortest) time that the Alaskan has to wait for his pet to return to him at Point A.

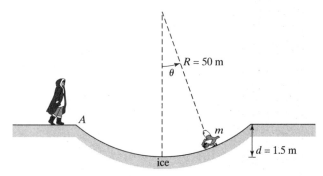

FIGURE 14–10

27. (II) In Section 14–5 of the text, the oscillation of a simple pendulum (Fig. 14–11) is viewed as linear motion along the arc length x and analyzed via $F = ma$. Alternately, the pendulum's movement can be regarded as rotational motion about its point of support and analyzed using $\tau = I\alpha$. Carry out this alternate analysis and show that $\theta(t) = \theta_{max} \cos\left(\sqrt{\frac{g}{L}} t + \phi\right)$, where $\theta(t)$ is the angular displacement of the pendulum from the vertical at time t, as long as its maximum value $\theta_{max} < 15°$.

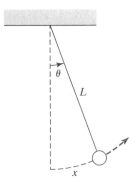

FIGURE 14–11

28. (II) A 2.0-kg mass is connected to a spring with spring constant $k = 200$ N/m and is placed on a frictionless table. The equilibrium position is at $x = 0$. The mass is displaced from its equilibrium position and starts to oscillate. If, at $t = 0$, $x = -5.0$ cm, and $v = +20$ cm/s, what is the speed of the mass each time it passes through $x = +3.0$ cm?
(a) Find the unknown speed using the results of Newton's second law applied to this situation; i.e., solve this problem using $x = A \cos(\omega t + \phi)$ and $v = -\omega A \sin(\omega t + \phi)$.
(b) Find the unknown speed using conservation of energy.

29. (III) A spring is hanging vertically, supported at its top end by a rigid support. A 2.0-kg mass is attached to the spring's lower end and is held vertically at rest when the spring is stretched 20 cm as shown in Fig. 14–12a. Next, this same mass and spring are placed on a horizontal frictionless table and the free end of the spring is anchored as shown in Fig. 14–12b. An x-axis is defined with its origin at the equilibrium position for this system and the positive direction to the left. Finally, the mass is displaced to $x = +5.0$ cm and, at that position and time $t = 0$, is released with a shove such that $v = -10$ cm/s. Find the amplitude A of the resulting simple harmonic motion by the following two methods:
(a) Applying Newton's second law (see Section 14–2 in the text), we know the position of the mass is given by $x(t) = A \cos(\omega t + \phi)$. Starting with this relation for $x(t)$, apply the initial conditions to find A.
(b) Determine A using conservation of energy.

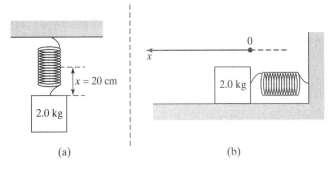

FIGURE 14–12

30. (III) Imagine a frictionless tunnel dug in a straight line through Earth, connecting two cities on its surface (Fig. 14–13a). Let's investigate the motion of a gravity-powered train of mass m moving within this tunnel.

(a) Consider the simplest possible model for Earth: a sphere of radius R with a uniform density ρ. Inside Earth, when it is a radius r away from the center, the train's weight mg is less than at Earth's surface. As described in Appendix C in the text, this reduction of mg results from the fact that only the portion of Earth's mass within the sphere of radius r contributes to the gravitation force at that location. Applying Eq. 6–4 in the text to the interior of Earth, the train's weight at a distance r from Earth's center is given by

$$mg(r) = G\frac{mM_{within}}{r^2} = \frac{Gm}{r^2}\left(\rho \frac{4}{3}\pi r^3\right) = \frac{4\pi Gm\rho r}{3}$$

where the mass within a sphere of radius r is $M_{within} = \rho \frac{4}{3}\pi r^3$, assuming the sphere has uniform density ρ (Fig. 14–13b). Define an x-axis along the direction of the tunnel with its origin at the tunnel's midpoint. Apply Newton's second law to the train in the x-direction and show that it executes SHM within the tunnel with a period $T = \sqrt{\frac{3\pi}{G\rho}}$.

(b) Taking $\rho = 5520$ kg/m³ (= average density of the real Earth), show that the travel time for the train between the two cities is 42 min.

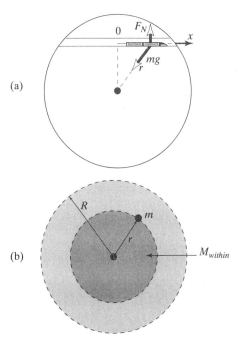

FIGURE 14–13

31. (III) A large mass M is ring-shaped with radius R and a small mass m is placed at a distance x along the ring's axis (Fig. 14–14a). Define the x-axis with its origin at the ring's center and positive to the right.

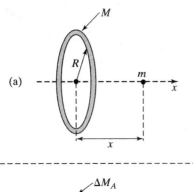

(a)

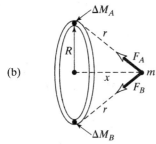

(b)

FIGURE 14–14

(a) Show that the gravitational force on the small mass due to the ring is directed along the axis and is given by

$$F = -\frac{GMmx}{(x^2 + R^2)^{3/2}}.$$

[*Hint*: Think of the ring as a collection of very small point masses ΔM, where $\sum_{\text{entire ring}} \Delta M = M$ (see Fig. 14–14b). Consider the net force ΔF that a particular pair of ΔM, located on the opposite sides of the ring, produces on the mass m.]

(b) In the limit $x \ll R$, show that the force relation in part (a) reduces to Hooke's law and determine the "spring constant" k for this system in terms of $M, m,$ and R.

(c) If the small mass moves along the x-axis such that its distance from the origin x is always much less than R, the result of part (b) tells us that it will execute simple harmonic motion. Determine the frequency f of this oscillatory motion.

CHAPTER 15

Wave Motion

Sections 15–1 and 15–2

1. (II) A vacationing man, standing on a beach, is trying to decide if he is a strong enough swimmer to swim to an offshore island. In a swimming pool, he knows he can swim 300 m, but is too tired to swim after that. To estimate the distance to the island, he observes that it takes an ocean wave 56 s to travel from the island to the beach. He estimates the distance between wave crests to be 20 m and notes that successive wave crests reach the beach every 8.0 s. Based on this information, can the man be justified in concluding that he is a strong enough swimmer to reach the island?

2. (II) An automated grid of seismometer stations could warn an operating nuclear power plant to shut down in the event of a large earthquake. Various types of seismic waves emanate from the epicenter of a large quake, including two types of "body" waves (P and S waves), the slowest of which travels at about 5.5 km/s, and the more damaging surface waves that move at 3.5 km/s. The difference in arrival times of the P and S waves at three seismograph stations can be used to locate the quake's epicenter. Suppose a grid of seismometer stations is dense enough so that there will always be 3 stations within 100 km of any epicenter and that, after obtaining both P and S wave arrival-time data from 3 stations, it takes 7.0 s for an automated computer program to locate the epicenter and generate a radio warning to the nuclear power station. Assuming a radio transmission time of 100 ms and that it takes 10 s to shut down the power plant once the warning is received, at what distance x from the epicenter can the plant be, if it is able to safely shut down before the arrival of the violent surface waves at its location?

3. (II) The elastic modulus E for a material can be measured using the pulse-echo technique. In this method, a piezoelectric transducer, which acts as both a source ("speaker") and receiver ("microphone") of longitudinal ultrasonic waves, is placed on one of the two flat faces of a cylindrical-shaped sample of length L and diameter d. The transducer emits a wave pulse that travels the length of the cylinder, reflects off the opposite face, and returns to the transducer. The transducer transforms the reflected pulse into an electrical signal, which is displayed on an oscilloscope, allowing the pulse's roundtrip transit time T in the sample to be measured. If this technique is applied to a 5.5-g uniform cylinder with $L = 3.0$ cm and $d = 0.93$ cm and T is found to be 11.8 μs, determine E for the material from which the sample is constructed. Using Table 12–1 in the text, identify the material.

4. (II) A particular sound endures in the brain as a memory for up to 0.1 s. Thus, if a reflected sound wave reaches a person within 0.1 s after the initial sound, rather than being heard as a distinct second sound ("echo"), it will be perceived by the person as a prolonging of the first sound ("reverberation"). Show why reverberation occurs in spaces (such as showers and tunnels) with highly reflective walls, whose height, width, and/or length dimensions are approximately 20 m or less.

5. (II) Except for the tiniest ripples, waves on the surface of the ocean do not depend on the properties of water such as density and surface tension. The primary "return force" for water piled up in the wave crests is due to the gravitational attraction of Earth. Thus, the speed v (m/s) of these surface water waves depends on the acceleration due to gravity g. It is reasonable to expect that v might also depend on water depth h and the wave's wavelength λ. Assume wave speed is given by the functional form $v = kg^\alpha h^\beta \lambda^\gamma$, where α, β, γ, and k are numbers without dimension.
(a) In deep water, well-submerged water does not affect the motion of waves at the surface. Thus, v becomes independent of depth h (i.e., $\beta = 0$). Using only dimensional analysis, determine the formula for the speed of surface water waves in deep water.
(b) In shallow water, the speed of surface water waves is found experimentally to be independent of the wavelength (i.e., $\gamma = 0$). Using only dimensional analysis, determine the formula for the speed of surface water waves in shallow water.

Section 15–3

6. (II) The Sun radiates light-wave energy equally in all directions. At Earth's location, spacecrafts above the atmosphere have measured the intensity of this energy to be 1350 W/m². How much light-wave energy does the Sun emit every second?

Section 15–4

7. (II) A sinusoidal traveling wave has frequency 1000 Hz and phase velocity 500 m/s.
(a) At a given time, find the distance between any two locations that correspond to a difference in phase of $\pi/6$ rad.
(b) At a fixed location, by how much does the phase change during a time interval of 1.0×10^{-4} s?

8. (II) The displacement of a bell-shaped wave pulse is described by the following relation that involves the exponential function

$$D(x, t) = A \exp[-\alpha(x - vt)^2]$$

where the constants $A = 10.0$ m, $\alpha = 2.0$ m^{-1}, and $v = 3.0$ m/s.

(a) Over the range $-10.0\text{ m} \leq x \leq +10.0\text{ m}$, use a graphing calculator or a computer to plot $D(x,t)$ at each of the three times $t = 0$, $t = 1.0\text{ s}$, and $t = 2.0\text{ s}$. Do these three plots demonstrate the wave-pulse shape shifting along the x-axis by the expected amount over the span of each one-second interval?
(b) Repeat part (a), but this time assume the displacement is given by

$$D(x,t) = A\exp[-\alpha(x+vt)^2].$$

Section 15–5

9. (II) In deriving the expression $v = \sqrt{F_T/\mu}$ for the velocity of a transverse wave on a string, it was assumed that the wave's amplitude D_M is much less than its wavelength λ. Assuming a sinusoidal wave shape $D = D_M \sin(kx - \omega t)$, show via the partial derivative $v' = \partial D/\partial t$ that the assumption $D_M \ll \lambda$ implies that the maximum transverse speed $v'_{\max}$ of the string itself is much less than the wave velocity. If $D_M = \lambda/100$, determine the ratio $v'_{\max}/v$. [Note: For functions of several variables (e.g., u is a function of the two variables x and t), a partial derivative tells how the function changes when one variable is varied, while the other variables are held constant. To compute a partial derivative, take the derivative of the function with respect to the chosen variable as usual and assume that the other variables are held fixed (i.e., they are constants).]

10. (III) Prove the displacement $D(x,t) = f(x \pm vt)$ satisfies the wave equation for any arbitrarily chosen function f. Begin by writing $D(x,t) = f(u)$, where $u = x \pm vt$ and use the chain rule for partial derivatives $\dfrac{\partial D(x,t)}{\partial \alpha} = \dfrac{df(u)}{du}\dfrac{\partial u}{\partial \alpha}$, where α can be, e.g., x or t.
[Note: For functions of several variables (e.g., u is a function of the two variables x and t), a partial derivative tells how the function changes when one variable is varied, while the other variables are held constant. To compute a partial derivative, take the derivative of the function with respect to the chosen variable as usual and assume that the other variables are held fixed (i.e., they are constants).]

Section 15–7

11. (II) Seismic reflection prospecting is commonly used to map deeply buried formations containing oil. In this technique, a seismic wave generated on Earth's surface (e.g., by an explosion or falling weight) reflects from the subsurface formation and is detected upon its return to ground level. By placing ground-level detectors at a variety of locations relative to the seismic-wave source and observing the variation in the source-to-detector travel times, the depth of the subsurface formation can be determined.
(a) Assume a ground-level detector is placed a distance x away from a seismic-wave source and that a horizontal boundary between overlaying rock and a subsurface formation exists at depth D (Fig. 15–1a). Determine an expression for the time T taken by the reflected wave to travel from source to detector, assuming the seismic wave propagates at constant speed v.
(b) Suppose several detectors are placed along a line at different distances x from the source as shown in Fig. 15–1b. Then, when a seismic wave is generated, the different travel times T for each detector are measured. Starting with your result from part (a), explain how a graph of T^2 vs. x^2 can be used to determine D.

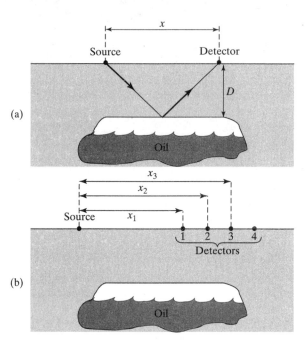

FIGURE 15–1

12. (III) Consider a right-directed traveling wave $D_I = A_I \sin(k_1 x - \omega t)$ with $A_I > 0$ incident from material 1 on its boundary with material 2, where the wave speeds in the two materials are $v_1 = \omega/k_1$ and $v_2 = \omega/k_2$. As the incident wave strikes the boundary, let the reflected wave back into material 1 and the transmitted wave into material 2 be given by $D_R = A_R \sin(k_1 x + \omega t)$ and $D_T = A_T \sin(k_2 x - \omega t)$, respectively. Define the location of the boundary as $x = 0$ (see Fig. 15–2).
(a) The total wave to the left of the origin is the sum of the incident and the reflected waves $D_{left}(x,t) = D_I + D_R$, while to the right of the origin there is only the transmitted wave $D_{right}(x,t) = D_T$. At the origin, the wave must be continuous at all times, i.e., $D_{left}(x=0,t) = D_{right}(x=0,t)$. Apply this condition to show $A_I - A_R = A_T$.
(b) At the origin, the slope of the wave (or equivalently, the partial derivative of D with respect to x) must also be continuous at all times, i.e., $\dfrac{\partial D_{left}}{\partial x} = \dfrac{\partial D_{right}}{\partial x}$ at $x = 0$ for all t. For a function of several variables (in our case, D is a function of the two variables x and t), a partial derivative tells how the function changes when one variable is varied, while the other variables are held constant. To compute a partial derivative, take the derivative of the function with respect to the chosen variable as usual and assume that the other variables are held fixed (i.e., they are constants). Apply this condition to show $\dfrac{v_2}{v_1}(A_I + A_R) = A_T$.

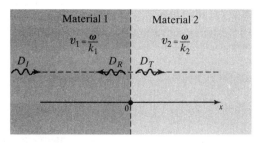

FIGURE 15–2

(c) Show that $A_R = \dfrac{v_1 - v_2}{v_1 + v_2} A_I$.

(d) Explain why, at $x = 0$, $D_{left} = (A_I - |A_R|)\sin(-\omega t)$ and $D_{left} = (A_I + |A_R|)\sin(-\omega t)$ when $v_1 > v_2$ and $v_1 < v_2$, respectively. Thus, the incident and reflected waves combine at that point as if the reflected wave (relative to the incident wave) is shifted by half a wavelength ("180° phase shift") or is unshifted in the former or latter case, respectively.

13. (III) A right-directed traveling wave $D_I = A_I \sin(k_1 x - \omega t)$, incident from material 1 on its boundary with material 2, produces a reflected wave back into the first material $D_R = A_R \sin(k_1 x + \omega t)$ with the reflection amplitude given by $A_R = \dfrac{v_1 - v_2}{v_1 + v_2} A_I$, where the wave speeds in the two materials are $v_1 = \omega/k_1$ and $v_2 = \omega/k_2$ (see Fig. 15–2).
(a) The reflection coefficient R is defined as the ratio of the reflected-to-incident wave intensities. Determine the expression for R.
(b) When you stand on a sidewalk and look at a store window, light from your face is incident upon the window glass. What percentage of this incident light intensity is reflected back to your eyes from the front face of the glass? In glass, the speed of a light wave is about 0.67 times its speed in air.
(c) If, when it is incident upon the interface between two materials, you want a large fraction of a wave's energy to be reflected back into the original material 1 (rather than transmitted into the new material 2), how should v_1 and v_2 be related?
(d) If, when it is incident upon the interface between two materials, you want none of a wave's energy to be reflected back into the original material 1 (i.e., all of the wave energy is transmitted into the new material 2), how should v_1 and v_2 be related? This important situation is called "impedance matching."

Section 15–9

14. (II) In music, two notes that have a frequency ratio of 2:1, 5:4, 4:3, and 3:2 are called an octave, third, fourth, and fifth, respectively. The six lowest allowed standing wave frequencies on a string with fixed ends are the $n = 1$, $n = 2$, $n = 3$, $n = 4$, $n = 5$, and $n = 6$ harmonics. List all pairs of these harmonics that are (a) octaves, (b) thirds, (c) fourths, and (d) fifths.

15. (II) Consider the possible standing waves on two strings, each fixed at both ends. The first string has a length of 90 cm and the second has a length of 60 cm.
(a) The 90-cm string is found to have a fundamental frequency of 400 Hz. At what velocity do waves travel on this string?
(b) The second string has the same tension F_T and mass per unit length μ as the first string. What is the lowest standing wave frequency that the two strings have in common?
(c) Assume the two strings begin with the same tension. By what factor would you have to increase the tension in the first string so that its fundamental frequency becomes the same as the fundamental frequency of the second string?

16. (II) On an electric guitar, a "pickup" under each string transforms the string vibration directly above it into an electrical signal. If a pickup is placed 16.25 cm from one of the fixed ends of a 65.00-cm electric guitar string, which of the harmonics from $n = 1$ to $n = 12$ will not be "picked up" by this pickup?

17. (II) A 65-cm guitar string is fixed at both ends. In the frequency range between 1.0 kHz and 2.0 kHz, the string is found to resonate only at frequencies 1.2 kHz, 1.5 kHz, and 1.8 kHz. What is the speed of traveling waves on this string?

18. (II) Your guitar teacher wants you to try an alternate way of tuning one of the strings on your instrument. He requests that you "tune" a 65.0-cm string, which is fixed at both ends and has a mass per unit length of 3.00 g/m, so that two of its harmonics have frequencies f of precisely 630 Hz and 4725 Hz, and additionally there are exactly 38 harmonics with frequencies in between these two (i.e., the string has 40 harmonics in the range 630 Hz $\leq f \leq$ 4725 Hz). To do this, what is the correct value for the string tension F_T?

19. (II) Two oppositely directed traveling waves given by $D_1 = (5.0 \text{ mm})\cos[(2.0 \text{ m}^{-1})x - (3.0 \text{ rad/s})t]$ and $D_2 = (5.0 \text{ mm})\cos[(2.0 \text{ m}^{-1})x + (3.0 \text{ rad/s})t]$ overlap and form a standing wave. Determine the position of nodes along the (positive and negative) x-axis.

20. (II) In an "open-cavity" helium-neon laser, a standing wave of light is created between two mirrors mounted a distance L apart on a flat optical tabletop. Since the mirrors used are almost perfectly reflecting, a node can be assumed at each end of the standing wave.
(a) The created standing wave is the nth harmonic of the laser cavity of length L. In laser terminology, this standing wave is called the nth laser mode. If $L = 0.500$ m, approximately which mode will be created between the mirrors (i.e., find the ballpark value of n).
(b) If, for some reason, the length between the mirrors increases, the laser mode will increase. Each change from n to $(n + 1)$ is called a "mode hop." If L is increased by 0.250 mm, how many mode hops will the laser experience? A helium-neon laser emits light of wavelength $\lambda = 632.8$ nm.

21. (III) Suppose we posit that the wave velocity v on a string with mass per unit length μ is related to the string tension F_T by the following power law $v = AF_T^N$, where A and N are constants. This relation can be verified using the experimental setup shown in Fig. 15–3. One end of a string is attached to an oscillator with fixed oscillation frequency f and very small oscillation amplitude, while the other end passes over a pulley and supports a hanging mass M at rest. The horizontal portion of the string with length L then oscillates at frequency f with (to a good approximation) its ends fixed and the string's tension $F_T = Mg$ can be varied by changing M. The collection of data then consists of finding the masses M_1, M_2, M_3, etc., which produce the $n = 1$, $n = 2$, $n = 3$, etc., harmonic standing waves on the string. These data can then be analyzed

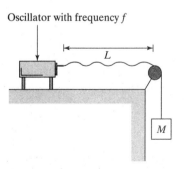

FIGURE 15–3

by plotting $\log(n)$ [y-axis] vs. $\log(Mg)$ [x-axis]. Show that, assuming our posited power law, this plot should yield a straight line and explain how N and A can be determined from the slope and y-intercept of this line. [Note: If, in this experimental process, the analysis of data yields $N = 1/2$ and $A = 1/\sqrt{\mu}$, then the experiment supports the theory given in Sections 15–2 and 15–5 of the text.]

22. (III) Consider a string of length $L = 65.0$ cm with both ends fixed by supports, vibrating at its fundamental frequency $f_1 = 330$ Hz. Now assume that the length of this string "grows" by a small amount ΔL, while at the same time the distance between its supports increases by ΔL, so that the string's tension and mass per unit length (and wave velocity v) remain unchanged (see Fig. 15–4). The fundamental frequency will change by Δf_1 during this process.

(a) Show that $\Delta L \approx -\dfrac{v}{2f_1^2}\Delta f_1$.

(b) If a certain instrument is able to measure a change in frequency as small as 0.1 Hz in this situation, what minimum ΔL could be detected? [The growth of thin film materials on a quartz disk is commonly determined by an analogous method.]

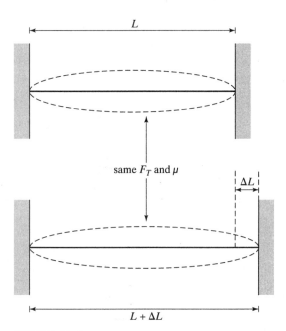

FIGURE 15–4

23. (III) A wire is composed of aluminum with length $L_1 = 0.600$ m and mass per unit length $\mu_1 = 2.70$ g/m joined to steel with length $L_2 = 0.882$ m and mass per unit length $\mu_2 = 7.80$ g/m. This composite wire is fixed at both its ends and is held at a uniform tension $F_T = 120$ N. Find the lowest frequency standing wave that can exist on this wire, assuming there is a node at the joint between aluminum and steel. How many nodes (including the two at the ends) does this standing wave possess?

Section 15–10

24. (II) Fishermen on the shore of a lake or bank of a stream often try to be quiet so as not to scare fish away. Use your knowledge of refraction to determine how far back from the shore a 1.7–m tall fisherman has to stand so that the sound of his voice cannot be heard by fish in the water (see Fig. 15–5). At this location, the sound, when incident from air, will be bent so that it travels along the water's surface. The speed of sound is about 343 m/s in air and 1440 m/s in water. Assume that the fisherman's voice does not propagate through the ground.

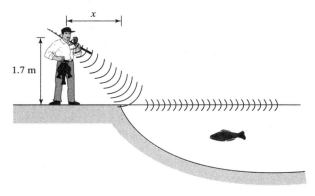

FIGURE 15–5

General Problems

25. (II) Manufacturers typically offer a particular guitar string in a choice of diameters so that a player can tune his or her instrument with a preferred string tension. For example, a nylon high-E string is available in a low- and a high-tension model with diameter 0.699 mm and 0.724 mm, respectively. Assuming the density ρ of nylon is the same for each model, compare (as a ratio) the tension in a tuned high-tension string with that in a tuned low-tension string. When mounted on a tuned guitar, a high-E string is fixed at both ends with length $L = 65.0$ cm and has a fundamental frequency $f_1 = 329.6$ Hz.

26. (II) The high-E string on a guitar is a string fixed at both ends with length $L = 65.0$ cm and fundamental frequency $f_1 = 329.6$ Hz. On an acoustic guitar, this string typically has a diameter of 0.33 mm and is commonly made of brass (7760 kg/m^3), while on an electric guitar it has a diameter of 0.25 mm and is made of nickel-coated steel (7990 kg/m^3). Compare (as a ratio) the high-E string tension on an acoustic guitar with that on an electric guitar.

27. (II) Two wave pulses $D_1(x - vt)$ and $D_2(x + vt)$ on a string are shown in Fig. 15–6 at an initial time $t = 0$. The middle of the string is defined as $x = 0$. The string is held at a tension $F_T = 100$ N and has a mass per unit length of 2.5 g/m. Sketch the shape of the rope at $t = 20$ ms, $t = 25$ ms, and $t = 30$ ms.

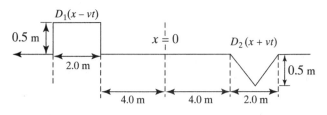

FIGURE 15–6

28. (II) Consider a sound wave, which travels at 340 m/s and has a frequency of 500 Hz, and a light wave, which travels at 3.0×10^8 m/s and has a frequency of 6.0×10^{14} Hz. Explain why the sound wave "bends around a corner" when passing through a 1.0-m wide door, while the light wave appears to travel in a straight line when passing through this same opening.

29. (III) Consider a string of length $L = 0.50$ m with its ends located at $x = 0.00$ and $x = 0.50$ m, both of which are fixed. The first five harmonics of this string have wavelengths $\lambda_1 = 1.00$ m, $\lambda_2 = \frac{1}{2}$ m, $\lambda_3 = \frac{1}{3}$ m, $\lambda_4 = \frac{1}{4}$ m, and $\lambda_5 = \frac{1}{5}$ m. According to Fourier's theorem, any shape of this string can be formed by a sum of its harmonics, with each harmonic having its own unique amplitude A. If we only include the first five harmonics in such a summation, the displacement D of the string at a time $t = 0$ is given by following expression

$$D(x, 0) = A_1 \sin\left(\frac{2\pi}{\lambda_1} x\right) + A_2 \sin\left(\frac{2\pi}{\lambda_2} x\right) + A_3 \sin\left(\frac{2\pi}{\lambda_3} x\right) + A_4 \sin\left(\frac{2\pi}{\lambda_4} x\right) + A_5 \sin\left(\frac{2\pi}{\lambda_5} x\right).$$

Imagine plucking this string at its midpoint, distorting it into a triangular shape as shown in Fig. 15–7 at $t = 0$. Using a graphing calculator or a computer,

(a) show that the above expression can fairly accurately represent the string's triangular shape at $t = 0$, if $A_1 = 1.00$, $A_2 = 0.00$, $A_3 = -0.11$, $A_4 = 0.00$, and $A_5 = 0.040$.

(b) show that, by including terms up to the eleventh harmonic in the summation for D, the expression becomes a (slightly)

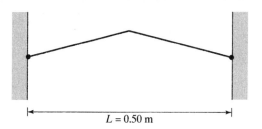

FIGURE 15–7

more accurate representation of the triangular shape (e.g., the "corner" becomes sharper) than in part (a), if for the nth harmonic $\lambda_n = (1.00 \text{ m})/n$ and $A_n = \dfrac{\sin(n\pi/2)}{n^2}$.

30. (III) Consider a right-directed traveling wave $D_I = A_I \sin(k_1 x - \omega t)$ with $A_I > 0$ incident from material 1 on its boundary with material 2, where the wave speeds in materials 1 and 2 are $v_1 = \omega/k_1$ and $v_2 = \omega/k_2$ and the densities are ρ_1 and ρ_2, respectively. As the incident wave strikes the boundary, the reflected wave back into material 1 and the transmitted wave into material 2 are given by $D_R = A_R \sin(k_1 x + \omega t)$ and $D_T = A_T \sin(k_2 x - \omega t)$, respectively. It can be shown that the amplitude of the reflected wave is $A_R = \dfrac{v_1 - v_2}{v_1 + v_2} A_I$. Apply conservation of energy to the incident, reflected, and transmitted waves and show, with the help of the given expression for A_R, that

$$A_T = \sqrt{\frac{\rho_1}{\rho_2} \frac{2v_1}{v_1 + v_2}} A_I.$$

CHAPTER 16

Sound

Section 16–1

1. (II) At many schools, a motion sensor is used to measure distance in instructional physics lab experiments. This device can accurately measure the distance d from itself to an object of interest via the "sonar" technique used in older autofocus cameras. A short ultrasonic pulse is emitted from the motion sensor, which reflects from any objects it encounters, creating "echo" pulses upon their arrival back at the sensor. The motion sensor measures the time interval T between the emission of the original pulse and the arrival of the first echo. Assume, as is typical, the 20°C velocity of sound $v = 343 \text{ m/s}$ is programmed into the motion sensor's software driver.
   (a) The time interval T can be measured with high precision only if it is at least 1.0 millisecond ($= 0.0010 \text{ s}$) long. What is the smallest distance (at 20°C) that can be measured with the motion sensor?
   (b) By default, the motion sensor makes 20 distance measurements every second (i.e., it emits 20 sound pulses per second at evenly spaced time intervals). Thus, the measurement of T must be completed within the time interval between the emissions of successive pulses. What is the largest distance (at 20°C) that can be measured with the motion sensor?
   (c) Assume during a lab period that the room's temperature increases from 20°C to 23°C. What percent error will this introduce into the motion sensor's distance measurements?

2. (II) On a 20°C day, you view a distant carpenter who is repeatedly hammering the head of a nail exactly once every second. With each hammer swing, you hear the sound of the hammer striking the nail at the exact same instant that you see the hammer striking the nail. What is your (minimum) distance from the carpenter?

Section 16–2

3. (II) The pressure change ΔP of a 400 Hz sound wave in air is given by

$$\Delta P = -\Delta P_M \cos\left[\frac{2\pi}{\lambda}x - 2\pi f t\right].$$

At a particular location x, assume that $\Delta P = +\frac{\Delta P_M}{3}$ at time t. Determine the two possible values for ΔP at that same location exactly 1.0 ms later.

Section 16–3

4. (II) At a painfully loud concert, a 120 dB sound wave travels away from a loudspeaker at 343 m/s. How much sound wave energy is contained in each 1.0 cm³ volume of air in the region near this speaker?

5. (II) If the pressure amplitude of a sound wave is increased by a factor of two, by how much does the sound level of the sound wave increase?

6. (II) At an aerial fireworks show, a shell explodes 100 m above the ground, creating a colorful display of sparks. How much greater is the sound level of the explosion for a person who is standing at a point directly below the explosion than for a person who is a horizontal distance of 200 m away (Fig. 16–1)?

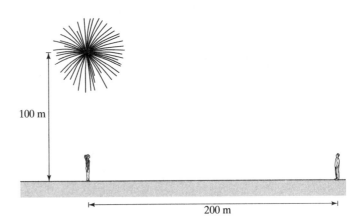

FIGURE 16–1

7. (II) A pair of inexpensive headphones can produce a sound level of up to 105 dB in a listener's ears. If somehow several such headphones, each playing a different song at 105 dB, could all be attached at a listener's ears, how many would be required to create a sound level equivalent to that in the front row at a loud concert (120 dB)? [*Hint:* Since each headphone plays a different song, the average total intensity will be the sum of the intensities from each headphone. Alternately, if they all played the same song, the amplitudes from each headphone would add up (i.e., there would be constructive interference).]

8. (II) The sound level of a lion's roar is measured to be about 90 dB at a distance from the lion of 10 m. If air did not absorb sound energy, how far away (km) could the roar be heard by a person? The roar consists mostly of frequencies in the range of 120 to 240 Hz, where the threshold of hearing is about 25 dB. [Because, in reality, air does absorb 200-Hz sound (at a rate of 1.8 dB/km), a lion's roar can be heard from a distance of 8 km on a savannah.]

9. (II) In your aisle seat in the first row at a loud concert, which is 4.0 m directly in front of a loudspeaker on stage, the sound level is 120 dB. If you think the music is too loud and want to move away from the stage to an aisle seat where the sound level is 100 dB, what row should you move back to? Assume the aisle seat in each successive row is 1.0 m farther away from the loudspeaker.

Section 16–4

10. (II) (a) On the planet Tatooine, lightning strikes 1080 meters from the Star Wars bar. Inside the bar, Luke Skywalker and Chewbacca hear the thunder 5.0 seconds after seeing the lightning flash. What is the speed of sound in the atmosphere of planet Tatooine?
    (b) Inspired by the romance of the storm, Chewbacca wants to capture the attention of a Wookiee standing across the room by emitting a mating call that will cause the atmospheric gases in its intracranial cavity to vibrate in the especially meaningful fifth-harmonic standing wave pattern. Chewbacca knows that the hole in a Wookiee's head is a tube 60-cm long and open at both ends. Chewbacca emits his call with his vocal cord that is a 40-cm long string, fixed at both ends, whose mass is 0.010 kg. Chewbacca decides to vibrate his string in the second-harmonic standing wave pattern. If the frequency of his call equals the frequency of the Wookiee's fifth harmonic, the gases will vibrate in the desired pattern and true love will follow. What tension should be in Chewbacca's vocal cord?

11. (II) A bugle is simply a tube of fixed length that behaves as if it is open at both ends. A bugler, by adjusting his lips correctly and blowing with proper air pressure, can cause a harmonic (usually other than the fundamental) of the air column within the tube to sound loudly. The standard military bugle, although unable to play all the notes in the musical scale, can be used to play famous tunes like *Taps* and *Reveille*, which only require the following four musical notes: G4 (392 Hz), C5 (523 Hz), E5 (659 Hz), and G5 (784 Hz).
    (a) For a certain length L, a bugle will have a sequence of four harmonics whose frequencies very nearly equal those associated with the notes G4, C5, E5, and G5. Determine this L.
    (b) Which harmonic is each of the (approximate) notes G4, C5, E5, and G5 for the bugle?

12. (II) A tube with a cap on one end, but open at the other end, produces a standing wave whose fundamental frequency is 130.8 Hz. The speed of sound is 343 m/s.
    (a) If the cap is removed, what is the new fundamental frequency?
    (b) How long is the tube?

13. (II) On a steel-string acoustic guitar, all of the six strings are fixed at both ends with length $L = 65$ cm and are under nearly the same tension $F_T = 130$ N. The fundamental frequency f_1 of each string is different because the mass per unit length μ of the strings differs over a range of 0.70 to 11 g/m. With this design, the instrument must be sturdy enough to withstand a total force due to the strings of 6×130 N $= 780$ N. Consider an alternate design for a guitar in which μ is the same for each string, say 5.0 g/m and tuning is achieved by varying the tension in each string. How much total force would the strings exert on the instrument in this case? The tuned frequencies of the six guitar strings are 329.6, 246.9, 196.0, 146.8, 110.0, and 82.4 Hz.

14. (II) On a nylon-string guitar, the three high-pitch strings are composed of pure nylon $(\rho = 1100 \text{ kg/m}^3)$, but the three low-pitch strings have a nylon core with metal wire wrapped over it. Let's explore why. When mounted on a tuned guitar, all of the six strings are fixed at both ends with length $L = 65$ cm and are under nearly the same tension $F_T = 70$ N. The fundamental frequency f_1 of each string is different. For the highest- and lowest-pitch string, $f_1 = 329.6$ Hz and 82.4 Hz, respectively.
    (a) Show that the required mass per unit length μ for a guitar string of given fundamental frequency is

$$\mu = \frac{F_T}{4f_1^2 L^2}.$$

   Determine the required μ-value for the nylon-string guitar's highest- and lowest-pitch string.
    (b) Guitar makers generally want the string diameters on their instruments to vary by, at most, a factor of two. Show that if all of the strings on a guitar were made of pure nylon, the ratio of diameters between the lowest- and highest-pitch strings would be unacceptably large.

15. (II) When a player's finger presses a guitar string down onto a fret, the length of the vibrating portion of the string is shortened, thereby increasing the string's fundamental frequency (see Fig. 16–2). The string's tension and mass per unit length remain unchanged. If the unfingered length of the string is $L = 65.0$ cm, determine the positions x of the first six frets, if each fret raises the pitch of the fundamental by one musical note in comparison with the neighboring fret. On the equal tempered musical scale, the ratio of frequencies of neighboring notes is $2^{1/12}$.

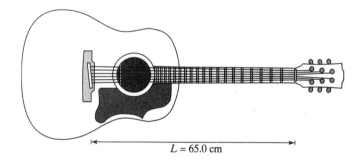

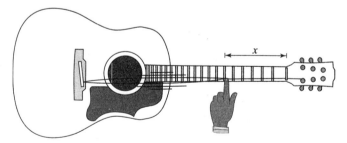

FIGURE 16–2

16. (II) A tube of length L, closed at one end, has nine overtones in the audible range. From this information you can determine a range of possible values for L (i.e., $L_{\min} < L < L_{\max}$). Find $L_{\min}$ and $L_{\max}$, assuming the velocity of sound is 340 m/sec and the audible range ends at 20,000 Hz.

17. (II) Determine the fundamental and first overtone frequencies for an 8.00-m long hallway with all doors closed. Model the hallway as a tube closed at both ends.

18. (II) In a quartz oscillator, a transverse (shear) standing sound wave is excited across the thickness t of a quartz disk and its frequency f is detected electronically. The parallel faces of the disk are unsupported and so behave as "free ends" when the sound wave reflects from them (see Fig. 16–3). If the oscillator is designed to operate with the first harmonic, determine the required disk thickness if $f = 10\,\text{MHz}$. The density and shear modulus of quartz are $\rho = 2650\,\text{kg/m}^3$ and $G = 2.95 \times 10^{10}\,\text{N/m}^2$. The velocity of a transverse sound wave in a solid is given by $v = \sqrt{G/\rho}$; see Section 15–2 in the text. [Quartz oscillators are used as stable clocks in electronic devices.]

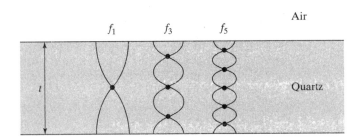

FIGURE 16–3

19. (II) In a quartz oscillator, a standing sound wave is excited across the thickness t of a quartz disk and its frequency f is detected electronically. The parallel faces of the disk are unsupported and so behave as "free ends" when the sound wave reflects from them (see Fig. 16–3). High-frequency oscillators are usually designed to operate in high-overtone modes because the fundamental mode would require such a small (and fragile) disk thickness. For $f = 100\,\text{MHz}$, determine t, if the quartz oscillator excites the fifth-harmonic standing wave. How many times larger is this thickness than that needed if the oscillator operated using the first harmonic? The speed of sound in quartz is $3340\,\text{m/s}$. [Quartz oscillators are used as stable clocks in electronic devices.]

20. (III) In a quartz crystal microbalance (QCM), the growth of a thin-film layer on the surface of a quartz disk is monitored by changes in a standing wave frequency. In this device, a standing sound wave is excited across the thickness of the quartz disk and its frequency f is detected electronically. The parallel faces of the disk are unsupported and so behave as "free ends" when the sound wave reflects from them (see Fig. 16–3). Assume that the disk has an initial thickness t, but that a vapor above the disk deposits a thin film of material on its upper face, increasing the disk thickness by a small amount Δt. Further assume that the thin film is rigidly attached to the disk and, for simplicity, that the speed of sound in it is the same as in quartz.
(a) Assume that prior to thin-film growth the third harmonic oscillation with frequency f_3 is excited across the thickness of the disk. If a thin film is then deposited, show that the resulting thickness change Δt is related to the change in third-harmonic frequency f_3 by

$$\Delta t = -\frac{3v}{2f_3^2}\Delta f_3$$

where $v = 3340\,\text{m/s}$ is the velocity of sound in quartz.

(b) For a typical QCM, $f_3 = 5\,\text{MHz}$ and a decrease in f_3 can be measured to an accuracy of about $0.1\,\text{Hz}$. Determine the minimum thin-film thickness (in nm) that can be detected by a QCM. [In the fabrication of layered semiconductor devices (such as diode lasers), QCMs are often used to monitor the growth of the thin-film layers.]

Section 16–5

21. (III) The manner in which a string is plucked determines the mixture of harmonic amplitudes in the resulting wave. Consider a $\frac{1}{2}$ m-long string that is fixed at both its ends located at $x = 0.0$ and $x = \frac{1}{2}\,\text{m}$. The first five harmonics of this string have wavelengths $\lambda_1 = 1.0\,\text{m}$, $\lambda_2 = \frac{1}{2}\,\text{m}$, $\lambda_3 = \frac{1}{3}\,\text{m}$, $\lambda_4 = \frac{1}{4}\,\text{m}$, and $\lambda_5 = \frac{1}{5}\,\text{m}$. According to Fourier's theorem, any shape of this string can be formed by a sum of its harmonics, with each harmonic having its own unique amplitude A, where we limit the sum to the first five harmonics in the following expression

$$D(x) = A_1 \sin\left(\frac{2\pi}{\lambda_1}x\right) + A_2 \sin\left(\frac{2\pi}{\lambda_2}x\right) + A_3 \sin\left(\frac{2\pi}{\lambda_3}x\right) + A_4 \sin\left(\frac{2\pi}{\lambda_4}x\right) + A_5 \sin\left(\frac{2\pi}{\lambda_5}x\right).$$

Here D is the displacement of the string at a time $t = 0$. Imagine plucking this string at its midpoint (Fig. 16–4a) or at a point two-thirds from the left end (Fig. 16–4b). Using a graphing calculator or a computer, show that the above expression can fairly accurately represent the shape in
(a) Fig. 16–4a if $A_1 = 1.00$, $A_2 = 0.00$, $A_3 = -0.11$, $A_4 = 0.00$, and $A_5 = 0.040$.
(b) Fig. 16–4b if $A_1 = 0.87$, $A_2 = -0.22$, $A_3 = 0.00$, $A_4 = 0.054$, and $A_5 = -0.035$.

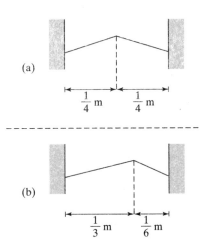

FIGURE 16–4

Section 16–6

22. (II) Jeremy is working on a "silence machine." He places two speakers 3.0 m apart and stands in front of one of the speakers at a distance of 4.0 m as shown in Fig. 16–5. Both speakers are

connected to the same source, so they emit sound waves of the same frequency and they are in phase. Jeremy starts listening when the frequency is 20 Hz. He gradually increases the frequency and eventually encounters a frequency at which he hears almost no sound. What is this frequency?

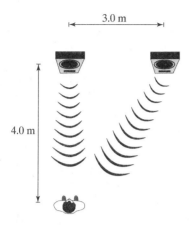

FIGURE 16–5

23. (II) When pure tones with frequencies 440 and 448 Hz are sounded together, what pitch is heard and what is the frequency of beats? Repeat the question for the frequencies 440 and 432 Hz sounded together.

Section 16–7

24. (II) If a speaker mounted on an automobile broadcasts a song, with what speed (km/h) does the automobile have to move toward a stationary listener so that the listener hears the song with each musical note shifted up by one note in comparison with the song heard by the automobile's driver? On the equal tempered musical scale, the ratio of frequencies of neighboring notes is $2^{1/12}$.

25. (II) Assume a sports car can attain speeds of up to 200 km/h. If a speaker mounted on this car broadcasts a musical note C (262 Hz) as heard by the car's driver, and the car is driven at its maximum speed toward a stationary listener, approximately what musical note does the listener hear (see Table 16–3 in the text)?

26. (II) A wave on the surface of the ocean has wavelength 40 m and is traveling east at a speed of 20 m/s relative to the ocean floor. If, on this stretch of ocean surface, a powerboat is moving at 15 m/s (relative to the ocean floor), how often does the boat encounter a wave crest, if the boat is traveling (a) west? (b) east?

General Problems

27. (II) A man is standing between a parked car and a car that is moving toward him with velocity v. The drivers in each car are honking their horns, trying to signal to the man that he is in the path of the moving car. The horns are identical and emit sound at a frequency of 500 Hz. Unfortunately, the man is a physicist and has found it interesting that when the horns are honking at him, he hears a beat frequency of 15 beats per second and is busy calculating the speed of the oncoming car. If he is quick at algebra, what value does he obtain for v?

28. (II) Two identical tubes, each closed at one end, have a fundamental frequency of 440 Hz at 25.0°C. In one tube the air temperature is increased to 30.0°C. If the two pipes are sounded together, what beat frequency results?

29. (II) A man stands between two fixed-end strings that are both vibrating in their fundamental standing-wave patterns. The strings have the same length $L = 65$ cm and the same mass per unit length $\mu = 0.010$ kg/m. When standing still, the man hears the same frequency coming from each string, but when he walks toward the string at his right with a speed $v = 4.0$ m/s, he hears a beat frequency of 6.0 beats/s. Assuming the speed of sound is 340 m/s, what is the tension in each string?

30. (III) Imagine an acoustical space consisting of two parallel walls separated by a distance $L = 10$ m, with room temperature air in between, as shown in Fig. 16–6. A person standing next to the left wall emits a short burst of sound with initial intensity I directed toward the right wall, and this sound wave subsequently bounces between the two walls. Assume that, due to transmission into the wall, at each reflection from a wall, the reflected intensity is only a fraction $\alpha = 0.70$ of the incident intensity (a value characteristic for bare concrete blocks). Also, assume that the sound wave travels only along the horizontal axis.
(a) Find the number of reflections after which the sound level is 60 dB less than its initial sound level.
(b) How long does it take for the sound level to be reduced by 60 dB from its initial value? This time can be defined as the "reverberation time" of the acoustical space.

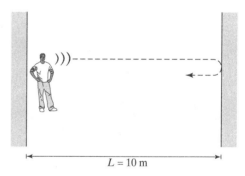

FIGURE 16–6

CHAPTER 17

Temperature, Thermal Expansion, and the Ideal Gas Law

Section 17–1

1. (II) Assume you have a 1-liter bottle filled with pure water (call this 1000 cm³ water sample the "original water"). You dump all of the original water into the ocean and wait for the amount of time necessary for the original water's molecules to be uniformly dispersed throughout the ocean. After this long wait, you refill the 1-liter bottle with ocean water. Estimate how many of the original water's molecules end up back in the bottle when it is refilled. Assume the ocean covers 70% of Earth's surface, has an average depth of 4 km, is composed completely of pure water, and no water evaporates from it.

Section 17–2

2. (II) The numerical value for a system's temperature on the Fahrenheit and Celsius scales is the same, if the system's temperature is $-40°$. Based on this fact, one can convert between the two scales using the following method: (1) add 40 to the given temperature regardless of whether you are converting from °F to °C or vice versa; (2) multiply that result by 5/9 or 9/5 if converting from °F to °C or °C to °F, respectively; and (3) subtract 40 from that result. Prove that this method works both when converting from °F to °C as well as from °C to °F.

Section 17–4

3. (II) At a given latitude, ocean water in the so-called "mixed layer" (from the surface to a depth of about 50 m) is at approximately the same temperature due to the mixing action of waves. Assume that because of global warming the temperature of the mixed layer at each latitude is increased by 0.5°C, while the temperature of the deeper portions of the ocean remains unchanged. Estimate the resulting rise in sea level. Recall that the ocean covers 70% of Earth's surface.

4. (II) Under normal operating conditions, a typical car has 17.0 L of liquid coolant ("antifreeze") circulating at a temperature of 93°C through its engine's cooling system. Assume that, in this normal condition, the coolant completely fills the 3.5-L volume within an aluminum radiator and the 13.5-L internal cavities within the steel engine. When a car overheats, the radiator, engine, and coolant expand, so a small reservoir is connected to the radiator to catch any resultant coolant overflow. How much coolant will overflow to the reservoir if this system is heated from 93°C to 105°C? Model the radiator and engine as a hollow thin shell of aluminum and steel, respectively. The coefficient of volume expansion for coolant is $\beta = 410 \times 10^{-6} \, (\text{C}°)^{-1}$.

5. (II) In our everyday experience, the thermal expansion of solid objects occurs at a level that is almost imperceptible. Instead of, say, meter-long objects expanding by a noticeable 1 cm, they instead expand by 0.1 mm or less because we do not commonly experience vast temperature changes. What does it take to get an obvious thermally induced increase in length? Consider using aluminum, which has one of the largest coefficients of linear expansion among the common metals, to fashion a rod of length 100 cm at room temperature. By how much would you have to increase the temperature of this rod in order for it to grow in length by 1 cm? Assume that aluminum's coefficient of linear expansion remains constant over this entire heating process. Would the rod melt before it reached a length of 101 cm? The melting point of aluminum is 660°C.

6. (II) In an "open-cavity" helium-neon laser, a standing wave of light with wavelength $\lambda = 633 \, \text{nm} \, (1 \, \text{nm} = 10^{-9} \, \text{m})$ is created between two mirrors mounted a distance L apart on a flat optical tabletop.
(a) The created standing wave is the nth harmonic of the laser cavity where $n = \dfrac{L}{(\lambda/2)}$ (see Eq. 15–17 in the text). In laser terminology, this standing wave is called the nth laser mode. Assume that in the early morning when this laser is first set up, the laboratory room is at a temperature T and the distance between the mirrors is exactly $L = \dfrac{1}{2}$ m. Which mode (to 3 significant figures) will be created between the mirrors (i.e., determine the approximate value of n)?
(b) If the length between the mirrors increases by a distance equal to $\dfrac{\lambda}{2}$, the laser mode will change from n to $(n + 1)$. This event is called a "mode hop." If, by late morning, the laboratory's room temperature has increased by 1C°, how many mode hops will the laser experience? Assume the optical tabletop is the common type, which is composed of stainless steel.
(c) Repeat part (b), but this time assume the optical tabletop is composed of the expensive alloy called Super Invar, which has the extremely low coefficient of linear expansion value of $0.2 \times 10^{-6} \, (\text{C}°)^{-1}$.

7. (II) Wine bottles are never completely filled. Instead, a small volume of air (called "headspace") is left in the glass bottle's cylindrically shaped neck (inner diameter $d = 18.5$ mm) to allow for wine's fairly large coefficient of thermal expansion. The distance H between the surface of the liquid contents and the bottom of the cork is called the "headspace height" (Fig. 17–1). A wine bottle contains 750 ml of liquid and, when

bottled at 20°C, vintners typically choose H to be 1.5 cm. Consider such a typical unopened bottle. Determine its resulting headspace height H (expressed as a percent of its 20°C value) if, after being purchased, this bottle is placed in a 30°C room. Due to its alcoholic content, wine's coefficient of volume expansion is typically double that of water; in comparison, the thermal expansion of glass can be neglected.

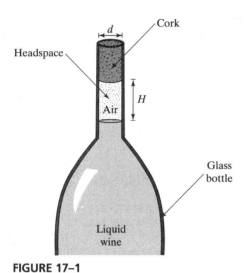

FIGURE 17–1

Section 17–5

8. (II) Assume a fluid is completely enclosed in an absolutely rigid container so that its volume cannot vary. Show that, if the temperature of the fluid is changed by an amount ΔT, then there will be a change in the fluid pressure $\Delta P = B\beta\Delta T$, where B is the bulk modulus for the fluid and β is the coefficient of volume expansion.

Sections 17–7 and 17–8

9. (II) In an airport shop, you buy an "air-tight" bag of potato chips, which was packaged in a processing plant at sea level, and place the chips in your carry-on luggage. You then board an airplane and when it reaches cruising altitude, you're ready for a snack. When you take the potato chips out of your luggage, before opening the bag you notice that it has noticeably "puffed up." Given that airplane cabins are typically pressurized at 0.75 atm, and assuming the absolute temperature inside an airplane is about the same as inside a potato chip processing plant, by what percentage has the bag "puffed up" in comparison with when it was packaged?

10. (II) A typical scuba tank, when fully charged, contains 12 L of air at 204 atm. Assume an "empty" tank contains air at 34 atm and is connected to an air compressor at sea level. The air compressor intakes air from the atmosphere, compresses it to high pressure, and then inputs this high-pressure air into the scuba tank. If the (average) flow rate of air from the atmosphere into the intake port of the air compressor is 300 L/min, how long will it take to fully charge the scuba tank? Assume the tank remains at the same temperature as the surrounding air during the filling process.

11. (II) If air supply were the only limiting factor, to what depth could a recreational diver descend with conventional scuba equipment? Let's answer this question by determining the depth at which a diver would exhaust all of his or her available air within a very short time, say, ten minutes. The regulator portion of a scuba apparatus takes high-pressure air from the tank and converts it to the pressure of the immediate environment. Assume that, at sea level, an active scuba diver breathes air at a rate of 10 L/min and that water temperature is the same at all depths that a diver can access.
    (a) A standard scuba tank, when fully charged, stores 12 L of air at a pressure of 204 atm. How long can this tank supply breathing air to a diver at the ocean surface?
    (b) Given that scuba apparatus supplies air to a diver at the pressure of the surrounding water and that a scuba diver's lungs maintain their same volume regardless of depth, at what depth would a diver use the entire tank's air supply in ten minutes? Assume it takes negligible time to descend to (and then ascend from) this depth.
    [In reality, recreational divers are limited to a maximum depth of 40 m due to the problem of nitrogen dissolving in the blood.]

12. (II) At a dive shop, a scuba tank is filled with air to a pressure of 204 atm when the air temperature is 29°C. A diver takes the tank out on a boat, puts it on, and jumps into the ocean. After a short time treading water on the ocean surface, the diver checks the tank's pressure and is surprised to find that it is only 184 atm. Assuming the diver has inhaled a negligible amount of air from the tank, what is the temperature of the ocean water?

13. (II) Assume that in an alternate universe the laws of physics are very different from ours. In particular, say that "ideal" gases behave as follows:
    i. At constant temperature, pressure is inversely proportional to the square of the volume per mole.
    ii. At constant pressure, the volume per mole varies directly with the 2/3 power of the temperature.
    iii. At 273.15 K and 1 atm pressure, 1 mole of an ideal gas is found to occupy 22.4 L.
    Obtain the form of the ideal gas law in this alternate universe, including the value of the gas constant R.

14. (II) A 3.0-L helium gas cylinder for sale at a party store is advertised as being capable of inflating approximately 35 latex balloons. If the cylinder is pressurized to 135 atm, what diameter balloons are assumed in the advertisement? Take the balloons to be perfectly spherical in shape. When inflated, the pressure inside a balloon is approximately equal to atmospheric pressure.

15. (II) In order to properly function, a temperature control system in your laboratory requires that nitrogen gas at slightly above atmospheric pressure be continuously flowed through it at a rate of 5.0 L/min (and then exhausted to the atmosphere). You decide to supply the nitrogen to your system from a nitrogen gas cylinder, which can be delivered to your lab in portable 49-L cylinders. When delivered, the nitrogen within the cylinder is at a pressure of 200 atm; using a regulator, this high-pressure nitrogen can be drawn from the cylinder and converted to slightly above atmospheric pressure. How often will a new nitrogen cylinder need to be delivered to your lab, given the planned rate of usage?
    Assume that your lab is at sea level, that it is maintained at a temperature of about 20°C, and that you run the temperature control system 8.0 hours per day.

16. (II) A typical scuba tank, when fully charged at 20°C, contains 12 L of air at a pressure of 204 atm. What is the total mass (kg)

of the pressurized air and how does this mass compare (express as a percentage) with the mass of the metal tank? When it contains no air, the metal tank has a mass of 14 kg. The average molecular mass of air is 29.

17. (II) At a depth of 30 m in the ocean, a balloon is filled with air from a scuba tank to a volume of 2.0 liters. The tank supplies air at a pressure equal to that of the surrounding seawater. The balloon is then tied so air cannot escape from it and it is taken slowly to the surface (i.e., slow enough that the balloon's air temperature always equals that of the surrounding water). Assuming that the absolute temperature of the seawater changes insignificantly with depth, what is the volume of air in the balloon at the ocean surface?

18. (II) A stoppered test tube traps 25 cm³ of air at a pressure of 1 atm and a temperature of 20°C. The cylindrically shaped stopper at the test tube's mouth has a diameter of 1.50 cm and will "pop off" the test tube if a net force of 10 N is applied to it. To what temperature would one have to heat the trapped air in order to "pop off" the stopper? Assume the air surrounding the test tube is always at a pressure of 1.0 atm. Neglect the stopper's weight.

19. (III) To open a bottle of wine, a net upward force F must be applied to the cork in order to extract it from the bottle's cylindrically shaped neck (inner diameter of 18.5 mm). One bottle-opening method takes advantage of the "headspace" in the unopened bottle's neck (i.e., the volume of trapped air between the bottom of the cork and the surface of the liquid contents). The headspace height H is typically 1.5 cm (see Fig. 17–1). In this method a syringe-like needle, which is attached to a small canister filled with a pressurized liquid/gas mixture of propellant, is pressed through the cork, then a burst of gas propellant is injected through the needle's tip into the headspace. For a typical wine bottle, when the headspace gauge pressure reaches about 4.0 atm, the cork will move upward. Assume the originally trapped air in the headspace, as well as the atmospheric air surrounding the bottle, have a pressure of 1.0 atm and a temperature of 20°C, and that the gas propellant obeys the ideal gas law.
(a) After the propellant is injected and the headspace is at a gauge pressure of 4.0 atm, what net force F acts on the cork? This force is sufficient to push the cork out of the neck.
(b) How many moles of gas propellant are introduced into the headspace in this injection process? Assume the temperature of the headspace remains at 20°C throughout. Also, it is valid to assume the mixed gas in the headspace is composed of air and propellant at absolute pressures of 1.0 atm and 4.0 atm, respectively.
(c) The propellant has an atomic mass of 102 and a liquid density of 1.220 g/cm³, and, when stored in its canister, most of its mass is in the liquid form. If a "full" canister is capable of opening 80 wine bottles, what volume of liquid propellant does it contain?

Section 17–9

20. (II) A gas is mostly empty space. Demonstrate the veracity of this statement by assuming that an ideal gas at STP fills a cubic volume $V = L^3$, where L is the length of one of the cube's sides. Then compute the fraction of this volume V that is occupied by matter. Assume that the spatial extent of common gas molecules is about $L_o = 0.3$ nm so one gas molecule occupies an approximate volume of L_o^3.

General Problems

21. (II) When packaging wine in a bottle, vintners always leave a "headspace" in the bottle's neck [i.e., a volume of trapped air between the bottom of the cork and the surface of the liquid contents (see Fig. 17–1)]. If a bottle of wine is taken from 20°C to 30°C, its headspace volume decreases to about 25% of its original value due to the large coefficient of volume expansion for alcohol. If air is contained in the headspace, determine whether there is any danger of "popping" the bottle's cork because of this rise in temperature. For the headspace, assume that its absolute pressure is 1.0 atm at 20°C, that a gauge pressure of about 4.0 atm is required to move the cork, and that no air molecules are lost by dissolving in the wine during the compression process.

22. (II) Assume an ideal gas of N molecules, initially at volume V_o and temperature T_o, expands to volume V and temperature T, so that its change in volume is $\Delta V = V - V_o$. Using the ideal gas law, show that $\Delta V = \beta V_o \Delta T$ (Eq.17–2 in the text) as long as $\beta = \dfrac{1}{T_o}$ and the expansion takes place at constant pressure P.

23. (II) Skin divers breathe through short tubular "snorkels" while swimming underwater very near the surface. One end of the snorkel attaches to the diver's mouth while the other end protrudes above the water surface. Unfortunately, snorkels cannot support breathing to any great depth. In fact, a typical diver below a water depth of only 30 cm cannot draw a breath through a snorkel. Based on this fact, what is the approximate fractional change in a typical person's lung volume when drawing a breath? Assume that in equilibrium the air pressure in a diver's lungs matches that of the surrounding water pressure.

24. (III) A steel ball floats in a bath of mercury at temperature $T = 20°C$ with a fraction X_o of its total volume submerged. The temperature of the ball and bath is then increased by $\Delta T = 50\,\text{C}°$ and the fraction of the ball's volume submerged changes to X.
(a) Show that $X_o = 0.57$ (see Section 13–6 in the text).
(b) Show that, if a material's density is initially ρ_o and then its temperature is changed by ΔT, the density ρ at the final temperature is

$$\rho = \frac{\rho_o}{1 + \beta \Delta T},$$

assuming the material's coefficient of volume expansion β remains constant over the given range of temperatures.
(c) Show that the change in the fractional submerged volume is approximately given by

$$X - X_o \approx X_o(\beta_M - \beta_S)\,\Delta T$$

where β_M and β_S are the coefficients of volume expansion for mercury and steel, respectively. [Hint: Use the binomial expansion and only keep the largest terms.]
(d) Show that the fractional submerged volume increases by about 1% during this heating process; i.e., compare $(X - X_o)$ with X_o.

25. (III) Wine bottles are never completely filled. Instead, a small volume of air (called "headspace") is left in the glass bottle's cylindrically shaped neck (inner diameter $d = 18.5$ mm) to allow for wine's fairly large coefficient of thermal expansion. The distance H between the surface of the liquid contents and the bottom of the cork is called the "headspace

height" (Fig. 17–1). If the headspace height is made too small during bottling, the bottle's cork may "pop" when subjected to the temperature variations of everyday life. A wine bottle contains 750 ml of liquid; when bottled at 20°C, assume the headspace air is at 1.0 atm absolute pressure with headspace height H_o. Assume further that H_o is minimally acceptable; i.e., when this bottle is placed in a 30°C room, the headspace height is reduced to H due to the thermal expansion of the bottled wine and the headspace gauge pressure reaches the level required (about 4.0 atm) to move the bottle's cork. Use this information to determine H_o. Assume that no air molecules are lost by dissolving in the wine during the compression of the headspace and that wine's coefficient of volume expansion is double that of water; in comparison, the thermal expansion of glass can be neglected. [Vintners typically choose a headspace height of 1.5 cm at 20°C.]

26. (III) From Earth's surface (defined as $y = 0$) up through the troposphere (which is the lowest 10-km thick layer of Earth's atmosphere), the atmosphere's absolute temperature T only varies by about 20%. Let's model this situation by assuming that the atmosphere's temperature T is constant and determine the resulting atmospheric pressure P as a function of altitude y above Earth's surface.

    (a) Starting with $\dfrac{dP}{dy} = -\rho g$ (Eq. 13–4 in the text), use the definition of density ρ and the ideal gas law to show

    $$P = P_o \exp\left(-\frac{mgy}{kT}\right)$$

    where P_o is atmospheric pressure at Earth's surface and m is the average mass of an air molecule. Assume the acceleration due to gravity is constant.

    (b) The result in part (a) can be written as

    $$P = P_o \exp\left(-\frac{y}{H}\right)$$

    where $H = \dfrac{kT}{mg}$ is called the scale height of the atmosphere. H is the height at which atmospheric pressure has decreased by a factor of $e^{-1} = 0.37$ of its value at Earth's surface. Show that $H = 7.6$ km, given that the average molecular mass of air is 29. For the assumed constant atmospheric temperature T, use the average temperature of the troposphere, which is about 260 K.

    (c) Commercial airliners fly at a cruising altitude of about 10 km. Use your findings from part (b), which give a fairly accurate description of the troposphere, to determine atmospheric pressure at this altitude. The small value you will obtain explains why airline cabins are pressurized; but perhaps surprisingly, the cabins are only pressurized to about 0.75 atm (which accounts for the "popping'" of passenger's ears during air travel). Why do you think airplanes are designed to be pressurized only to 0.75 atm, rather than 1.00 atm?

27. (III) (a) In Problem 26, a constant-temperature model of Earth's atmosphere was shown to give the following atmospheric pressure P as a function of altitude y above Earth's surface

    $$P = P_o \exp\left(-\frac{y}{H}\right)$$

    where $H = \dfrac{kT}{mg}$ is called the scale height, T is the assumed constant temperature, and m is the average mass of an air molecule. Using this model, which provides a fairly accurate description of Earth's troposphere, determine the altitude $y_{1/2}$ (as a multiple of H) at which P is one-half its value at sea level.

    (b) Show that, for this atmospheric model, the relation in part (a) also implies that the air density ρ as a function of altitude is

    $$\rho = \rho_o \exp\left(-\frac{y}{H}\right).$$

    (c) Consider a rectangular vertical column of cross-sectional area A that extends from sea level up to the farthest reaches of the atmosphere. Assume the acceleration due to gravity is constant. Show via integration that the weight w of air contained within this column is

    $$w = \rho_o g H A.$$

    (d) One-half of the weight w found in part (c) is contained within the column's bottom portion of height h. Show, via evaluating an appropriate integral, that $h = y_{1/2}$.

28. (III) From Earth's surface (defined as $y = 0$) up through the troposphere (which is the lowest 10-km thick layer of Earth's atmosphere), the atmosphere's absolute temperature T only varies by about 20%. This variation of atmospheric temperature T as a function of altitude y in Earth's troposphere is accurately described by the following linear relation:

    $$T = T_o - \Gamma y$$

    where T_o is the temperature at Earth's surface and the constant Γ is called the lapse rate. For the "standard atmosphere" used by meteorologists, $T_o = 288$ K (15°C) and $\Gamma = 6.50$ C°/km.

    (a) Starting with $\dfrac{dP}{dy} = -\rho g$ (Eq. 13–4 in the text), use the definition of density ρ, the ideal gas law, and the given temperature variation of Earth's troposphere to show that the resulting atmospheric pressure P as a function of altitude y above Earth's surface is

    $$P = P_o \left(\frac{T_o - \Gamma y}{T_0}\right)^\alpha$$

    where $\alpha = \dfrac{mg}{k\Gamma}$, P_o is atmospheric pressure at Earth's surface, and m is the average mass of an air molecule. The acceleration due to gravity is assumed constant.

    [Hint: $\displaystyle\int \dfrac{dx}{a + bx} = \dfrac{1}{b}\ln(a + bx)$, where a and b are constants.]

    (b) For this atmospheric model, the atmosphere has a finite thickness. Determine this thickness (km).

    (c) Commercial airliners fly at a cruising altitude of about 10 km. First show that $\alpha = 5.26$ and then use the expression from part (a) to determine atmospheric pressure at this altitude. The average molecular mass of air is 29.

29. (III) Altimeters are used by pilots, mountain climbers, and skydivers to determine altitudes above Earth's surface of up to 10 km. These devices do not measure altitude y directly, but instead accurately measure atmospheric pressure P and then use the relation $P = P_o\left(\dfrac{T_o - \Gamma y}{T_o}\right)^\alpha$, where the constants $\alpha = 5.26$, $T_o = 288$ K, $\Gamma = 6.50$ C°/km, and P_o is atmospheric pressure at Earth's surface (see Problem 28), to convert this measured quantity P to an associated altitude y.

(a) Using the given relation between P and y, show that altitude as a function of atmospheric pressure is given by

$$y = \frac{T_o}{\Gamma}\left[1 - \left(\frac{P}{P_o}\right)^{1/\alpha}\right].$$

(b) Starting with $\Delta y \approx \left|\dfrac{dy}{dP}\right|\Delta P$, show that if an altimeter can measure atmospheric pressure P to an accuracy of ΔP, the resulting accuracy Δy in altitude will be given by

$$\Delta y \approx \frac{1}{\alpha}\left[\frac{T_o}{\Gamma} - y\right]\frac{\Delta P}{P}$$

and go on to show that $\left[\dfrac{T_o}{\Gamma} - y\right]$ can be approximated as $\dfrac{T_o}{\Gamma}$ for all relevant altitudes of altimeter use and so

$$\Delta y \approx \frac{T_o}{\alpha\Gamma}\frac{\Delta P}{P}.$$

(c) A high-quality altimeter can measure atmospheric pressure to an accuracy of about 0.1%. Show that this precision in pressure measurement results in an altitude determination accurate to about 8 m.

(d) Determining altitude via atmospheric pressure is not free from difficulties as atmospheric pressure can be affected by changing weather patterns. On a day full of weather changes, an approaching cold front, e.g., can change air pressure by as much as 1%. What apparent altitude change would an altimeter indicate when subjected to this weather-induced pressure fluctuation?

30. (III) It has been observed that, if released from sea level, helium-filled latex balloons usually burst when they reach an altitude of about 8 km. When filled to its recommended size, the helium pressure within a latex balloon can be accurately assumed to equal atmospheric pressure (it is actually slightly greater because of the stretched latex). If a spherical balloon is filled to its recommended diameter d at sea level, then released, by what fraction has its diameter increased just before it bursts at an altitude of 8 km? In working this problem, assume that the balloon rises slow enough for it to equilibrate to the temperature of the surrounding air at all times and that atmospheric temperature T and pressure P as a function of altitude y above sea level are accurately modeled by $T = T_o - \Gamma y$ and $P = P_o\left(\dfrac{T_o - \Gamma y}{T_o}\right)^\alpha$, respectively, where the constants $\alpha = 5.26$, $T_o = 288$ K, $\Gamma = 6.50$ C°/km, and P_o is atmospheric pressure at sea level.

CHAPTER 18

Kinetic Theory of Gases

Section 18–1

1. (II) Can pockets of vacuum persist in an ideal gas? Assume that a room is filled with air at 20°C and that somehow a small spherical region of radius 1.0 cm within the room becomes devoid of air molecules. Estimate how long it will take for air to refill this region of vacuum. Assume the atomic mass of air is 29.

2. (II) Air at a given temperature T consists chiefly of a mixture of nitrogen (N_2) molecules and oxygen (O_2) molecules. What is the ratio of the rms speeds of these two molecules in air?

3. (II) One assumption in the kinetic theory of a gas is that the separation between gas molecules is much greater than the diameter of each molecule. Does air in an average-sized room conform to this criterion? One way to answer this question is to compare the volume V_G occupied by air in the gaseous state with the volume V_L when the air molecules are closely packed together in the liquid state. If the ratio V_G/V_L is much larger than one, then the gas fulfills the stated criterion of kinetic theory. Assume a room with typical dimensions of 4.0 m × 4.0 m × 3.0 m, which is initially filled with air at 20°C and 1.0 atm. If all of the air in this room is condensed into 77 K liquid air, show that $V_G/V_L \gg 1$. When air (atomic mass 29) is cooled to 77 K, all of its major components $(N_2, O_2,$ and Ar) condense into a liquid mixture with a density of 875 kg/m³.

4. (II) (a) For an ideal gas at temperature T show that

$$\frac{dv_{\text{rms}}}{dT} = \frac{1}{2}\frac{v_{\text{rms}}}{T}.$$

(b) Using the approximation $\Delta v_{\text{rms}} \approx \dfrac{dv_{\text{rms}}}{dT}\Delta T$, show that

$$\frac{\Delta v_{\text{rms}}}{v_{\text{rms}}} \approx \frac{1}{2}\frac{\Delta T}{T}.$$

(c) If the average air temperature changes from −5°C in winter to 25°C in summer, determine the percent change in the root-mean-square velocity of air molecules between these seasons.

5. (III) In a typical room, how many air molecules rebound from each wall every second? Estimate the frequency of molecular rebound from a particular wall (i.e., the number of molecules rebounding per second) as follows:
(a) Assume an ideal gas of N molecules is contained within a cubic room with sides of length L at temperature T and pressure P. Prove that the frequency f with which gas molecules strike the wall as shown in Fig. 18–1 is

$$f = \frac{\bar{v}_x}{2}\frac{P}{kT}L^2$$

where $\bar{v}_x$ is the average x-component of a molecule's velocity.
(b) Approximate $\bar{v}_x$ as $\bar{v}_x \approx \sqrt{\overline{v_x^2}}$, and go on to show that the above expression can then be written as

$$f \approx \frac{PL^2}{\sqrt{4mkT}}$$

where m is the mass of a gas molecule.
(c) Assume a cubic air-filled room is at sea level, has a temperature 20°C, and has sides of length $L = 3$ m. The atomic mass of air is 29. Determine f.

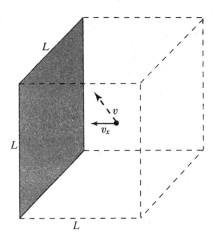

FIGURE 18–1

Section 18–2

6. (II) At room temperature, the speed distribution of conduction electrons in a semiconductor, such as silicon, is the Maxwell distribution of speeds (this is not true for conduction electrons in metals). What is the approximate speed for a typical conduction electron in silicon?

99

7. (II) A gas consisting of 15,200 molecules, each of mass 2.00×10^{-26} kg, has the distribution of speeds given in Table 18-1, which crudely mimics the Maxwell distribution.

TABLE 18-1

Number of Atoms	Speed (m/s)
1600	200
4100	400
4700	600
3100	800
1300	1000
400	1200

(a) Determine the rms speed v_{rms} for this distribution of speeds.
(b) Given your value for v_{rms}, what (effective) temperature would you assign to this gas?
(c) Determine the mean speed $\bar{v}$ of this distribution and use this value to assign an (effective) temperature to the gas. Is the temperature you find here consistent with what you determined in part (b)?

8. (III) A gas consisting of 15,200 molecules, each of mass 2.00×10^{-26} kg, has the distribution of speeds given in Table 18-1. Demonstrate that these data are consistent with the Maxwell distribution of speeds as follows:
(a) Determine the rms speed v_{rms} for the distribution of speeds given in Table 18-1 and use this value to assign an (effective) temperature to the gas.
(b) If the given distribution is consistent with the Maxwell distribution of speeds, the number ΔN of gas molecules with speeds in the finite range from v to $v + \Delta v$ will be approximately

$$\Delta N \approx 4\pi N \left(\frac{m}{2\pi kT}\right)^{3/2} v^2 \exp\left(-\frac{mv^2}{2kT}\right) \Delta v.$$

Using the temperature found in part (a), evaluate this expression for $v = 200, 400, 600, 800, 1000,$ and 1200 m/s with $\Delta v = 200$ m/s and show that the predicted values for ΔN are consistent with the distribution given for the gas. For each speed v, show that the percent difference between the predicted and the given value for ΔN is always less than 10%.

9. (III) Equation 18–6 in the text is Maxwell's expression for the distribution of molecular speeds within a gas; i.e., it allows one to determine the number of gas molecules with a speed in the range v to $v + dv$, where v is the magnitude of a molecule's velocity vector **v**. Maxwell also gave the following expression describing the number of molecules with a given x-component of the velocity vector

$$f(v_x) = N\sqrt{\frac{m}{2\pi kT}} \exp\left(-\frac{mv_x^2}{2kT}\right) \qquad -\infty \le v_x \le +\infty.$$

The number dN_x of molecules with an x-component velocity in the range v_x to $v_x + dv_x$ is given by $dN_x = f(v_x)\, dv_x$.

(a) Show that if one averages over all the possible positive and negative values for v_x, $\bar{v}_x = 0$. Give a physical interpretation of this result.
(b) Show that if one only averages over all the possible positive values for v_x, then

$$\bar{v}_x = \sqrt{\frac{2kT}{\pi m}}.$$

(c) Use your result from part (b) to show that $\bar{v}_x = \frac{1}{2}\bar{v}$ and $\bar{v}_x = \sqrt{\frac{2}{\pi}}\sqrt{\overline{v_x^2}}.$

Section 18–4

10. (II) Show that humid air, which contains water vapor, is less dense than dry air, if both types of air are at the same temperature and pressure.

11. (II) A pressure cooker is a sealed pot designed to cook food with the steam produced by boiling water typically at 120°C. These cookers are especially useful at higher altitudes where water boils at temperatures well below 100°C. One (older) method for maintaining the proper pressure within the pressure cooker involves simply using a weight with mass m to plug a small opening in the cooker's lid (Fig. 18–2). As water in the cooker is heated, the internal pressure of the resulting steam increases until the weight is lifted slightly. Some steam then exits the cooker, decreasing the internal pressure to the point where the weight falls into its original position and again prevents steam from exiting the cooker. If the diameter d of the opening in the cooker's lid is 3 mm, what should m be in order to cook food at 120°C? Assume that atmospheric pressure outside the cooker is 1.01×10^5 Pa.

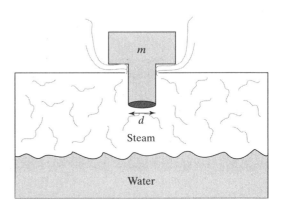

FIGURE 18–2

12. (II) When using a mercury barometer (see pp. 338–339 in the text), the vapor pressure of mercury is usually assumed to be zero. While very small, mercury's vapor pressure at room temperature has the nonzero value of about 0.0015 mmHg. At sea level, the height h of mercury in a barometer is typically about 760 mm. If the vapor pressure of mercury is neglected when using a mercury barometer to measure atmospheric pressure,
(a) is the true atmospheric pressure (slightly) greater or less than the value determined from the barometer?
(b) what percent error is introduced to the determined value for atmospheric pressure by this neglect?

13. (II) On a backpacking trip in the mountains, you take along a thermometer, and while cooking dinner one evening, you find that at the location of your campsite water boils at 90°C. Use Table 18–2 in the text to determine the elevation of your campsite. Assume, as is common among meteorologists, that atmospheric pressure P and altitude y above sea level are related by the following expression

$$y = \frac{T_o}{\Gamma}\left[1 - \left(\frac{P}{P_o}\right)^{1/\alpha}\right]$$

where $T_o = 288$ K, $\Gamma = 6.5$ C°/km, $\alpha = 5.26$, and P_o is atmospheric pressure at sea level.

14. (II) In the analysis of a barometer (see pp. 338–339 in the text), the vapor pressure of its fluid is usually assumed to be zero. For a barometer that uses water, however, this approximation may introduce a significant error due to water's fairly large vapor pressure at room temperature. At sea level, the height h of water in a barometer is typically about 10 m. If the vapor pressure of water is neglected when using this barometer at sea level and 20°C to determine atmospheric pressure,
    (a) is the true atmospheric pressure greater or less than the determined value?
    (b) what percent error is introduced to the determined value for atmospheric pressure by this neglect?

15. (III) A naphthalene moth ball evaporates over time t at a rate proportional to its surface area A. That is, if M is the mass of the moth ball

$$\frac{dM}{dt} = -kA$$

where k is some constant. Assume a spherical mothball starts out at time $t = 0$ with a radius of $r = 0.50$ cm and after 6.0 months its radius is $r = 0.40$ cm. After what time will the moth ball be completely evaporated? Assume the moth ball is of uniform construction so its density is constant.

16. (III) What is the mathematical relation between water's boiling temperature and atmospheric pressure?
    (a) Using the data from Table 18-2 in the text, in the temperature range from 50°C to 150°C, plot $\ln(P)$ versus $(1/T)$, where P is water's saturated vapor pressure (Pa) and T is temperature on the Kelvin scale. [Note: you must convert each temperature in the table to Kelvin.] Show that a straight-line plot results and determine the slope and the y-intercept of the line.
    (b) Your graph demonstrates that a linear relation exists between the plotted quantities. Show that this result implies that P and T obey the following exponential relation

$$P = B\exp\left(-\frac{A}{T}\right)$$

where A and B are constants. Use the slope and the y-intercept found from your plot to show that $A \approx 5000$ K and $B \approx 7 \times 10^{10}$ Pa.

17. (III) How does the boiling temperature of water change with altitude?
    (a) The variation of atmospheric pressure P with altitude y above sea level is found to fairly accurately obey the following relation

$$P = P_o \exp\left(-\frac{y}{H}\right)$$

where the constants $H = 7.6$ km and $P_o = 1.0$ atm. Equate this relation with the exponential relation between atmospheric pressure P and water's boiling temperature T obtained in Problem 16 to show

$$T = A\left(\frac{y}{H} - C\right)^{-1}$$

where the constant $C = \ln(P_o/B)$.
(b) At altitude $y = 0$, water's boiling temperature is $T_o = 373$ K. Use this fact to show

$$T = T_o\left(1 + \frac{y}{D}\right)^{-1}$$

where the constant $D = AH/T_o$.
(c) Show that for altitudes up to 10 km, the quantity y/D will always be less than 0.10. Apply the binomial expansion to the result of part (b) and go on to show that

$$\frac{dT}{dy} \approx -\frac{T_o}{D}.$$

Finally, show that this expression implies that water's boiling temperature decreases with altitude gain by about 3.7°C/km.

Section 18–5

18. (III) How well does the ideal gas law describe the pressurized air in a scuba tank?
    (a) To fill a typical scuba tank, an air compressor takes in about 2300 L of air at 1.0 atm and compresses this gas into the tank's 12-L internal volume. If the filling process occurs at 20°C, show that a scuba tank holds about 96 moles of air.
    (b) Assume a 12-L scuba tank holds 96 moles of air at 20°C. Use the ideal gas law to predict the air's pressure within the tank.
    (c) Assume a 12-L scuba tank holds 96 moles of air at 20°C. Use the van der Waals equation of state to predict the air's pressure within the tank. For air, the van der Waals constants are $a = 0.1373$ N·m⁴/mol² and $b = 3.72 \times 10^{-5}$ m³/mol.
    (d) Taking the van der Waals pressure as the true air pressure, show that the ideal gas law predicts a pressure that is in error by only about 3%.

Section 18–6

19. (II) Vacuum chambers used in scientific research laboratories typically have spatial dimensions on the scale of 1 m. Below a certain threshold pressure, the air molecules (0.3 nm diameter) within such chambers are in the "collision-free regime." In this regime, it is highly unlikely that a particular air molecule will collide with another air molecule as it traverses from one chamber wall to the next. Find the threshold pressure for the collision-free regime for a typical-sized vacuum chamber in a room temperature (20°C) lab.

20. (II) Vacuum gauges are commonly calibrated in the pressure unit of torr or millitorr. A rule of thumb commonly used by those working with vacuum systems is that the mean free path of air molecules (0.3 nm diameter) at room temperature (20°C) is given by

$$l_M \text{ (in cm)} \approx \frac{5}{P(\text{in millitorr})}.$$

Starting with Eq. 18–10b in the text, verify that this relation is roughly valid.
[Hint: For vacuum researchers, numbers other than multiples of five are too hard to remember.]

General Problems

21. (II) To associate a meaningful temperature with a sample of ideal gas, the sample must contain a sufficient number N of molecules so that the Maxwell distribution of speeds results from an adequately large number of random intermolecular collisions. Random process theory shows that N must be at least 1×10^6 in order for the Maxwell distribution to be a valid description of the gas. For an ideal gas at STP, what is the smallest length scale L over which a valid temperature can be assigned? That is, for a sample of the gas with $N = 1 \times 10^6$ and cubic volume $V = L^3$, where L is the length of cube's side, what is L?

22. (II) A sample of liquid cesium is heated in an oven to 400°C and the resulting vapor is used to produce an atomic beam. The volume of the oven is 50 cm³, the vapor pressure of Cs at 400°C is 17 mmHg, and the diameter of cesium atoms in the vapor is 0.33 nm.
    (a) Calculate the mean speed of cesium atoms in the vapor.
    (b) For a single Cs atom in the vapor, determine the number of collisions it undergoes with other cesium atoms per second.
    (c) What is the total number of collisions per second between all of the cesium atoms in the vapor? Note that a collision involves two Cs atoms; assume the ideal gas law holds.

23. (II) When the speed v of an object exceeds about one-tenth the speed of light (i.e., $v > 0.1c$), where the speed of light $c = 3 \times 10^8$ m/s, the expression $K = \frac{1}{2}mv^2$ used in kinetic theory is no longer valid and must be replaced by the more general expression for relativistic kinetic energy (see Eq. 7–12 in the text). For the Maxwell distribution of speeds, only 1% of the molecules have speeds that exceed 4 times v_{rms}, so we expect that a gas will be on the "relativistic threshold" when its rms speed is $\frac{0.1c}{4} = 0.025c$.
    (a) For a gas of helium, what temperature corresponds to the relativistic threshold (i.e., above what temperature must relativistic effects be taken into account)?
    (b) Given that the temperature at one of the hottest places in the universe—the center of the Sun—is about 15×10^6 K, is relativity a necessary ingredient in the kinetic-theory description of gases that occur in nature?

24. (II) Sound waves in a gas can only propagate if the gas molecules collide with each other on the time scale of the sound wave's period. Thus, the highest possible frequency f_{max} for a sound wave in a gas is approximately equal to the inverse of the average collision time between molecules. Assume a gas, composed of molecules with mass m and radius r, is at pressure P and temperature T.
    (a) Show that
    $$f_{\text{max}} \approx 16\, Pr^2 \sqrt{\frac{\pi}{mkT}}.$$
    (b) Determine f_{max} for 20°C air ($r = 1.5 \times 10^{-10}$ m) at sea level. How many times larger is f_{max} than the highest frequency in the human audio range (20 kHz)? The atomic mass of air is 29.

25. (III) In order to study the surface of a given material, surface science researchers must perform their experiments in a time less than the "monolayer time" t_M, which is the time it takes to adsorb a single layer of contaminant molecules (e.g., air) on the "clean" surface under investigation. Let's estimate the monolayer time as follows:
    (a) Assume an ideal gas of N "contaminant" molecules is contained within a cubic vacuum chamber with sides of length L at temperature T and pressure P. Prove that the frequency f with which gas molecules strike the chamber wall shown in Fig. 18–1 is
    $$f = \frac{\bar{v}_x}{2}\frac{P}{kT}L^2$$
    where $\bar{v}_x$ is the average x-component of a molecule's velocity.
    (b) Approximate $\bar{v}_x$ as $\bar{v}_x \approx \sqrt{\overline{v_x^2}}$ and go on to show that the above expression can then be written as
    $$f \approx \frac{PL^2}{\sqrt{4mkT}}$$
    where m is the mass of a gas molecule.
    (c) Assume now that the chamber wall's surface starts out "clean" [i.e., it consists of M "chamber wall" atoms (e.g., iron if the chamber is made out of iron) distributed over an area of L^2] and that as the contaminant gas molecules strike the wall, rather than rebounding, they stick to whichever "chamber wall" atom they have collided with. Show that the time required to cover the entire wall with a monolayer of contaminant molecules is approximately
    $$t_M \approx \frac{M}{L^2}\frac{\sqrt{4mkT}}{P}.$$
    (d) For a typical solid surface, M/L^2 is about 1×10^{19} m^{-2}. Assume that a surface science researcher needs to keep a surface clean of atmospheric (e.g., nitrogen, oxygen) contaminant molecules ($m = 5 \times 10^{-26}$ kg) for 1 hour in order to carry out a planned investigation in a room temperature (20°C) lab. First, show that if the chamber pressure is $P = 1$ atm, the monolayer time is unacceptably short; then determine the required chamber pressure for this investigation. Compare (as a ratio) this required pressure with standard atmospheric pressure.

26. (III) In an atom trap, a laser cools a sample of atoms to a very low temperature and confines them to a small spatial region. The atoms in the trap move in accordance with the Maxwell distribution of speeds and the confinement region is so small it can be considered to be a point. The atoms' most probable speed v_p (and thus their temperature) may be determined by the following "time-of-flight" procedure. As shown in Fig. 18–3, a plane that is a distance d below the trap is the "detection region." At time $t = 0$, the trap's laser is turned off and the atoms, each with a particular initial speed v_o and launch angle $-90° \leq \theta_o \leq +90°$ (relative to the horizontal, positive is upward), follow projectile trajectories under the influence of gravity to the detection plane.
    (a) Since atoms with speeds near v_p are most plentiful, it is easy to identify when these atoms pass through the detection plane (because of the large detection signal produced). Considering only atoms with speed v_p at $t = 0$, the first and last of these atoms to arrive at the detection plane are the downward- ($\theta_o = -90°$) and upward-directed ($\theta_o = +90°$) atoms, respectively. In the time-of-flight experiment, the arrival times t_F and t_L of the (first) downward- and (last) upward-directed

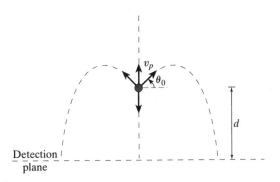

FIGURE 18-3

atoms with speed v_p are measured. Show that the speed of the atoms is given by

$$v_p = \frac{g(t_L - t_F)}{2}.$$

(b) Show that, in a time-of-flight experiment, the temperature T of trapped atoms on the Kelvin scale is given by

$$T = \frac{m}{8k}\left[g(t_L - t_F)\right]^2$$

where m is the mass of an atom in the trap (commonly, rubidium).
(c) If, in a time-of-flight experiment with Rb-87, one measures $t_L - t_F = 80$ ms, what is T?

27. (III) In the game of paintball, players use guns powered by pressurized gas supplied from storage tanks to propel paintballs at opposing team members. Carbon dioxide is commonly employed as the pressurized gas because of its favorable phase diagram. As shown in Fig. 18–7 in the text (p. 474), CO_2 coexists as a liquid and a gas at 20°C when at a pressure of 56 atm. In Fig. 18–4, a CO_2 tank for paintball is initially filled so that 34% of its volume contains liquid CO_2 at 56 atm, while the remaining 66% of the volume contains CO_2 vapor at 56 atm. A paintball gun is fired by drawing some of the 56-atm vapor from the tank. Some of the liquid will then quickly evaporate and restore the vapor to its original 56-atm pressure. A typical paintball CO_2 tank has a volume of 500 cm³. Show that when it is initially filled, a large fraction of the CO_2 is compactly stored in the dense liquid state. That is, compare the number of moles n_L of CO_2 in the liquid state with the number of moles n_V in the vapor state by determining the ratio $x = \frac{n_L}{n_V}$. The density of liquid CO_2 at 56 atm is 1.977 g/cm³. Assume the ideal gas law approximately holds for CO_2 vapor at 56 atm and 20°C. The atomic mass of CO_2 is 44.

28. (III) In reality, carbon dioxide vapor is not described very well by the ideal gas law at high pressures.
(a) Rework Problem 27 using the van der Waals equation (rather than the ideal gas law) to model the CO_2 vapor at 56 atm. For CO_2, the van der Waals constants are $a = 0.3643$ N·m⁴/mol² and $b = 4.27 \times 10^{-5}$ m³/mol.
(b) In this high-pressure situation, what percent error was made in Problem 27 in determining x by using the ideal gas law rather than the more accurate van der Waals equation?
[*Hint*: Find the root of the cubic equation of the form $An^3 + Bn^2 + Cn + D = 0$ using a calculator or a computer.]

29. (III) Let's explore why, although its molecules are in rapid motion, Earth's atmosphere does not escape into outer space.
(a) Recall that, if starting from Earth's surface, in order to escape Earth's gravitational attraction a mass must have a speed of at least the escape velocity $v_{\text{esc}} = 11{,}200$ m/s (Eq. 8–20 in the text). Assume there are $N = 1 \times 10^{44}$ total molecules in Earth's atmosphere (see chapter 17, Problem 63 in the text) and define N_{esc} to be the number of these molecules that have speed sufficient to escape Earth (i.e., have speed $v \geq v_{\text{esc}}$). Show that N_{esc} is given by

$$N_{\text{esc}} = 4\pi N \left(\frac{m}{2\pi kT}\right)^{\frac{3}{2}} \int_{v_{\text{esc}}}^{\infty} v^2 \exp\left(-\frac{1}{2}\frac{mv^2}{kT}\right) dv.$$

(b) By making the substitution $x^2 = \frac{1}{2}\frac{mv^2}{kT}$, show

$$N_{\text{esc}} = \frac{4N}{\sqrt{\pi}} \int_{x_o}^{\infty} x^2 e^{-x^2} dx$$

where $x_o = \frac{v_{\text{esc}}}{v_p}$ and $v_p = \sqrt{\frac{2kT}{m}}$ is the most probable speed of the Maxwell distribution of speeds for the atmospheric gas.

(c) For $x_o \gg 1$, $\int_{x_o}^{\infty} x^2 e^{-x^2} dx \approx \frac{x_o}{2} e^{-x_o^2}$ (see Problem 30). Thus, in this limit, show that

$$N_{\text{esc}} \approx \frac{2Nx_o}{\sqrt{\pi}} e^{-x_o^2} = \frac{2Nx_o}{\sqrt{\pi}} 10^{-x_o^2(\log_{10} e)}.$$

(d) Given that Earth's atmosphere is composed of air with an (average) molecular mass of 29 and a typical temperature of $T = 250$ K, verify that x_o for air is much greater than one and then show that N_{esc} is negligibly small.

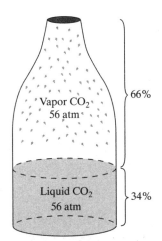

FIGURE 18-4

30. (III) For $x_o \gg 1$, show that $\int_{x_o}^{\infty} x^2 e^{-x^2}\, dx \approx \dfrac{x_o}{2} e^{-x_o^2}$ as follows

(a) Starting with $\int_{x_o}^{\infty} x^2 e^{-x^2}\, dx$, integrate twice by parts to show that

$$\int_{x_o}^{\infty} x^2 e^{-x^2}\, dx = \left(\dfrac{x_o}{2} + \dfrac{1}{4x_o}\right) e^{-x_o^2} - \dfrac{1}{4} \int_{x_o}^{\infty} \dfrac{e^{-x^2}}{x^2}\, dx$$

which, for $x_o \gg 1$, reduces to

$$\int_{x_o}^{\infty} x^2 e^{-x^2}\, dx \approx \dfrac{x_o}{2} e^{-x_o^2} - \dfrac{1}{4} \int_{x_o}^{\infty} \dfrac{e^{-x^2}}{x^2}\, dx.$$

(b) Noting that $\int_{x_o}^{\infty} \dfrac{e^{-x^2}}{x^2}\, dx < \int_{x_o}^{\infty} \dfrac{e^{-x_o^2}}{x^2}\, dx$, go on to show that, for $x_o \gg 1$,

$$\dfrac{1}{4} \int_{x_o}^{\infty} \dfrac{e^{-x^2}}{x^2}\, dx \ll \dfrac{x_o}{2} e^{-x_o^2}$$

and therefore, in this limit,

$$\int_{x_o}^{\infty} x^2 e^{-x^2}\, dx \approx \dfrac{x_o}{2} e^{-x_o^2}.$$

CHAPTER 19

Heat and the First Law of Thermodynamics

Section 19–1

1. (II) How much energy does it take to "warm up" a wetsuit? A wetsuit is a form-fitting suit made of neoprene worn by scuba divers to keep warm while immersed in cold water. When a wetsuit-wearing diver first jumps into the ocean, water leaks into the gap region between the diver's skin and the fabric of the suit, forming a water layer of typical thickness 0.50 mm throughout the suit. This water layer is then warmed by the diver's body. Assuming the total surface area of the wetsuit covering the diver is about 1.0 m², and that ocean water enters the suit at 10°C and is warmed by the diver to skin temperature 35°C, how much energy (in units of candy bars) is required by this heating process? One candy bar has 300 kcal.

Sections 19–3 and 19–4

2. (II) How much energy and money are required to bring a pot of water to boiling temperature? Assume the pot contains 1.0 L of liquid water, which is heated from 20°C to 100°C, and that water's specific heat remains at the constant value of $c = 1.0 \, \text{cal/g} \cdot \text{C}°$ over this entire temperature range.
   (a) Show that the energy required to boil this water could run a 100-W light bulb for almost an hour.
   (b) Determine the cost of this required energy. Assume an electric stove is used to boil the water and that electrical energy costs 10 cents per 3.6×10^6 J. One kilowatt-hour, the unit that power companies use to measure amounts of energy, equals 3.6×10^6 J.

3. (II) A soft rubber ball, with mass m and specific heat 1850 J/kg·C°, is initially held at rest at a height of 2.7 m above a hard concrete floor. The ball is released and, after bouncing from the floor, returns to rest at a height of 1.3 m. Assuming that 90% of the lost potential energy heats the ball, what temperature increase does it experience as a result of this single bounce?

4. (II) The large quantity of heat produced by high-power lasers is commonly exhausted using a "flow-through" water-cooling system. In such a system, tap water flowing in the cooling system's pipes passes through the laser system where it absorbs the laser's waste heat, after which the heated water flows out of the system and is discharged into a drain. In one real-life flow-through system, a power of 10 kW can be discharged to the drain, if 20 L of water flow through the system per minute. If the tap water is initially at 20°C, at what temperature is the water discharged to the drain?

5. (III) Two assistant professors, Jacques and Katherine, invite the president, dean of faculty, and their spouses over for dinner. Fifteen minutes before the guests are to arrive, the hosts realize that there is only one 0.75-L bottle of white wine in their 2.0°C refrigerator. Jacques goes to the store and returns home with an identical bottle of white wine, but it is unchilled at 21.0°C. Jacques insists that white wine must be drunk at a proper serving temperature in the range of 9.0 to 12.0°C and so suggests the following plan to Katherine: Pour the contents of both the 2.0°C and 21.0°C bottles into a large plastic pitcher and quickly stir them together; then use this equilibrated mixture to refill both the 2.0°C and 21.0°C wine bottles. Assume the mass of a glass wine bottle is 0.54 kg and the specific heat and density of wine are the same as for water. Also assume that plastic is such a poor heat conductor that the pitcher in which the wine is mixed does not change its temperature during the rapid mixing process. If Jacques implements his plan, when placed on the dinner table minutes later, will the wine in both bottles be within the proper range of serving temperature?

6. (III) Katherine quickly figures out that Jacques plan will not chill both wine bottles to the proper serving range (see Problem 5). So she proposes the following scheme: Pour the entire contents of the 21.0°C bottle into a large plastic pitcher; then pour 55% of the contents of the 2.0°C bottle into the empty 21.0°C bottle. Next use the 21.0°C wine from the plastic pitcher to refill both bottles to their original levels (i.e., 55% and 45% of the pitcher's contents go into the 2.0°C and 21.0°C bottles, respectively). Assume the mass of a glass wine bottle is 0.54 kg and the specific heat and density of wine are both the same values as for water. Using Katherine's plan, when placed on the dinner table minutes later, will the wine in both bottles be within the proper range of serving temperature?

Section 19–5

7. (II) You have 340 g of a soda drink $(c = 1.0 \, \text{cal/g} \cdot \text{C}°)$ at room temperature of 22°C. How many 25-g ice cubes at −10°C should you add to the drink, so that after the ice melts, the mixture is at a refreshing 2°C?

8. (II) A dictum of high-altitude mountain climbing is: *Do Not Eat Snow*. Rather, one should first melt snow using a stove, then drink the resulting liquid water in order to hydrate one's body. To understand the physics behind this dictum, consider the following two scenarios as a means toward hydrating a body with 1.0 L of water:
   (a) Reinhold eats 1000 g of −10°C snow and allows his body to warm it to his body temperature of 37°C. How much energy is required by Reinhold's body to power this process?

105

How many candy bars would Reinhold have to eat to obtain the required amount of energy? One candy bar has 300 kcal. (b) Trina acquires 1000 g of −10°C snow and melts it using her stove, producing 1000 g of 2°C liquid water. She then drinks the 2°C water and allows her body to warm it to her body temperature of 37°C. How much energy is required by Trina's body to power this process? How many candy bars would Trina have to eat to obtain the required amount of energy?

9. (II) How much longer does it take to vaporize water than to freeze it?
(a) Consider a 0.30-kg sample of water initially at 50°C. Assume a heater is turned on, causing heat to flow into the sample at a rate of 7.0 kW. When the sample is completely converted into 100°C steam, the heater is turned off. What is the heater's total "on-time" and what percentage of this on-time is spent converting 100°C liquid water into 100°C steam?
(b) Consider again the 0.30-kg sample of water initially at 50°C. Assume a cooler is turned on, causing heat to flow out of the sample at a rate of 7.0 kW. When the sample is completely converted into 0°C ice, the cooler is turned off. What is the cooler's total on-time and what percentage of this on-time is spent converting 0°C liquid water into 0°C ice?
(c) How many times longer is the heater's on-time than the cooler's on-time?

Sections 19–6 and 19–7

10. (II) One mole of helium gas is contained in a cylinder-piston arrangement. The cylinder is submerged in a 0°C ice-water bath as shown in Fig. 19–1. The piston is slowly pushed down until the volume of the helium is one-half of its original value. Heat is free to flow through the cylinder so that the helium's temperature is a constant 0°C during the compression. How much ice (in the bath) melts during the compression?

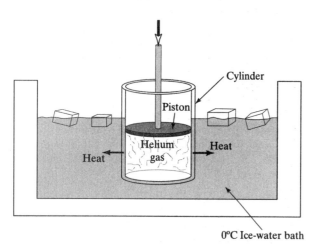

FIGURE 19–1

Section 19–9

11. (II) Starting with $PV^\gamma = $ constant, show that $P^{1-\gamma}T^\gamma = $ constant for an ideal gas undergoing an adiabatic process.

12. (II) A bicycle pump consists of a cylinder whose internal dimensions are 20 cm long by 3.0 cm diameter when the pump handle is pulled all the way out (Fig. 19–2a). The pump contains (diatomic) air at 20.0°C and 1.0 atm. If the outlet of the pump is blocked and the handle is pushed in quickly until the internal length of the pump cylinder is only 10 cm (Fig. 19–2b), to what temperature does the air within the pump rise? Assume that heat has no time to flow as the air is compressed quasistatically.

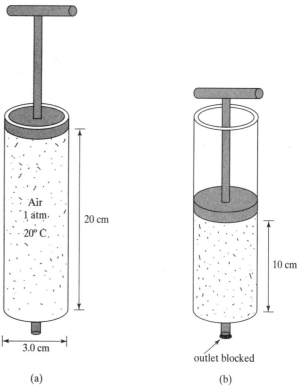

FIGURE 19–2

13. (II) When wine is being bottled, a cork cylinder of length $L = 3.8$ cm is quickly pressed into the top of the bottle's cylindrically shaped neck of inner diameter 18.5 mm (Fig. 19–3a). As the cork is being inserted, it compresses the (diatomic) air in the bottle's neck, so that the bottle's resulting "headspace," which is the air-filled volume between the bottom of the fully inserted cork and the surface of the liquid contents, is at high pressure (Fig. 19–3b). Determine the headspace air pressure just after a cork has been inserted. Assume that the air originally in the bottle's neck is at 1.0 atm, that the cork insertion takes place so quickly that this trapped air is compressed adiabatically, that the top of the cork is flush with the bottle's top after insertion, and that the resulting headspace height H is the typical value of 1.5 cm. [Within a few minutes after corking, much of this high-pressure air dissolves into the wine and the headspace pressure is returned to about 1.0 atm.]

14. (II) The pressurized air in a commercial airliner's passenger cabin is usually supplied from its jet engines. Airliners cruise at an altitude of around 10 km, where the outside air temperature is about −50°C and the air pressure is 0.25 atm. Because the air at this altitude is so "thin," the first stage of a jet engine intakes the high-altitude (diatomic) air and compresses it by a factor of about 30. Because the engine compresses air so quickly, assume the compression process occurs adiabatically. This pressurized air is then mixed with fuel and combusted in order to turn a turbine. Prior to fuel mixing, however, some of the compressed air is diverted and used to

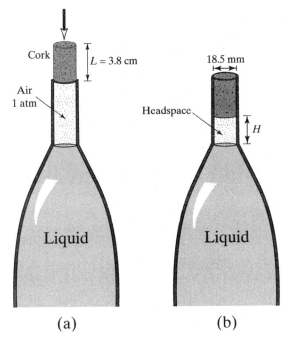

FIGURE 19–3

(a) Use this model to estimate the rate of heat flow $\Delta Q/\Delta t$ from the diver's skin to the cold water in units of kcal/h.
(b) How many candy bars would a diver have to eat in order to provide the energy needed for loss of "body heat" during one hour of diving? One candy bar has 300 kcal.

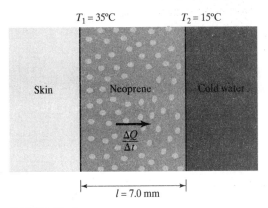

FIGURE 19–4

pressurize the passenger cabin. Show that this air, however, is very hot (i.e., determine its temperature) and so cannot be directly introduced into the cabin. [In fact, this air must be run through an efficient air conditioning system before being sent to passengers.]

15. (III) When bottling wine, a small volume of trapped (diatomic) air called the "headspace" is left between the bottom of the inserted cork and the top of the liquid contents (Fig. 19–3). If insufficient headspace height is left after the cork is inserted, the headspace pressure will simply eject the cork. Assuming a cylindrical cork of length $L = 3.8 \text{ cm}$ is quickly pressed into the top of the bottle's cylindrically shaped neck (inner diameter of 18.5 mm), derive an expression for the resulting headspace pressure P as a function of headspace height H, where H can vary from 0 (bottom of cork flush with liquid surface) to 2.0 cm. Plot P vs. H in the range $0.1 \text{ cm} \leq H \leq 2.0 \text{ cm}$ and show that P becomes very large for $H < 1.5 \text{ cm}$. Assume that the liquid level is adjusted for each choice of H so that, after insertion, the top of the cork is always flush with the top of the bottle. Also assume that the air is compressed adiabatically during the insertion process. [As a result of this plot, vintners usually choose $H = 1.5 \text{ cm}$, so that the cork will not be immediately ejected by the high headspace pressure.]

Section 19–10

16. (II) A wetsuit is a form-fitting suit made of neoprene worn by scuba divers to keep warm while immersed in cold water. When wearing this suit, how serious is the loss of "body heat" to the surrounding cold water for a scuba diver? As a simple model for this situation, assume only a neoprene layer of thickness $l = 7.0 \text{ mm}$ sandwiched between a diver's skin and the ocean's cold water at temperatures $T_1 = 35°C$ and $T_2 = 15°C$, respectively (Fig. 19–4). For neoprene, $k = 0.14 \text{ J/(s·m·C°)}$, and the surface area for a diver is $A \approx 1.0 \text{ m}^2$.

17. (II) Where is the habitable zone for humans in other solar systems? That is, at what orbital distance about some star other than the Sun would a planet be found with a temperature similar to that on Earth?
(a) Consider a planet with radius R and average temperature T that orbits at distance d about a star (Fig. 19–5). Assume the star outputs radiation uniformly into its surrounding space at a rate L (called the star's "luminosity" with units of W) and that the planet (with an effective cross-sectional area πR^2) absorbs a portion of this radiation and then, in order to maintain thermal equilibrium at temperature T, reradiates (uniformly in every direction) all of this energy back into space. Assuming the planet is a perfect blackbody with $e = 1$, show that its orbital distance and temperature are related by

$$d = \frac{1}{\sqrt{16\pi\sigma}} \frac{\sqrt{L}}{T^2}$$

where σ is the Stefan-Boltzmann constant.

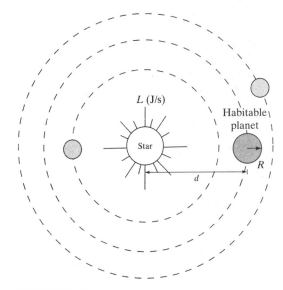

FIGURE 19–5

(b) Consider a planet orbiting a star with luminosity L. Show that if this planet has the same average temperature as Earth, its orbital distance d is

$$d = d_E \sqrt{\frac{L}{L_S}}$$

where d_E is the Earth-Sun distance and L_S is the luminosity of the Sun.

(c) The nearby star Vega has a luminosity 54 times greater than that of the Sun. If Vega has a planetary system, at what orbital distance (expressed as a multiple of d_E) would a planet habitable for humans be located?

18. (III) Many insulating materials derive their low value of thermal conductivity as the result of trapped air in small pores within the material. Consider the following simple model for a porous material (Fig. 19–6): A composite with cross-sectional area A and thickness l consists of a sequence of N layers, each of thickness Δx. Of these layers, N_A consists of trapped air and N_H consists of host material with thermal conductivities k_A and k_H, respectively. Note $N = l/\Delta x = N_A + N_H$. Also define $p = N_A/N$, where the porosity p is the fraction of the material's volume that is filled with trapped air.

(a) Assume a temperature difference ΔT is maintained across the composite of thickness l. Noting that the rate of heat flow must be the same through each layer, show that the thermal conductivity of this composite is given by

$$\frac{1}{k} = \frac{p}{k_A} + \frac{(1-p)}{k_H}.$$

(b) Using the fact that the thermal conductivity of air is much smaller than that of any (host) solid material, show that the relation from part (a) reduces to

$$k \approx \frac{k_A}{p}.$$

For styrofoam, $k = 0.025 \, \text{J}/(\text{s} \cdot \text{m} \cdot \text{C}°)$. According to this model, what is the porosity of styrofoam?

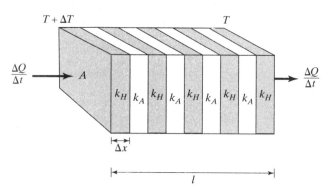

FIGURE 19–6

19. (III) A wetsuit is a form-fitting suit made of neoprene worn by scuba divers to keep warm while immersed in cold water. The suit allows a thin layer of water to enter and be trapped between the neoprene fabric and the diver's skin, where the diver's body warms it. It is commonly claimed that this thin layer of trapped water is responsible for the insulating property of a wetsuit; in this problem, we will explore the veracity of this claim. Consider the following model for a wetsuit: A diver's skin and cold water at temperatures T_1 and T_2, respectively, are separated by dual layers of body-warmed trapped water and neoprene of thicknesses l_1 and l_2, respectively (Fig. 19–7). T_x is the temperature at the trapped water-neoprene interface and the thermal conductivities of warm (30–35°C) water and neoprene are $k_1 = 0.62$ and $k_2 = 0.14 \, \text{J}/(\text{s} \cdot \text{m} \cdot \text{C}°)$, respectively.

(a) Show that the rate of heat flow $\Delta Q/\Delta t$ from the diver's skin to the cold water is given by

$$\frac{\Delta Q}{\Delta t} = A \frac{T_1 - T_2}{l_1/k_1 + l_2/k_2}$$

where A is the area of the diver's skin covered by the wetsuit. [Hint: $(T_1 - T_2) = (T_1 - T_x) + (T_x - T_2)$.]

(b) For a typical diver, the area of covered skin and skin temperature are $A \approx 1.0 \, \text{m}^2$ and $T_1 = 35°C$, respectively. When diving in water of temperature $T_2 = 15°C$, a neoprene thickness of $l_2 = 7.0 \, \text{mm}$ is commonly used. For a properly fitted suit, the layer in which water is trapped might have thickness $l_1 = 0.50 \, \text{mm}$. Using this information, determine the rate of heat flow $\Delta Q/\Delta t$, assuming a wetsuit filled with trapped water so that the layer with thickness l_1 contains warm water.

(c) Now assume a wetsuit without the "benefit" of trapped water. That is, recalculate $\Delta Q/\Delta t$, assuming a wetsuit where the layer of thickness l_1 has not been filled with water, but rather contains air. For air, $k = 0.023 \, \text{J}/(\text{s} \cdot \text{m} \cdot \text{C}°)$.

(d) Finally, recalculate $\Delta Q/\Delta t$, assuming a "perfect-fit" wetsuit, where both layers l_1 and l_2 are both filled with neoprene.

(e) Comparing the three values you obtained for $\Delta Q/\Delta t$, comment on the claim that a layer of trapped water enhances the insulating property of a wetsuit. [In fact, divers use "leaky" wetsuits, not due to some desirable property of water, but because they are inexpensive in comparison with leak-proof drysuits.]

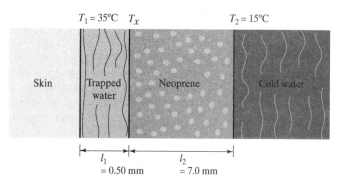

FIGURE 19–7

General Problems

20. (II) A helium-containing latex balloon, with 15-L volume and 8°C temperature, is brought into a very large 20°C room. If left there long enough so that thermal equilibrium is achieved, determine the heat (J) that flows from the room into the balloon. Assume the internal pressure of the balloon is always 1.0 atm, the mass of the (deflated) latex balloon itself is 2.8 gm, and the specific heat of latex at 1.0 atm pressure is 1840 J/kg·C°.

21. (II) A 10.0 m × 10.0 m × 5.0 m classroom is filled with 50 students taking the final exam for their introductory physics course.
    (a) By using appropriate values for pressure and temperature, approximate the number of air molecules in the classroom.
    (b) On the average, a person consumes about 2400 kcal per day and transfers 1/24 of this amount as heat to the surrounding environment every hour. Assuming the door is closed so the classroom's air is kept (fairly) static, by how much will the room's temperature increase during the hour-long exam? Assume that all of the heat flow from the students is transferred to the classroom air. [Given this final assumption, the calculation determines the upper bound of the room's temperature increase.]

22. (II) What determines the value of the latent heat of vaporization for a substance? The latent heat provides energy for two processes: changing the substance's internal energy by breaking chemical bonds that bind molecules together in the liquid state and doing work as the freed molecules expand into a gas at atmospheric pressure. Consider a 1.0-kg sample of liquid water with volume V_L at 100°C and 1.0 atm to which 2.26×10^5 J (= 539 cal) of heat is added, changing the sample into steam with volume V_S at 100°C and 1.0 atm. The molecular mass of water is 18 and, at 100°C, the density of liquid water is 958.4 kg/m³.
    (a) Determine the work done as 1.0 kg of liquid water expands at 1.0 atm into 1.0 kg of steam.
    (b) Using the known value of latent heat and your result from part (a), determine the amount of energy required to break chemical bonds when 1.0 kg of water is changed into 1.0 kg of steam at 100°C. Assuming each liquid water molecule has the same bonding energy, what is the bonding energy per molecule in liquid water at 100°C?
    (c) What percentage of the latent heat of vaporization goes into doing work and what percentage goes into breaking chemical bonds?

23. (II) A microwave oven is used to heat 200 g of water. On its maximum setting, the oven can raise the temperature of the liquid water from 20°C to 100°C in 1 min 45 s (= 105 s).
    (a) At what rate does the oven input energy to the liquid water?
    (b) If the power input from the oven to the water remains constant, determine how many grams of water will boil away if the oven is operated for 2.00 min (rather than just 1 min 45 s).

24. (II) Suppose an average-sized room holds 100 kg of supersaturated air at 20°C and 500 g of water vapor suddenly condenses into a cloud which fills the room. Assuming all of the heat flow from the condensing vapor goes into warming the surrounding air and that this process occurs at constant pressure, determine the room's air-temperature increase ΔT. The latent heat of vaporization for water at 20°C is 585 cal/g and the molecular mass of air is 29 g/mol. [In reality, as the air warms, some energy would be used to re-vaporize a portion of the cloud, so our value ΔT is an overestimate.]

25. (II) If one mole of monatomic ideal gas at pressure P_1, volume V_1, and temperature $T_1 = 300$ K expands until its volume has increased by 10%, how much more work is done if the expansion is done isothermally rather than adiabatically?

26. (II) A 20 cm³ air bubble ($\gamma = 1.4$) is at the bottom of a 40 m deep lake where the temperature is 4.0°C. The bubble rises to the lake's surface where the temperature is 20.0°C.
    (a) Assuming the pressure at the lake's surface is 1.0 atm, find the pressure at a depth of 40 m.
    (b) Find the volume of the air bubble just as it reaches the surface, assuming that it rises slowly enough so that its temperature is the same as that of the surrounding water at all times.
    (c) Find the volume of the air bubble just as it reaches the surface, assuming that it rises so fast that no heat has time to flow into it from the surrounding water.

27. (III) Consider a sample of air (called a parcel), which is moving to a different altitude y in Earth's atmosphere (Fig. 19–8). As the parcel changes altitude it always assumes the pressure P of the surrounding air at its present altitude and thus, from Eq. 13–4 in the text,

$$\frac{dP}{dy} = -\rho g$$

where ρ is the parcel's altitude-dependent mass density. During this motion, the parcel's volume will change and, because air is a poor heat conductor, we assume this expansion or contraction will take place adiabatically.
(a) As an ideal gas undergoes an adiabatic process, $P^{1-\gamma}T^\gamma =$ constant (see Problem 11). Using this relation, show that the parcel's pressure and temperature are related by

$$(1-\gamma)\frac{dP}{dy} + \gamma\frac{P}{T}\frac{dT}{dy} = 0$$

and thus

$$(1-\gamma)(-\rho g) + \gamma\frac{P}{T}\frac{dT}{dy} = 0.$$

(b) Use the ideal gas law with the result from part (a) to show that the change in the parcel's temperature with change in altitude is given by

$$\frac{dT}{dy} = \frac{1-\gamma}{\gamma}\frac{mg}{k}$$

where m is the average mass of an air molecule.
(c) Given that air is a diatomic gas with an average molecular mass of 29, show that $\frac{dT}{dy} = -9.8$ C°/km. This value is called the "adiabatic lapse rate for dry air."
(d) In California, the prevailing westerly winds descend from one of the highest elevations (the 4000-m Sierra Nevada

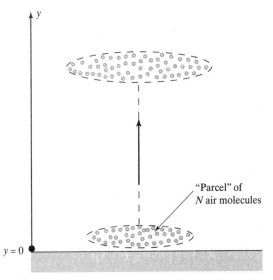

FIGURE 19–8

Mountains) to one of the lowest elevations (sea level Death Valley) in the continental United States. If a dry wind has a temperature of $-5°C$ at the top of the Sierra Nevadas, what is the wind's temperature after it has descended to Death Valley?

28. (III) Assume a given amount of ideal gas has an initial pressure P_o and volume V_o. Consider the following two possible processes by which this gas might be expanded—isothermal and adiabatic, which obey the relations $PV = $ constant and $PV^\gamma = $ constant, respectively. By evaluating the derivative $\frac{dP}{dV}$ at the point (P_o, V_o) for each of these processes, show that the adiabatic curve is steeper than the isothermal curve, as long as $\gamma > 1$. Is γ always greater than one for all types (e.g., monatomic, diatomic) of ideal gases?

29. (III) Starting with the first law of thermodynamics in differential form and noting that for N molecules of a monatomic ideal gas the internal energy is given by $U = \frac{3}{2}NkT$, show that, for an adiabatic process, the temperature and volume of a monatomic gas are related by $T^{3/2}V = $ constant.

30. (III) Isaac Newton knew that the speed of a sound wave in a gas of density ρ is given by $v = \sqrt{B/\rho}$ (Eq. 15–4 in the text), where the bulk modulus B is defined by $B = -V\frac{dP}{dV}$ (Eq. 12–7b in the text). Consider a sound wave moving through an ideal gas of temperature T.

(a) Newton believed that the regions of compressions and expansions that make up a sound wave must be in thermal equilibrium with the surrounding gas and so these processes take place at constant temperature T. Show that, based on this reasoning, the speed of sound is predicted to be

$$v = \sqrt{\frac{RT}{M}}$$

where M is the molecular mass of the gas.

(b) Because gases are poor heat conductors and sound wave compressions and expansions occur very quickly, one might conjecture that these are adiabatic (rather than isothermal) processes. Assuming pressure and volume are related by $PV^\gamma = $ constant, show that the speed of sound is predicted to be

$$v = \sqrt{\frac{\gamma RT}{M}}.$$

(c) At $20°C$, the speed of sound in the diatomic gas air is observed to be 343 m/s. Which of the two models given above predicts the correct value for the speed of sound? For air, the average molecular mass is 29 g/mol.

CHAPTER 20

Second Law of Thermodynamics

Section 20–3

1. (II) Assume a Carnot engine is operating between temperatures T_H and T_L and that you have the ability to either increase the temperature T_H by an amount ΔT while holding T_L constant or decrease the temperature T_L by an amount ΔT while holding T_H constant. Which of these two possible actions will result in the greatest increase of efficiency for the engine?

2. (II) A Carnot engine has an efficiency $e = 0.60$. At the end of each adiabatic compression, the monatomic ideal gas in its cylinder has a volume of 50 cm^3, while during each isothermal expansion, the gas pressure decreases by a factor of one-half. Determine the volume of the gas at the end of each adiabatic expansion.

3. (II) An engine, working between the temperatures of 600 K and 400 K, takes 1.0 mole of monatomic ideal gas around a Carnot cycle such that its smallest volume (at the end of the adiabatic compression) and its largest volume (at the conclusion of the adiabatic expansion) are 0.050 cm^3 and 0.40 cm^3, respectively.
   (a) Determine the change, if any, in efficiency of this engine when its monatomic gas is replaced by 1.0 mole of diatomic gas.
   (b) Determine the change, if any, in both the volume at the end of the isothermal compression and the volume at the conclusion of the isothermal expansion when the engine's monatomic gas is replaced by 1.0 mole of diatomic gas.

4. (II) An engine, working between the temperatures of 600 K and 400 K, takes 1.0 mole of gas around a Carnot cycle such that its smallest volume (at the end of the adiabatic compression) and its largest volume (at the conclusion of the adiabatic expansion) are 0.050 cm^3 and 0.40 cm^3, respectively. By what factor is the net work done per cycle changed if the gas used in this engine is diatomic rather than monatomic?

5. (II) In a properly functioning internal combustion engine, whose performance can be approximated by the Otto cycle, gasoline vapor must be ignited at the end of the cylinder's adiabatic compression by the spark from a spark plug rather than by the elevated temperature of the engine's working gas (air-gasoline mixture). Given that the ignition temperature of (87-octane) gasoline vapor is 430°C and assuming that the working gas is diatomic and enters the cylinder at 25°C, determine the maximum compression ratio for an internal combustion engine.

6. (II) The high- and low-temperature reservoirs for a Carnot engine are a large vat of boiling water at a pressure of 1.0 atm and a large block of aluminum at some unknown temperature. If, during one cycle of the engine, it is found that the engine absorbs 20.0 J of heat from the boiling water and does 12.0 J of net work, what is the temperature of the aluminum block?

7. (III) The working substance of a certain Carnot engine is 1.0 mole of an ideal monatomic gas. During the isothermal expansion portion of this engine's cycle, the volume of the gas doubles, while during the adiabatic expansion the volume increases by a factor of 5.70. The work output of the engine is 900 J in each cycle. Compute the temperatures of the two reservoirs between which this engine operates.

Section 20–4

8. (II) During winter, the monthly electricity bill to operate a heat pump that heats a certain house is $50. The coefficient of performance for the heat pump is 3.0. If the house were heated using electric heaters rather than the heat pump, determine the monthly electric bill. Assume electric heaters can transform electricity into heat with 100% efficiency.

9. (II) Heat pumps are suitable for warming houses only when the outside temperature is no colder than about −5°C. To demonstrate this fact, assume a (real) heat pump with a coefficient of performance (CP) that is 20% that of an ideal heat pump is used to maintain the inside temperature of a house at 22°C.
   (a) Show that the coefficient of performance for an ideal heat pump is given by $\dfrac{T_H}{T_H - T_L}$.
   (b) By what factor does the CP of the (real) heat pump decrease when the outside temperature is changed from 10°C to 0°C?
   (c) For what outside temperature does the (real) heat pump's CP = 2.0? Below this CP-value, heat pumps are usually aided by supplementary heaters (e.g., electric heaters) in house heating.

10. (II) A refrigeration system, designed to condense nitrogen vapor already at its boiling point of 77 K into a 77 K liquid, is placed in a small 295 K room. When operating, the system expels its exhaust heat into the room's air.
    (a) What is the minimum amount of heat Q_{min} this system can exhaust to the room per mole of condensed nitrogen (heat of vaporization 2793 J/mol)?
    (b) If the room's dimensions are $3.0\text{ m} \times 4.0\text{ m} \times 3.0\text{ m}$, by approximately how much would Q_{min} increase the temperature of the room's air? The constant-pressure specific heat and density of room-temperature air are $1000\text{ J/kg}\cdot\text{K}$ and 1.23 kg/m^3.

11. (II) As a holdover from when commercial refrigeration was used mostly to make ice to sell to people who had iceboxes, refrigeration units are commonly rated in "tons." By definition, a

111

one-ton air conditioning system can remove the amount of heat from a low-temperature area necessary to freeze one British ton (2000 pounds = 909 kg) of 0°C water into 0°C ice in one 24-hr day. If, on a 35°C day, the interior of a house is maintained at 22°C by the continuous operation of a five-ton air conditioning system, how much does this cooling cost the homeowner per hour? Assume the work done by the refrigeration unit is powered by electricity that costs $0.10 per kWh and that the unit's coefficient of performance is 15% that of an ideal refrigerator. 1 kWh = 3.60×10^6 J.

12. (III) An ideal heat pump is used to maintain the inside temperature of a house at $T_I = 22°C$ when the outside temperature is T_O. Assume that when it is operating, the heat pump does work at a rate of 1500 W. Also assume that the house loses heat via conduction through its walls and other surfaces at a rate given by $\left(650 \dfrac{W}{C°}\right)(T_I - T_O)$.

(a) Show that the coefficient of performance for an ideal heat pump is given by $\dfrac{T_H}{T_H - T_L}$.

(b) For what outside temperature would the heat pump have to operate at all times in order to maintain the house at an inside temperature of 22°C?

(c) If the outside temperature is 8°C, what percentage of the time does the heat pump have to operate in order to maintain the house at an inside temperature of 22°C?

Sections 20–5 and 20–6

13. (II) Two 1.0-kg blocks (A and B) of copper are in thermal contact and initially at 20°C. Assume that heat then flows from A to B until block A has been cooled to 10°C and block B has been warmed to 30°C (see Fig. 20–1). Show that this process will not occur spontaneously in nature.

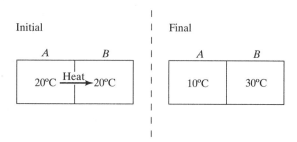

FIGURE 20–1

14. (II) At Earth's orbital radius, the intensity of sunlight (above the atmosphere) is 1350 W/m². Given that the surface temperature of the Sun is 5800 K, determine the change in entropy of the Sun per second.

15. (II) A certain wax has these thermodynamic properties: heat of fusion 100 cal/g, melting point 300 K, and specific heat at constant pressure 0.50 cal/g·C° when in the liquid state. A 30-g sample of this wax, initially in solid form at 300 K, having been placed in an open vessel over a heater, is first melted and then warmed to the final temperature 400 K. What is the change ΔS in the entropy of the sample?

16. (II) An ideal gas of n moles undergoes the reversible process ab shown in the PV diagram of Fig. 20–2. The temperature T of the gas is the same at points a and b. Determine the gas's change in entropy due to this process.

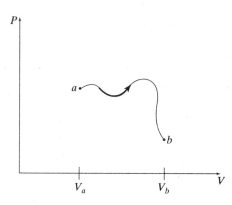

FIGURE 20–2

17. (II) Thermodynamic processes are sometimes represented on TS (temperature-entropy) diagrams, rather than PV diagrams. Determine the slope of a constant-volume process on a TS diagram when a system with n moles and constant-volume molar specific heat C_V is at temperature T.

18. (II) A monatomic ideal gas of n moles is initially at pressure P_o, volume V_o, and temperature T_o. If the gas expands at constant pressure to volume $2V_o$, determine its change in entropy ΔS.

19. (III) (a) A 100-g ice cube at 0°C is placed in a large 20°C room. Heat flows (from the room to the ice cube) such that the ice cube melts and the liquid water warms to 20°C. The room is so large that its temperature remains (more-or-less) 20°C at all times. Calculate the change in entropy for the (water + room) system due to this process. Will this process occur naturally?

(b) 100 grams of liquid water at 20°C is placed in a large 20°C room. Heat flows (from the water to the room) such that the liquid water cools to 0°C then freezes into a 0°C ice cube. The room is so large that its temperature remains 20°C at all times. Calculate the change in entropy for the (water + room) system due to this process. Will this process occur naturally?

(c) 100 grams of liquid water at 20°C is placed in a large 0°C room. Heat flows (from the water to the room) such that the liquid water cools to 0°C then freezes into a 0°C ice cube. The room is so large that its temperature remains 0°C at all times. Calculate the change in entropy for the (water + room) system due to this process. Will this process occur naturally?

20. (III) Two insulated containers A and B with total volume $V = V_A + V_B$ are each filled with a different ideal gas such that their pressures and temperatures are equal. The containers are then connected so that the two gases are allowed to mix completely. Define x and N to be the fraction of V possessed by container A and the total number of gas molecules, respectively. Determine the value of x that maximizes the change in entropy due to this mixing process.

21. (III) Consider an ideal gas of n moles and constant-volume molar specific heat C_V taken via a reversible process from an initial state to a final state with pressure, volume, and temperature of P_1, V_1, T_1 and P_2, V_2, T_2, respectively.

(a) Starting with the first law, show that the change in entropy for this gas is given by

$$\Delta S = nC_V \ln\left(\dfrac{T_2}{T_1}\right) + nR \ln\left(\dfrac{V_2}{V_1}\right).$$

(b) If the reversible process is adiabatic, then $T_1 V_1^{\gamma-1} = T_2 V_2^{\gamma-1}$. Using the result from part (a), show that $\Delta S = 0$ for an adiabatic process.

22. (III) Consider an ideal gas of n moles with constant-volume and constant-pressure molar specific heats C_V and C_P, respectively.
(a) Starting with the first law, show that when temperature and volume of this gas are changed by a reversible process, its change in entropy is given by

$$dS = nC_V \frac{dT}{T} + nR \frac{dV}{V}.$$

(b) Show that the expression in part (a) can be written as

$$dS = nC_V \frac{dP}{P} + nC_P \frac{dV}{V}.$$

(c) Using the expression from part (b), show that if $dS = 0$ for the reversible process (i.e., the process is adiabatic), then $PV^\gamma = $ constant, where $\gamma = C_P/C_V$.

23. (III) Assume a system consists of two regions of ideal gas, each with N molecules at temperature T, separated by a frictionless piston as shown in Fig. 20–3a, where initially the left- and right-hand regions are at pressures and volumes P_1, V_1 and P_2, V_2, respectively. The gases then cause the piston to move in such a way that the temperature of each gas is maintained always at T, while the volume of each region changes in magnitude by ΔV. At the conclusion of this reversible isothermal process, the final pressures and volumes of the two regions are P'_1, V'_1 and P'_2, V'_2, where $V'_1 = V_1 + \Delta V$ and $V'_2 = V_2 - \Delta V$ as shown in Fig. 20–3b.
(a) Show that the total change in entropy for the system due to this isothermal process is

$$\Delta S = Nk\left[\ln\left(1 + \frac{\Delta V}{V_1}\right) + \ln\left(1 - \frac{\Delta V}{V_2}\right)\right].$$

(b) Show that ΔS is maximized, if $\Delta V = \dfrac{V_2 - V_1}{2}$.

(c) Show that, if $\Delta V = \dfrac{V_2 - V_1}{2}$, then $P'_1 = P'_2$. [Note: $P'_1 = P'_2$ is the condition of equilibrium for this system, so the result of this problem suggests that the system's equilibrium is a state of maximum entropy.]

24. (III) A system consists of two regions of ideal gas, each with N molecules at temperature T, separated by a frictionless piston as shown in Fig. 20–4a, where initially the left- and right-hand regions have the same volume V. Imagine the gases then cause the piston to move in such a way that the temperature of each gas is maintained always at T, while the volume of each region changes in magnitude by ΔV. At the conclusion of this reversible isothermal process, the volumes of the left- and right-hand regions are $V'_L = V + \Delta V$ and $V'_R = V - \Delta V$ as shown in Fig. 20–4b.
(a) Show that the total change in entropy for the system due to this isothermal process is

$$\Delta S = Nk[\ln(1 + x) + \ln(1 - x)]$$

where $x = \Delta V/V$. The domain of x is $-1.0 \leq x \leq +1.0$.
(b) Plot $\Delta S/Nk$ vs. x over the range $-1.0 < x < +1.0$. Assuming the system is initially at $x = 0$ (i.e., the left- and right-hand regions have equal volumes), use your plot to explain why the system will not spontaneously move the piston as time progresses onward.

FIGURE 20–4

25. (III) Consider two different reversible processes, each of which takes n moles of a monatomic ideal gas from point a to point b on a PV diagram (Fig 20–5). The temperature of the gas is the same at point b as it is at point a. In Process I, the gas expands isothermally at temperature T from point a to point b. In Process II, the gas first expands at constant pressure from volume V_a to volume V_b, and then is taken from pressure P_a to pressure P_b at constant volume. Calculate the change in entropy ΔS of the gas for each process and show that $\Delta S_\mathrm{I} = \Delta S_\mathrm{II}$.

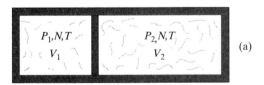

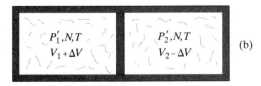

FIGURE 20–3

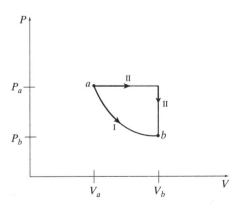

FIGURE 20–5

26. (III) An isolated system consists of a small sample (mass m, constant-volume specific heat c_V) and a heat reservoir at constant temperature T_R (see Fig. 20–6). The sample, which is initially separated from the reservoir and is at temperature T_S, is then placed in thermal contact with the reservoir. After a flow of heat during which it remains at constant volume, the sample is in thermal equilibrium with the reservoir. Show that the total entropy for this system increases independent of whether the sample is heated $(T_S < T_R)$ or cooled $(T_S > T_R)$ during this process as follows:

(a) Prove that the system's change in entropy ΔS during this process is

$$\Delta S = mc_V\left(\frac{1}{x} - 1 + \ln x\right)$$

where $x = T_R/T_S$.

(b) Plot $\Delta S/mc_V$ vs. x for $0 < x < 5$. Identify the range of x-values that corresponds to the sample being heated by the reservoir as well as the range associated with the sample being cooled. Then explain why your plot demonstrates that ΔS increases in both cases.

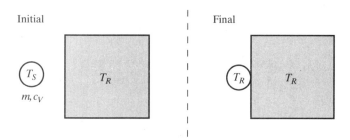

FIGURE 20–6

Section 20–9

27. (III) Consider an isolated gas-like system consisting of a box that contains $N = 10$ distinguishable atoms, each moving at the same speed v. The number of unique ways that these atoms can be arranged so that N_L atoms are within the left-hand half of the box and N_R atoms are within the right-hand half of the box is given by $\dfrac{N!}{N_L!N_R!}$, where, e.g., the factorial $4! = 4 \cdot 3 \cdot 2 \cdot 1$ (the only exception to this pattern is $0! = 1$). Define each unique arrangement of atoms within the box to be a microstate of this system. Now imagine the following two possible microstates: state A where all of the atoms are within the left-hand half of the box and none are within the right-hand half; and state B where the atoms are distributed uniformly throughout the box (i.e., there is the same number in the left-hand and right-hand halves). See Fig. 20–7.

(a) Assume the system is initially in state A and, at a later time, is found to be in state B. Determine the system's change in entropy. Will this process occur naturally?

(b) Assume the system is initially in state B and, at a later time, is found to be in state A. Determine the system's change in entropy. Will this process occur naturally?

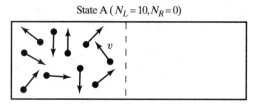

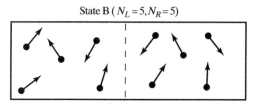

FIGURE 20–7

General Problems

28. (II) A heat engine takes a diatomic gas around the cycle shown in Fig. 20–8. The process from points a to b is an isothermal compression, while the process from points c to a is an adiabatic expansion.

(a) Using the ideal gas law, find how many moles of gas are in this engine.
(b) Determine the temperature at point c.
(c) Find the heat input into the gas during the process from points b to c.
(d) Find the work done by the gas during the process from points a to b.
(e) Find the work done by the gas during the process from points c to a.
(f) Find the engine's efficiency.
(g) What is the maximum efficiency possible for an engine working between T_a and T_c?

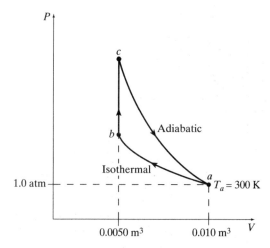

FIGURE 20–8

29. (II) The operation of a certain heat engine takes an ideal monatomic gas through a cycle, which appears on a PV diagram as the rectangle shown in Fig. 20–9.

(a) Determine the efficiency of this engine.
[*Hint*: $|Q_H|$ and $|Q_L|$ are the total heat input and total heat exhausted during one cycle of this engine, respectively.]

(b) Compare (as a ratio) the efficiency of this engine with that of a Carnot engine operating between T_H and T_L, where T_H and T_L are the highest and lowest temperatures achieved by the ideal gas in the cycle shown in Fig. 20–9.

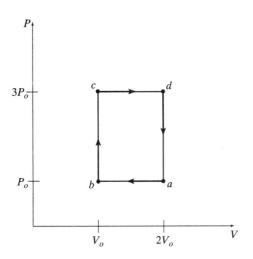

FIGURE 20–9

30. (III) A 70-kg block of copper, initially at 1000°C, is used as the high-temperature reservoir of a Carnot engine. The engine dumps its exhaust heat into a very large tub of ice and water at 0°C and it operates until the temperature of the copper block drops to that of the ice-water mixture. Assume the specific heat of copper remains constant (at its 20°C value) over the course of the engine's operation.
(a) What is the change in entropy of the copper block due to the engine's operation? Based on the fact that a Carnot engine is reversible, what is the change in entropy ΔS of the ice-water mixture due to the engine's operation?
(b) Determine the total amount of heat exhausted into the tub of ice and water during the engine's operation.
(c) Determine the total amount of work done during the engine's operation.
(d) Defining the effective efficiency e_{eff} as the ratio of the total work done to the total heat input from the high-temperature reservoir, determine e_{eff} for the operation of this engine. Compare e_{eff} with the efficiency this engine would have if the temperature of the copper block were somehow (e.g., with an embedded heater) maintained always at 1000°C.

CHAPTER 21

Electric Charge and Electric Field

Section 21–5

1. (II) An insulating brick has rectangular faces of dimensions 7 cm × 10 cm, separated by the brick's 20 cm length. If a positive point charge $+Q$ is firmly attached to the midpoint of one face of the brick, and a negative point charge $-Q$ is attached to the midpoint of the other face, for what minimum value of Q will the brick be fractured under compressive stress? For the compressive strength of brick, see Table 12–2 in the text. Ignore any polarization of charge that may be induced within the brick.

2. (II) A one-dimensional lattice of positive ions, each with charge $+Q$ and separated from its neighbors by a distance d, occupies the right-hand half of the x-axis. That is, there is a single ion at $x = 0$, $x = +d$, $x = +2d$, $x = +3d$, etc., out to infinity.
   (a) If an electron is placed at the position $x = -d$, determine F, the magnitude of force that the lattice exerts on the electron.
   (b) If the electron is instead placed at $x = -3d$, what is the value of F?
   [Hint: The infinite sum $\sum_{n=1}^{n=\infty} \frac{1}{n^2} = \frac{\pi^2}{6}$, where the sum is evaluated with n taking on all the positive integer values.]

3. (II) The constant of proportionality ϵ_o in Coulomb's law at first glance appears to be a fundamental constant of nature that must be obtained from experiment (similar to e.g., the Gravitational constant G). However, in the SI system of units, ϵ_o is a theoretically defined quantity with the following arbitrary definition: $\epsilon_o = \frac{1}{c^2 \mu_o}$ $C^2/N \cdot m^2$, where the numerical values of the constants c and μ_o are defined to be 299,792,458 and $4\pi \times 10^{-7}$, respectively.
   (a) Show that this definition for ϵ_o does indeed yield the numerical value quoted for this quantity in the text.
   (b) After defining ϵ_o in this way, one might posit a definition for the unit of charge using Coulomb's law, which makes use of the previously defined units for distance and force. Consider the following possible definition: If two identical positive point-like objects are placed exactly one meter apart and it is observed that they repel each other with a force of exactly one Newton, then the amount of charge on each object is defined to be $+Q$ Coulombs. If this definition is to be consistent with the currently used Coulomb unit, what is the proper value for Q? [In SI units, the Coulomb is actually defined in terms of a given amount of charge flow in an electric circuit.]

4. (II) Theoretical physicists commonly unclutter their equations by setting constants of proportionality equal to unity. In one such example, the cgs (or "Gaussian") system of units arbitrarily defines the constant $k = 1$ so that Coulomb's law reads $F = \frac{Q_1 Q_2}{r^2}$. Then the Gaussian unit of charge is defined as follows: If two positive charges of equal magnitude Q are placed 1 cm apart and found to repel each other with a force of 1 dyne, then Q is 1 electrostatic unit (esu). Use this definition to determine how many Coulombs (the SI unit of charge) are equivalent to 1 esu.

5. (II) Is the electric force strong enough to hold an atom together? Consider the simplest atom, hydrogen, in which an electron (mass m) orbits a stationary proton. It is known from experiment that the radius r of the electron's orbit is about 0.5×10^{-10} m. A relation from quantum theory called the Heisenberg Uncertainty Principle tells us that, when confined to a spatial region of 1×10^{-10} m, an electron must have a speed of at least $v_{\min} = 6 \times 10^5$ m/s.
   (a) Assume that an electron with speed v is held in a circular orbit about a stationary proton by the electrical attraction between these two charged objects. Show that for this model of the hydrogen atom to be consistent with the Uncertainty Principle the constant k in Coulomb's law must fulfill the criterion
   $$k \geq \frac{m v_{\min}^2 r}{e^2}.$$
   (b) Determine the minimum required value for k if the atom is held together by the electric force. Does the actual value for k exceed this minimal value?

6. (III) Two positive charges $+Q$ are affixed rigidly to the x-axis: one at $x = +d$ and the other at $x = -d$. A third charge $+q$ of mass m, which is constrained to move only along the x-axis, is displaced from the origin by a small distance $A \ll d$ and then released from rest.
   (a) Show that (to a good approximation) $+q$ will execute simple harmonic motion and determine an expression for its oscillation period T.
   (b) If these three charges are each singly ionized (i.e., $q = Q = +e$) sodium atoms at the equilibrium spacing $d = 3 \times 10^{-10}$ m typical of the atomic spacing in a solid, find T in picoseconds (ps).

Sections 21–6 to 21–8

7. (II) Two line-charges, each with charge $+Q$ uniformly distributed over a length L, are separated by a distance $2a$. Point P

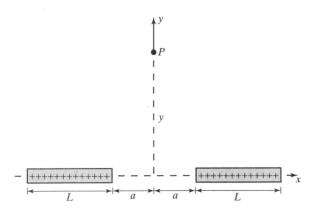

FIGURE 21–1

is a distance y along the line charges' perpendicular bisector, as shown in Fig. 21–1.

(a) Find the magnitude and direction of the electric field at point P.

(b) Assume a positive charge $+q$ with mass m is positioned at point P, then is released from rest and that you want to predict this charge's velocity v after it has been repelled to a distance infinitely far away from the line charges. If you decided to work this problem using Newton's force method, what differential equation would you need to solve in order to obtain the charge's position $y(t)$ as a function of time (and hence its velocity from $v = dy/dt$)? Just determine the appropriate differential equation to be solved (i.e., the equation involving y and its derivatives), but do not try to solve it (that's too hard!).

8. (II) (a) A line of charge possesses a uniform charge per length λ over its length L. Assume the line lies on the y-axis with its center (point O) at the origin and its ends at $y = \pm L/2$. Modify the solution of Example 21–10 in the text to show that the electric field at point P, a distance x along the axis (i.e., the line charge's perpendicular bisector), is given by

$$\mathbf{E} = \frac{\lambda}{2\pi\epsilon_o} \frac{L}{x(L^2 + 4x)^{1/2}} \mathbf{i}.$$

(b) Use your result from part (a) to find the electric field (magnitude and direction) a distance z above the center of a square loop with side length L and uniform charge per length λ (Fig. 21–2).

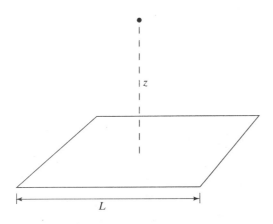

FIGURE 21–2

9. (II) A very thin line of charge lies along the x-axis from $x = -\infty$ to $x = +\infty$. Another similar line of charge lies along the y-axis from $y = -\infty$ to $y = +\infty$. Both lines have a uniform charge per length λ. Determine the resulting electric field magnitude and direction relative to the x-axis at a point (x,y) in the first quadrant of the x-y plane.

10. (II) A point-like charged ball (mass m, charge q) is hovering at rest at height h over a large horizontal uniformly charged plane (charge per unit area σ). What will be the acceleration of the ball (expressed as a multiple of g) if a disk of radius $R = h$ is removed from the plane directly underneath the ball?

11. (III) A thin hollow tube of radius a and length L has a charge $+Q$ distributed uniformly over its surface (Fig. 21–3). The x-axis lies on the axis of the tube such that the tube's left- and right-hand ends align with $x = -L/2$ and $x = +L/2$, respectively. Determine the electric field (magnitude and direction) on the x-axis at the position $x = x_o$, where $-L/2 < x_o < +L/2$.

[Hint: Imagine the tube as a collection of rings.]

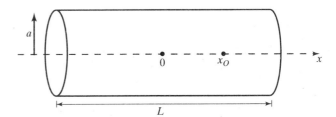

FIGURE 21–3

12. (III) In Example 21–11 in the text, it is shown that the electric field on the axis of a uniformly charged disk, a distance z from its center, is $E = \frac{\sigma}{2\epsilon_o}\left[1 - \frac{z}{(z^2 + R^2)^{1/2}}\right]$. It is then shown that, when $z \ll R$, the electric field can be approximated as $E \approx \frac{\sigma}{2\epsilon_o}$. Let's explore by how much the disk radius R must exceed the distance z in order for this approximation to be valid. In many cases, an approximation is satisfactory if it deviates from the true value by at most 1.0%. Determine how much larger R is than z (i.e., find the ratio R/z) when the approximate E deviates from the true E by a fractional amount of 0.010 (1.0% deviation).

13. (III) (a) A circular annulus of inner radius R_1 and outer radius R_2 has a total charge Q uniformly distributed over its surface. Appropriately modify the work of Example 21–11 in the text to show that the electric field on the axis of the annulus at a distance z from its center is given by

$$E = \frac{Qz}{2\pi\epsilon_o(R_2^2 - R_1^2)}\left[(z^2 + R_1^2)^{-1/2} - (z^2 + R_2^2)^{-1/2}\right].$$

(b) Investigate the prediction of your expression for E in the case of an annulus whose inner radius is almost equal to its outer radius (i.e., a charged annulus that approximates the shape of a thin ring). Take $R_1 = R$ and $R_2 = R + \Delta R$ in your result from part (a). Then, assuming $\Delta R/R \ll 1$, show that in this limit the above expression reduces to $E \approx \frac{1}{4\pi\epsilon_o}\frac{Qz}{(z^2 + R^2)^{3/2}}$, the result expected for the axial field of a charged thin ring with radius R.

14. (III) A thin glass rod is bent into a semicircle of radius R. A charge is nonuniformly distributed along the rod with a linear charge density given by

$$\lambda = \lambda_0 \sin \theta$$

where λ_0 is a positive constant and the rod's upper and lower halves subtend the angles $0 \le \theta \le +\frac{\pi}{2}$ and $0 \ge \theta \ge -\frac{\pi}{2}$, respectively. Point P is at the center of the semicircle (Fig. 21–4).
(a) Find the electric field **E** (magnitude and direction) at point P.
[*Hint*: Remember that $\sin(-\theta) = -\sin(\theta)$, so the upper half of the rod is positively charged, while the bottom half is negatively charged.]
(b) Assume $R = 1.0$ cm and $\lambda_0 = 1.0\ \mu\text{C/m}$. If an electron is placed at point P, determine its acceleration (magnitude and direction). Ignore the electron's weight.

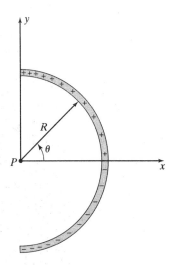

FIGURE 21–4

Section 21–10

15. (II) Breakdown of air ("sparking") occurs when an electric field with magnitude exceeding $E_B = 3 \times 10^6$ N/C is applied (under standard atmospheric conditions) within the air's volume. It is known that the approximate amount of energy required to liberate an electron from ("ionize") an air molecule is $E_I = 2 \times 10^{-18}$ J. Consider the following simple model to explain this breakdown effect: Free electrons in air have a mean free path (average distance traversed between subsequent collisions with molecules) of

$$l_M = \frac{1}{\pi\sqrt{2}r^2\,(N/V)}$$ (see Eq. 18–10b in the text; a factor of four is suppressed for collisions between molecules and point-like electrons). Here, $r = 1 \times 10^{-10}$ m is the radius of an air molecule and N/V is the number density of molecules in air. Initially, a small quantity of free electrons exists in air because they have been liberated from molecules by cosmic rays, light, etc. If an electric field of magnitude E_B is then applied within the air, each of these free electrons, on average, will gain enough kinetic energy as the field accelerates it over the distance l_M to liberate an additional electron when it collides with an air molecule. As this "chain-reaction" process is repeated, an avalanche of liberated electrons is produced and the air becomes electrically conductive. Show that this model offers a plausible explanation for the breakdown phenomenon as follows:

(a) Determine l_M for electrons in air at a pressure and temperature of 1 atm and 20°C.
(b) If, after having traveled a distance l_M, an electron is to acquire the final velocity needed to ionize an air molecule, determine the required ("ionizing") acceleration a_I. Assume that the electron starts from rest and that the acceleration is constant.
(c) To experience the acceleration a_I, to what electric field magnitude E must an electron be subjected? Is your predicted value for E consistent with (i.e., of the correct order of magnitude) the experimentally observed value for E_B? [We overestimate E_B in our calculation because breakdown first occurs due to the electrons that travel further than the average distance l_M between collisions.]

16. (II) Breakdown of an ideal gas ("sparking") occurs when an electric field with magnitude exceeding E_B is applied within the gas volume. Consider the following simple model to describe the breakdown phenomenon: Let E_I be the amount of energy required to liberate an electron from ("ionize") a gas molecule and l_M be the mean free path (average distance traversed between subsequent collisions with molecules) of a free electron within the gas. For free electrons,

$$l_M = \frac{1}{\pi\sqrt{2}r^2\,(N/V)}$$ (see Eq. 18–10b in the text; a factor of four is suppressed for collisions between molecules and point-like electrons). Here, r is the radius of a gas molecule and N/V is the number density of gas molecules. Initially, a small quantity of free electrons exists in the gas because they have been liberated from molecules by cosmic rays, light, etc. If an electric field of magnitude E_B is then applied within the gas, each of these free electrons, on average, will gain enough kinetic energy as the field accelerates it over the distance l_M to liberate an additional electron when it collides with a gas molecule. As this "chain-reaction" process is repeated, an avalanche of liberated electrons is produced and the gas becomes electrically conductive.

(a) Use this model to predict how E_B depends on the gas pressure P. The relation you derive approximates a result from plasma physics called Paschen's law.
(b) In order to decrease E_B by a factor of three, by what factor should the pressure of an ideal gas be reduced? [Plasma globes produce colorful displays of sparks using fairly small electric fields because the gas they contain is at low pressure.]

Section 21–11

17. (II) Consider the following crude model to describe the polarization of charge in an atom when it is placed within an electric field. Imagine a central positive point charge $+Q$ attached via springs to a spherical shell with a negative charge $-Q$ uniformly spread across its surface (Fig. 21–5). Assume the springs are neither stretched nor compressed when the shell is centered on $+Q$, and that k is the effective spring constant for this system when the center of the shell is displaced a distance x in a particular direction. Now assume that this system is placed in a region of space within which a uniform electric field $\mathbf{E} = E\mathbf{i}$, where E is a constant, exists at all points. Show that the electric field produces an electric dipole in this system, which is directly proportional to its magnitude E. That is, show $p = \alpha E$, where α is called the polarizability of the system. Determine α in terms of Q and k.

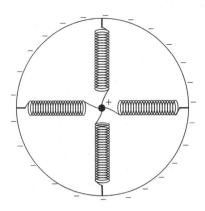

FIGURE 21–5

18. (II) A water molecule must be at what distance r_{align} from a singly charged $(Q = +e)$ ion in order for its dipole moment **p** to be aligned to the ion's electric field? Carry out the following "back-of-the-envelope" calculation to answer this question:
    (a) If a dipole moment p is aligned $(\theta = 0°)$ with an electric field E, how much energy is required to "misalign" it to an angle of $\theta = 90°$? Define this energy to be the "misalignment energy" U_{mis}.
    (b) From Section 19–8 in the text, at temperature T, a molecule possesses an average energy on the order of kT due to its random thermal motion (in this "order-of-magnitude" calculation we neglect multiplicative terms near unity like 3/2, 5/2, etc.). We posit that a dipole will remain aligned with an electric field if the energy it can potentially gain in a random "thermal" collision with another molecule ($\approx kT$) is less than U_{mis}. Use this idea to estimate the maximum value for r_{align} (in nm) at room temperature (20°C). For water, $p = 6.1 \times 10^{-30}\,\text{C}\cdot\text{m}$.

19. (III) A thin rod lies on the x-axis so that its geometric center is at $x = 0$ and its two ends are at $x = \pm L$. The rod's charge per length is given by $\lambda = \alpha x$, where α is a constant. Assume that a uniform electric field $\mathbf{E} = E\mathbf{j}$, with constant magnitude E, exists at all points along the x-axis.
    (a) The field exerts a force $dF = dQ\, E$ on an infinitesimal slice of the rod whose charge is dQ (Fig. 21–6a). Use this fact to compute the net torque τ (magnitude and direction) that the field exerts on the rod about $x = 0$.
    (b) Consider the infinitesimal electric dipole $d\mathbf{p}$ consisting of a positive and negative pair of infinitesimal slices of the rod, each with charge magnitude $|dQ|$ and at distance $|x|$ from the origin, as shown in Fig. 21–6b. Use the definition of dipole moment to write an expression for $d\mathbf{p}$, and then sum (via integration) the $d\mathbf{p}$ for all such pairs in order to determine $\mathbf{p}$, the rod's dipole moment.
    (c) Using $\boldsymbol{\tau} = \mathbf{p} \times \mathbf{E}$, compute the net torque $\boldsymbol{\tau}$ (magnitude and direction) that the field exerts on the rod. You should, of course, obtain the same result as in part (a).

20. (III) The left- and right-hand semicircular halves of a nonconducting hoop of radius R have uniform charge per arclengths $-\lambda$ and $+\lambda$, respectively, where λ is a positive constant (Fig. 21–7). Find an expression for the infinitesimal dipole moment $d\mathbf{p}$ of the pair of charges $-dQ$ and $+dQ$ at angle θ, and then sum (via integration) all of the $d\mathbf{p}$ to determine the dipole moment of the hoop.

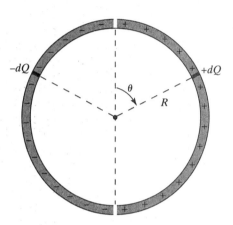

FIGURE 21–7

General Problems

21. (II) As demonstrated in this problem, an electric dipole placed in a nonuniform electric field will experience a net force. Assume an electric dipole, composed of charges $\pm Q$ separated by distance l, is aligned with its dipole moment **p** directed perpendicularly outward from an infinite line of charge with uniform positive charge per length $+\lambda$ (Fig. 21–8). If the midpoint of the dipole is a perpendicular distance y from the line charge, determine the force (magnitude and direction) exerted on the dipole by the line of charge. [In dielectrophoresis, a nonuniform electric field is used to separate electrically polarized particles (such as bacteria) from a mixture.]

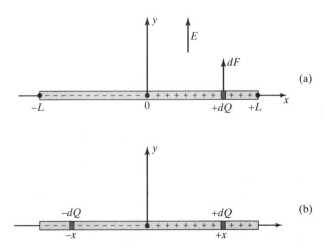

FIGURE 21–6

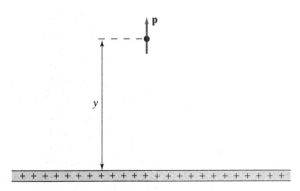

FIGURE 21–8

22. (II) Given that the human body is mostly made of water, estimate the total amount of positive charge in a 60-kg person.

23. (II) In the "international" system of units, which predated the currently used SI units, the definition for the unit of charge was based on the silver electroplating process. In this process, charge flow from a negative electrode to a positive electrode immersed in an aqueous solution of silver nitrate results in silver being deposited on the negative electrode. By definition, one "international coulomb" was defined as the amount of charge that must be emitted from the negative electrodes in order to deposit 1.11800 milligrams of silver. From chemistry it is known that, for each silver atom deposited, a single electron is emitted from the negative electrode. Using the best modern-day values (as given in the text) for the atomic mass of silver, Avogadro's number, and the charge of an electron, determine how many SI Coulombs are equivalent to one international coulomb.

24. (II) Five identical copper spheres, labeled A, B, C, D, and F, are each mounted on their own insulating stands and are separated from each other. Initially, all of the spheres are uncharged, except F, which has charge $+Q$. One by one, starting with A, followed by B, then C, and finally D, each sphere is subjected to the following procedure: The sphere is brought over to F, the sphere is placed in contact with F for a few moments, and then the sphere is returned to its original location. After this sequence using the four spheres is completed, what is the charge on each sphere (as a fraction of $+Q$)? That is, find the final values Q_A, Q_B, Q_C, Q_D, and Q_F. [This process was used by Coulomb to obtain known proportions of charge in his experiment.]

25. (III) A positive point charge q (mass m) moving at high speed v passes by a very massive positive point charge Q. Assume that during this pass, the charges are always far enough apart so that the electrical interaction between the two only produces a small transverse deflection of q from a straight-line path. Model this situation as follows: Assume the path of q can be approximated as a straight line defined to be the x-axis and that the speed of q along this path is always approximately v. Further assume that Q remains stationary at all times on the y-axis at the point $y = -b$ (see Fig. 21–9).
(a) The transverse impulse delivered to q during its pass is
$$J_y = \int_{t=-\infty}^{t=+\infty} F_y \, dt = \int_{x=-\infty}^{x=+\infty} (qE_y)\left(\frac{dx}{v}\right)$$
where E_y is the electric field produced by Q in the direction transverse to the path of q. Use our model to show that $J_y \approx 2kqQ/vb$.
(b) Use the expression for J_y to show that, at the conclusion of its pass, the angular deflection (in radians) of q from a straight-line path is $\theta \approx 2kqQ/mv^2b$.
(c) Determine θ when a proton with a speed $v = 1 \times 10^7$ m/s passes within distance $b = 1 \times 10^{-12}$ m of a lead nucleus.

26. (III) (a) An insulating rod, which lies on the x-axis, carries a charge $+Q$ uniformly distributed along its length L. If the right-hand end of this rod is positioned at $x = 0$, determine $E(x)$ for $x > 0$ (i.e., the electric field produced by this rod along the positive x-axis).
(b) A second identical uniformly charged rod is aligned with the x-axis so that its left-hand end is positioned at $x = +d$. Find the magnitude of the electric force F with which the first rod repels this second rod.

27. (III) Two identical insulating rods each carry a charge $+Q$ uniformly distributed over its length L. The rods are aligned parallel to each other, separated by a distance d (see Fig. 21–10). Find the repulsive electric force $\mathbf{F}$ that these rods exert on each other as follows:
(a) Explain, based on symmetry, why $\mathbf{F}$ is directed in the y-direction only and thus $\mathbf{F} = F_y\mathbf{j}$.
(b) Find the y-component of the electric field E_y created by the lower rod at the location of the upper rod through the following procedure: Assume the lower rod lies on the x-axis with the left-hand end of this rod at the origin as shown in

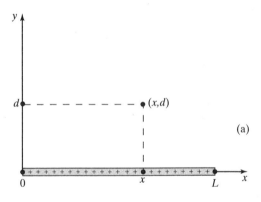

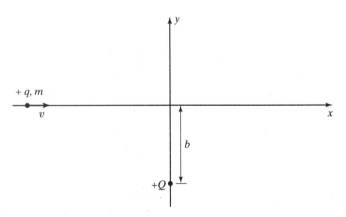

FIGURE 21–9

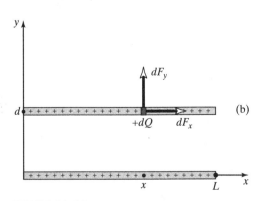

FIGURE 21–10

Fig. 21–10a. Define the rod's charge per length as λ, then determine E_y produced by this rod at the location (x, d).
(c) Assume that the upper rod is now placed on the line $y = +d$ with its left-hand end at the location $(0, d)$ as in Fig. 21–10b. By integrating $dF_y = dQ\, E_y$, calculate the force F_y.

28. (III) Two identical insulating rods each carry a charge $+Q$ uniformly distributed over its length L. The rods are aligned parallel to each other, separated by a distance d (see Fig. 21–10). Let **F** be the repulsive electric force that these rods exert on each other.
(a) Based on symmetry, explain why F_x, the x-component of **F**, must be zero.
Show explicitly that F_x is indeed zero through the following procedure:
(b) Assume that the lower rod lies on the x-axis with the left-hand end of this rod at the origin as shown in Fig. 21–10a. Define the rod's charge per length as λ, then determine E_x at the location (x, d), where E_x is the x-component of the electric field created by the lower rod.
(c) Assume the upper rod is now placed on the line $y = +d$ with its left-hand end at the location $(0, d)$ as in Fig. 21–10b. By integrating $dF_x = dQ\, E_x$, confirm that $F_x = 0$, as deduced by your symmetry argument in part (a).

29. (III) A 2.0-m long thin rod lies on the x-axis so that its geometric center is at $x = 0$ and its two ends are at $x = \pm L$, where $L = 1.0$ m. The rod's charge per length is given by $\lambda = \alpha x$, where the constant $\alpha = 3.0\ \mu C/m^2$ (see Fig. 21–11).
(a) Prove that the net charge on the rod is zero (i.e., it is neutral).
(b) Assume that a uniform electric field $\mathbf{E} = E\mathbf{i}$, with constant magnitude E, exists at all points along the x-axis. Prove that net force on the rod due to this field is zero.
[*Hint*: The field exerts a force $dF = dQ\, E$ on an infinitesimal slice of the rod whose charge is dQ.]
(c) Assume that a nonuniform electric field $\mathbf{E} = E\mathbf{i}$, with $E = ax + b$, exists along the x-axis, where $a = 100\ \text{N/C}\cdot\text{m}$ and $b = 500\ \text{N/C}$. Determine the net force F on the rod due to this field. Also explain qualitatively why the force is nonzero in this case.

30. (III) If an uncharged metal sphere of radius R is placed in a uniform electric field of magnitude E, a nonuniform charge per unit area $\sigma(\theta)$ will be induced on the sphere's surface, where the angle θ is measured relative to the uniform field's direction (Fig. 21–12). Advanced analysis techniques show that this surface charge is distributed according to the relation $\sigma(\theta) = 3\varepsilon_0 E \cos \theta$.
(a) Sketch σ vs. θ for the allowed range $0 \leq \theta \leq \pi$. Explain qualitatively why the largest magnitude positive and negative values of σ occur at the angles that they do.
(b) In Appendix C of the text, it is shown that, on the surface of a sphere, the infinitesimal area located between the angles θ and $\theta + d\theta$ is $dA = (2\pi R \sin \theta)(R\, d\theta) = 2\pi R^2 \sin \theta\, d\theta$. Use this fact to prove that, even though $\sigma(\theta)$ represents a separation of positive and negative charges on the sphere's surface, it also implies that the sphere's net charge is zero.

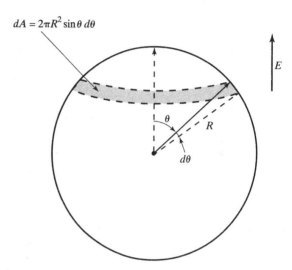

FIGURE 21–12

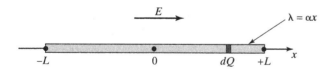

FIGURE 21–11

CHAPTER 22

Gauss's Law

Section 22–2

1. (II) Assume that in a certain region of space, the electric field is parallel to the x-axis and that its magnitude is independent of y and z; i.e., $E_y = 0$, $E_z = 0$, and $E_x(x) \neq 0$. Prove that, if there is no charge in this region, $E_x(x)$ = constant.

2. (II) A physicist needs to calculate the electric flux through the open surface S. As shown in Fig. 22–1, the oddly shaped S has an open "mouth" of radius R and a net charge $+Q$ is within its interior. Although the electric field enters the mouth "nicely," with a constant magnitude E directed perpendicular to the plane of the mouth, once inside S the electric field diverges in such a way that it has a variety of magnitudes and is directed at varying angles to the normal as it exits S at different locations, making the flux calculation through this surface difficult. Help the physicist by showing an easy method to determine the flux through S.

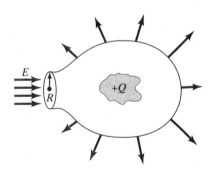

FIGURE 22–1

3. (II) A point charge $+Q$ sits a tiny distance Δx below the plane of a circle of radius R, aligned with the circle's center as shown in Fig. 22–2. Find the (approximate) electric flux through the circle, assuming Δx is negligibly small in comparison with R.
[Hint: Apply Gauss's law, exploiting the radial natural of the point charge's field.]

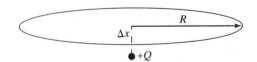

FIGURE 22–2

4. (II) In a region devoid of charge, consider two hypothetical electric field **E** patterns surrounding a point P: (A) From all directions, **E** converges toward P, or (B) **E** converges toward P from some directions and diverges from P in other directions. See Fig. 22–3. Assume no other fields (e.g., gravity) exist in this region of space.
(a) If an object with positive charge $+Q$ is to be placed at P and, once there, we wish it to be in stable equilibrium, explain why A is the desired electric field pattern to accomplish this task, rather than B.
(b) Prove that electric field pattern A cannot exist in our universe. Can pattern B exist? [This problem probes the physics underlying Earnshaw's theorem, which states that a charge cannot be held in stable equilibrium by electrostatic forces alone.]

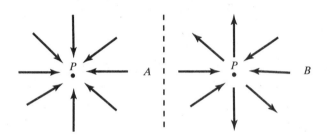

FIGURE 22–3

5. (II) (a) Prove that if **E** is uniform (i.e., constant direction and magnitude) in a given volume of space, then the volume charge density ρ_E is zero throughout that volume.
(b) In a given volume of space, if the volume charge density ρ_E is zero everywhere, does that necessarily imply that **E** is uniform throughout that volume? If so, offer a proof. If not, offer a counter example.

Section 22–3

6. (II) Let's demonstrate the veracity of Gauss's law from a numerical standpoint. In Example 21–10 in the text, using only Coulomb's law, it is shown that an infinitely long line of charge with uniform charge per length $+\lambda$ produces a radially outward directed electric field at point P of magnitude $E = \dfrac{1}{2\pi\epsilon_o} \dfrac{\lambda}{r}$, where r is the perpendicular distance from the line charge to P.

(a) Assuming $\lambda = 8.85 \, \mu\text{C/m}$, determine E_1 and E_2 (N/m) at distances $r_1 = 10.0 \, \text{cm}$ and $r_2 = 20.0 \, \text{cm}$, respectively, from the line.

(b) Imagine two cylindrically shaped (closed) gaussian surfaces S_1 and S_2 with radii $r_1 = 10.0$ cm and $r_2 = 20.0$ cm, respectively. Additionally, let both S_1 and S_2 have length $L = 50$ cm and be centered on the line of charge (Fig. 22–4). Find the surface area A_1 and A_2 (m²) of the cylindrically curved portions of S_1 and S_2, respectively (i.e., do not include the area of the two flat ends of each surface).
(c) Determine the electric flux Φ_E (N·m²/C) through both S_1 and S_2. Explain how it is that, even though $E_2 < E_1$, $\Phi_{E_1} = \Phi_{E_2}$.
(d) Determine the amount of charge Q_{encl} (C) enclosed by both S_1 and S_2. Then demonstrate numerically that Q_{encl}/ϵ_o equals the Φ_E value found in part (c).

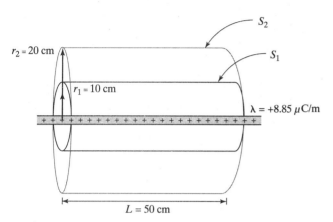

FIGURE 22–4

7. (II) An electric charge $+Q$ is distributed uniformly throughout a nonconducting sphere of radius r_o. With the goal of determining the electric field **E** produced by this sphere at locations outside its surface (i.e., at radial distances $r > r_o$ from the sphere's center), choose the (closed) gaussian surface shown in Fig. 22–5, which consists of a circle of radius r and a hemisphere of radius r, both of which are centered on the sphere.
(a) Let E be the electric field magnitude produced by the sphere at radial distance r. Determine the flux through the chosen gaussian surface in terms of E and r.
(b) What is the net charge enclosed within the chosen gaussian surface?
(c) Determine E using Gauss's law.

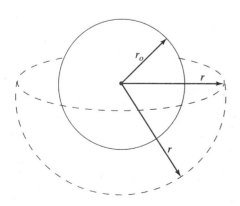

FIGURE 22–5

8. (II) A nonconducting sphere of radius r_o is uniformly charged with volume charge density ρ_E. It is surrounded by a concentric metal (conducting) spherical shell of inner radius r_1 and outer radius r_2, which carries a net charge $+Q$. Determine the resulting electric field in the regions (a) $r < r_o$, (b) $r_o < r < r_1$, (c) $r_1 < r < r_2$, and (d) $r > r_2$ where the radial distance r is measured from the center of the nonconducting sphere.

9. (II) A very long, solid nonconducting cylinder of radius R_1 is uniformly charged with volume charge density ρ_E (cross-sectional view shown in Fig. 22–6). It is surrounded by a cylindrical metal (conducting) tube of inner radius R_2 and outer radius R_3, which has no net charge. If the axes of the two cylinders are parallel, but displaced from each other by a distance d, determine the resulting electric field in the region $r > R_3$, where the radial distance r is measured from the metal cylinder's axis. Assume $d < (R_2 - R_1)$.

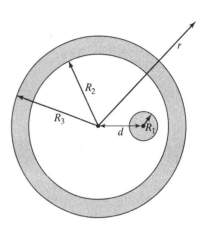

FIGURE 22–6

10. (II) A nonconducting sphere of radius r_o carries total charge $+Q$, where this charge is distributed in accordance with a spherically symmetric volume charge density $\rho_E = Cr^n$. Here r is the radial distance from the sphere's center, C is a constant, and n is an integer greater than -3.
(a) Determine C (in terms of $Q, r_o,$ and n).
(b) Prove that the magnitude of the electric field in the interior region of the sphere ($r < r_o$) is given by

$$E = \frac{Q}{4\pi\epsilon_o r_o^{n+3}} r^{n+1}.$$

(c) Let $Q = 1.0 \,\mu$C and $r_o = 10$ cm. Using a plotting calculator or a computer, plot E vs. r in the region $0 \leq r \leq r_o$, first for the case of uniform charge density ($n = 0$) and then for the case of the highly nonuniform charge density produced with $n = 50$.
(d) Offer a qualitative explanation for why the electric field is "excluded" from the sphere's interior in the nonuniform case with $n = 50$.

11. (II) Imagine that the region of space surrounding a point P is filled with a spherically symmetric volume charge density $\rho_E(r)$, where r is the radial distance from P. In addition, suppose it is found experimentally that the electric field in this region is directed radially outward from P with a magnitude E that is proportional to r^3 (i.e., $E = Cr^3$, where C is an empirically determined positive constant). With these givens, find the expression for $\rho_E(r)$.

12. (II) Consider a simple model for the phenomenon of charge screening in a plasma. A plasma is a neutral state of matter, consisting of positive ionized atoms with volume charge density $+\rho_E$ coexisting with an equal magnitude charge density of negative electrons. The ionized atoms are much more massive than the electrons and so, on short time scales, they appear stationary in comparison with the fast moving electrons. If an electric field appears within a plasma, after a brief time interval, these electrons will have moved in such a way as to cancel it. Thus (as in a conductor), the electric field in the bulk of a plasma can be taken to be zero. Assume an uncharged conducting sphere of radius r_o is placed in a plasma. If charge $-Q$ is then placed on the sphere at time $t = 0$, the electrons in the neighboring plasma will very quickly be repelled into the outer reaches of the plasma (and so, for our purposes, can be assumed to disappear), leaving behind a spherical shell of positive atoms with thickness L. During the moments immediately following $t = 0$, the positive charge in this shell can be assumed stationary with the shell surrounded by neutral plasma (Fig. 22–7). For this model,
(a) determine the screening length L.
(b) determine the electric field E as a function of distance r from the sphere's center in the screening region $r_o < r < r_o + L$.

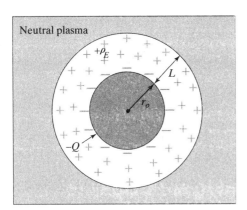

FIGURE 22–7

13. (II) A diode (an electronic component commonly used in electrical devices) can block the flow of electricity within itself by creating a "depletion region," which consists of neighboring volumes of positive and negative charges with volume charge densities ρ_+ and $-\rho_-$ and thicknesses t_+ and t_-, respectively. The depletion region is surrounded by neutral semiconductor material on both sides within which the electric field is zero as shown in Fig. 22–8. Assuming the cross-sectional area A of the diode is large, prove $\rho_+ t_+ = \rho_- t_-$.

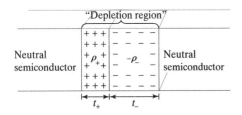

FIGURE 22–8

14. (II) Consider a metal layer in contact with a semiconductor (Fig. 22–9). Let a charge $-Q$ be placed on the surface area A of the metal, so that a uniform surface charge density $-\sigma$ exists at the metal-semiconductor boundary. If the semiconductor has positively charged impurity atoms embedded uniformly in its lattice structure, the movement of free electrons in the semiconductor will produce a positively charged layer of thickness W and a uniform volume charge density $+\rho_E$ (called the depletion region) in the region of the semiconductor nearest the metal. Beyond the depletion region, the semiconductor is neutral and the effect of $-Q$ is not felt (i.e., the electric field is zero here). This effect is called dielectric screening and W is the screening length.
(a) Using Gauss's law, prove $\sigma = \rho_E W$.
(b) Using the gaussian surface shown in Fig. 22–9, prove that the electric field at a distance x into the depletion region is given by

$$E = \frac{\rho_E}{\epsilon_o} x$$

where $x = 0$ is at the edge of the depletion region.
(c) Using your results from parts (a) and (b), show that the maximum electric field E_{max} within the depletion region occurs at the metal-semiconductor boundary and has a magnitude of σ/ϵ_o. For the typical values of $W = 0.20\,\mu m$ and $\rho_E = 1600\,C/m^3$, determine E_{max}. [In this problem, we ignore the effect of the semiconductor's dielectric constant K (see Section 24–5 in the text). To include this effect, replace ϵ_o by $K\epsilon_o$ in each expression. For the semiconductor silicon, $K = 11.7$.]

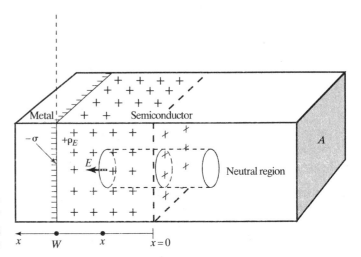

FIGURE 22–9

15. (II) A flat slab of nonconducting material has a thickness d, a very large (i.e., can be assumed infinite) cross-sectional area, and a uniform volume charge density $+\rho_E$.
(a) Prove that a uniform electric field exists outside of this slab. Determine its magnitude E and its direction (relative to the slab's surface).
(b) As shown in Fig. 22–10, the slab is aligned so that one of its surfaces lies on the line $y = x$. At time $t = 0$, a point-like particle (mass m, charge $+q$) is located at position $\mathbf{r} = +y_o \mathbf{j}$ and has velocity $\mathbf{v} = v_o \mathbf{i}$. Prove that, if $v_o \geq \sqrt{\sqrt{2}q y_o \rho d/\epsilon_o m}$, the particle will collide with the slab.

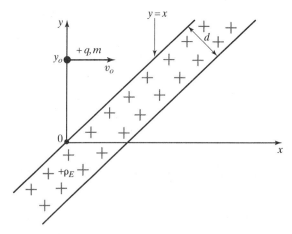

FIGURE 22–10

16. (II) A flat slab of nonconducting material has a thickness $2d$, which is small compared with its height and breadth. Define the x-axis to be along the direction of the slab's thickness with the origin at the center of the slab (Fig. 22–11). If the slab carries a volume charge density such that $\rho_E(x) = -\rho_o$ and $\rho_E(x) = +\rho_o$ in the regions $-d \leq x < 0$ and $0 < x \leq +d$, respectively, where ρ_o is a positive constant, determine the electric field **E** as a function of x in the regions (a) outside the slab, (b) $0 < x \leq +d$, and (c) $-d \leq x < 0$. [*Hint*: First, determine the electric field outside of a slab of thickness d that carries a uniform volume charge density ρ_o.]

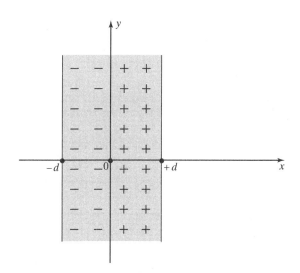

FIGURE 22–11

17. (II) A flat slab of nonconducting material has a thickness $2d$, which is small compared with its height and breadth. Define the x-axis to be along the direction of the slab's thickness with the origin at the center of the slab (Fig. 22–11). If the slab carries a volume charge density $\rho(x) = \rho_o\left(\dfrac{x}{d}\right)$, where ρ_o is a constant, determine the electric field **E** as a function of x in the regions (a) outside the slab and (b) inside the slab. [*Hint*: The value of **E** can be deduced qualitatively for a particular value of x.]

18. (II) A spherically symmetric volume charge density $\rho(r) = Ar^n$ exists in a region of space about the point P, where r is the radial distance from P to a location within that region, and A and n are constants. Further, in this region, it is found that an electric field of constant magnitude E_o emanates radially from P (i.e., $\mathbf{E} = E_o\hat{\mathbf{r}}$, where $\hat{\mathbf{r}}$ is the unit vector pointing radially outward from P). Determine the values of A and n.

19. (II) In devices such as air cleaners and photocopiers, the strong nonuniform electric field region near a thin, charged cylindrical wire is used to create ionized air molecules (for use in charging dust particles or a photocopier drum). Under standard atmospheric conditions, if air is subjected to an electric field magnitude that exceeds 2.7×10^6 N/C, its molecules will be dissociated into positively charged ions and free electrons. Typically, the region within which air is ionized (called the "corona discharge" region) occupies a cylindrical volume centered on the charged wire with a radius that is five times that of the wire. If the wire's charge per length $\lambda = +0.40$ nC/cm, what diameter wire (mm) should be used to create the corona discharge region? [Besides dissociating air, the charged wire repels the resulting positive ions from the corona discharge region, where they are put to use in charging objects.]

20. (II) An uncharged, solid conducting sphere of radius r_o contains two spherical cavities of radii r_1 and r_2, respectively. Point charge Q_1 is then placed within the cavity of radius r_1 and point charge Q_2 is placed within the cavity of radius r_2 (Fig. 22–12). Determine the resulting electric field (magnitude and direction) at locations outside the solid sphere (i.e., for $r > r_o$, where r is the radial distance from its center).

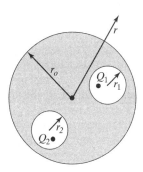

FIGURE 22–12

General Problems

21. (II) An insulating cube with sides of length a possesses a uniform volume charge density ρ_E. Consider a cube-shaped gaussian surface with sides of length $L > a$, whose center coincides with the center of the charged cube.
(a) Determine the electric flux through the given gaussian surface.
(b) Explain why, even though it has the same shape as the enclosed charge, the given gaussian surface cannot be used to determine the electric field produced by the charged cube. Is there some other choice for a gaussian surface that will allow one to determine the electric field created by the charged cube using Gauss's Law?

22. (II) In the primitive, and now discarded, "plum pudding" model, the hydrogen atom is imagined to be a nonconducting sphere of radius r_o carrying a uniformly distributed charge $+e$ within which a single point-like electron is embedded (Fig. 22–13).
(a) Taking the sphere's center as the equilibrium position of the electron, show that if the electron is displaced a distance $r < r_o$ from this equilibrium position, and then released, it will subsequently execute simple harmonic motion with a frequency given by $f = \dfrac{1}{2\pi}\sqrt{\dfrac{e^2}{4\pi\epsilon_o m r_o^3}}$. Assume the electron moves without friction within the nonconducting sphere.
(b) According to the plum-pudding model, this oscillating electron will emit light whose frequency matches that of its oscillation frequency f. Using hydrogen's known radius $r_o = 0.53 \times 10^{-10}$ m, what single frequency of light does this model predict the H atom to emit? [In contrast with this prediction, hydrogen emits many different frequencies of light, the highest of which is 3.3×10^{15} Hz. This flawed prediction is one of the reasons that this atomic model was discarded.]

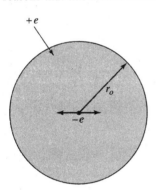

FIGURE 22–13

23. (II) After being struck by lightning, a large, flat metal roof carries a uniform surface charge density $+\sigma$. A ping-pong ball (mass m), attached to the roof by a nylon string of length L, picks up a charge $+Q$ (uniformly spread over its outer surface) and levitates in the electric field of the roof. What is the frequency of small oscillations of this upside-down pendulum? Assume that the electric force on the ping-pong ball is much greater than the gravitational force on it.

24. (II) A hemisphere of radius R is placed in a charge-free region of space within which exists a uniform electric field of magnitude E directed perpendicular to the hemisphere's circular base (Fig. 22–14).

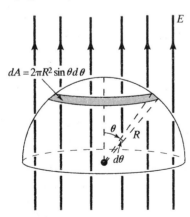

FIGURE 22–14

(a) Using the definition of Φ_E through an "open" surface, calculate (via explicit integration) the electric flux through the hemisphere. [Hint: In Appendix C of the text, it is shown that, on the surface of a sphere, the infinitesimal area located between the angles θ and $\theta + d\theta$ is $dA = (2\pi R \sin\theta)(Rd\theta) = 2\pi R^2 \sin\theta d\theta$.]
(b) Choose an appropriate gaussian surface, then use Gauss's law to much more easily obtain the same result as found in part (a) for the electric flux through the hemisphere.

25. (II) Consider the following simple model to describe the polarization of charge in a hydrogen atom when it is placed within an electric field. Imagine the atom to be a central positive point charge $+e$ embedded in a uniformly charged sphere of total charge $-e$ and radius R (Fig. 22–15). When not acted upon by external influences, the positive charge will be located at the center of the negatively charged sphere. Assume that the negative charge remains uniformly distributed over the sphere of radius R even when it is subjected to a field or when the positive charge is displaced from its center.
(a) Focusing on just the uniformly charged sphere, determine the electric field (magnitude and direction) created by this charge distribution at a distance d from its center, where $d < R$.
(b) Now assume a hydrogen atom is placed in an external electric field E_{ext}, which causes it to become "polarized" (i.e., the atom's opposite charges are displaced from one another). In the presence of this external field, use the given model to predict the hydrogen atom's electric dipole moment $p = ed$, where d is the equilibrium distance between the positive point charge and the center of the negative sphere.
[Hint: In equilibrium, the forces on the proton due to the external field and due to the charged sphere cancel each other out.]
(c) The ease with which an external field can induce a dipole moment in an atom is described by the relation $p = \alpha E_{ext}$, where α is defined as the atom's "atomic polarizability." Show that, according to the given model, the atomic polarizability for hydrogen is $\alpha_H = 4\pi\epsilon_o R^3$.
(d) From experiment, the radius of the hydrogen atom is 5×10^{-11} m and its atomic polarizability is 7×10^{-41} C$^2\cdot$m/N. Show that the model predicts the correct order of magnitude for α_H.

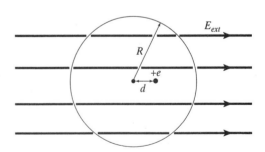

FIGURE 22–15

26. (II)(a) An electric dipole consists of two point **charges** $+Q$ and $-Q$ located at $z = -d$ and $z = +d$, respectively (Fig. 22–16a). Show that at a point in the x-y plane, which is a distance r from the origin, the electric field is
$$\mathbf{E} = \frac{1}{2\pi\epsilon_o}\frac{Qd}{(d^2+r^2)^{3/2}}\mathbf{k}.$$
[Hint: The point in question is on the perpendicular bisector of the dipole.]

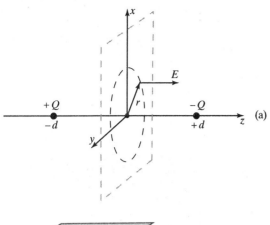

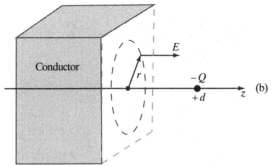

FIGURE 22–16

(b) Now assume that the surface of a conductor occupies the x-y plane and a point charge $-Q$ is located $z = +d$ as shown in Fig. 22–16b. Assume further that the conducting surface is positively charged in such a way that the electric field produced at the conductor's surface is exactly the same as that produced by the dipole in part (a). Determine $\sigma(r)$, the conductor's charge per unit area as a function of distance r from the origin.
(c) Prove that the total charge on the conductor's surface is $+Q$.

27. (II) (a) An infinitely long, solid nonconducting cylinder of radius r_o possesses a uniform volume charge density ρ_E. Show that the electric field at a point P inside this cylinder at a distance $r < r_o$ from its axis is $\mathbf{E} = \dfrac{\rho_E}{2\epsilon_o}\mathbf{r}$, where $\mathbf{r}$ spans the perpendicular distance from the cylinder's axis to point P.
(b) An infinitely long cylindrical cavity of radius $r'_o < r_o$ is cut out of the charged cylinder and the material is discarded (cross-sectional view shown in Fig. 22–17). The axis of the cavity is parallel to, and displaced by distance $\mathbf{d}$ from, the charged cylinder's axis. Show that a uniform electric field exists within this cavity given by $\mathbf{E} = \dfrac{\rho_E}{2\epsilon_o}\mathbf{d}$.
[*Hint*: Imagine the hollow cylinder as the superposition of two oppositely charged cylinders. Determine **E** at point P, noting $\mathbf{r} = \mathbf{d} + \mathbf{r}'$.]

28. (II) The electric field inside a sphere of radius r_o that carries a charge Q uniformly distributed within its interior can be written as $\mathbf{E} = \dfrac{Q}{4\pi\epsilon_o r_o^3}\mathbf{r}$ (see Example 22–4 in the text). Two diffuse spherical clouds have the same radius r_o, but are oppositely charged. One cloud has a charge $+Q$ and the other has a charge $-Q$; each charge is distributed uniformly throughout its sphere's interior. The two clouds interpenetrate each other as shown in Fig. 22–18, where **D** is the displacement from the center of the positive cloud to the center of the negative cloud.

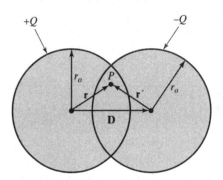

FIGURE 22–18

(a) Assuming the uniform charge distribution within each cloud is maintained in this situation, show that the electric field **E** in the cloud-overlap region is uniform, and determine this field's constant magnitude E and its direction. [*Hint*: Determine **E** at point P, noting $\mathbf{r} = \mathbf{D} + \mathbf{r}'$.]
(b) Using your expression for E, determine the maximum possible value for this quantity. Then offer a qualitative explanation that justifies the veracity of this result.

29. (II) Let's investigate what motivates excess charge to be driven to the surface of a conductor.
(a) Consider the hollow object O, which consists of a spherical shell of charge with radius r_o and a small hole of radius a drilled in its surface (Fig. 22–19a). Assume the surface charge density of this object has the constant value σ everywhere,

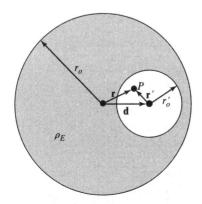

FIGURE 22–17

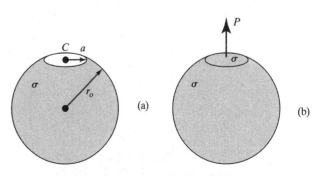

FIGURE 22–19

GENERAL PROBLEMS 127

except on the hole where it is zero. Determine the electric field E_C at the point C located at the center of the hole. [*Hint*: Consider O to be the composite object composed of a complete shell with charge $Q = \sigma A$ uniformly distributed over its surface area A and a small circular disk of radius a centered at C with surface charge density $-\sigma$.]

(*b*) If a charged disk of radius a and uniform surface charge density σ is placed in the hole on its surface (Fig. 22–19b), find the "electrostatic pressure" P (= electric force per unit area) that O will exert on this disk. Assume E_C provides a good estimate of the electric field at all locations on the hole.

(*c*) Suppose 100 excess electrons were placed within the interior of a metallic object and positioned uniformly on a spherical surface of radius $1.0\,\mu\text{m}$. Determine the pressure P that each electron would experience due to its repulsion from the other 99 electrons (ignore any effects due to other charges within the metal). This repulsive pressure drives the excess electrons to the metallic object's surface.

30. (II) A uniform electric field of magnitude E is parallel to the z-axis. Consider the half-cylinder gaussian surface whose rectangular base of length L and width $2R$ lies in the x-y plane and assume the gaussian surface encloses no charge (Fig. 22–20).

(*a*) From Gauss's law, what is the net electric flux through this gaussian surface?

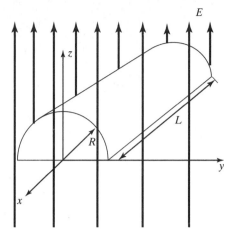

FIGURE 22–20

(*b*) Starting with the definition $\Phi_E = \int \mathbf{E} \cdot d\mathbf{A}$, explicitly calculate the electric flux through the closed half-cylinder's rectangular base, the cylindrically curved surface, and the two half-circular ends. Then, summing these contributions, verify that Gauss's law yielded the correct value for the net flux in this situation.

CHAPTER 23

Electric Potential

Section 23–2

1. (I) Thunderclouds can develop voltage differences of approximately 1×10^8 V. Given that an electric field of 3×10^6 V/m is required to produce an electrical spark within a volume of air, show that a thundercloud is only capable of producing a lightning bolt of several tens of meters in length. [Thus, when lightning strikes from a cloud to the ground, the bolt actually has to propagate as a sequence of steps, with each step several tens of meters long.]

2. (II) Suppose that as an electron with speed $v_o = 1.0 \times 10^7$ m/s passes point A, it enters into a region of space where it is traveling along the direction of a constant electric field with magnitude $E = 1000$ V/m. Define the electric potential at A to be zero.
(a) As the electron moves in this region of space, it is slowed by the field. Let B be the point at which the electron is finally stopped. What is the electric potential at B?
(b) Determine the electron's stopping distance (i.e., the distance between points A and B).

3. (II) In a region where a horizontal electric field of magnitude 15 V/m exists, the points A, B, and C are at the vertices of an equilateral triangle with side length 0.25 m (Fig. 23–1). Use Eq. 23–3 in the text to determine
(a) V_{BA}, the potential difference across the triangle leg AB.
(b) V_{CB}, the potential difference across the triangle leg BC.
(c) V_{CA}, the potential difference across the triangle leg AC.
(d) Is the potential difference one experiences when traveling along the path ABC the same as when traveling along the path AC?

4. (II) Define the x-axis to be along the axis of a uniformly charged disk of radius R with charge per unit area σ. In Example 21–11 of the text, it is shown that the electric field along this axis is given by $\mathbf{E} = \dfrac{\sigma}{2\epsilon_o} \left[1 - \dfrac{x}{\sqrt{x^2 + R^2}} \right] \mathbf{i}$.
Put this expression into Eq. 23–3 of the text to determine the electric potential $V(x)$ on the axis of the disk at a distance x from its center. Define $V = 0$ at $x = \infty$. [You should obtain the same result as that in Example 23–9 of the text, where $V(x)$ was determined by another method.]

5. (II) An electric field acts along the x-axis according to $\mathbf{E}(x) = -\dfrac{C}{x^3}\mathbf{i}$, where $C = 32$ V·m^2. Find the electric potential $V(x)$, if V is arbitrarily defined to be zero at $x = +2.0$ m.

6. (III) Two lines of charge on the x-axis are symmetrically placed about the origin, separated by a distance $2a$. Each line has charge $+Q$ uniformly distributed over its length L. The y-axis is the perpendicular bisector of this charge distribution; point P is located at $y = +b$ (see Fig. 23–2).
(a) Determine the electric field $\mathbf{E}$ at an arbitrary distance y along the y-axis.
(b) Using Eq. 23–3 in the text, find the electric potential at point P. Choose $V = 0$ at $y = \infty$.
(c) If a positive point charge $+q$ of mass m is released from rest at point P, what is its velocity after it has been repelled to a distance infinitely far away from the line charges?

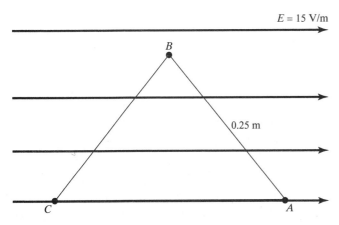

FIGURE 23–1

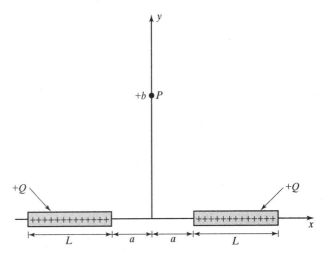

FIGURE 23–2

7. (III) This problem demonstrates that, while the functional form of the electric potential $V(r)$ does depend on the choice for "zero potential," potential *differences* between two points in space are independent of this choice. Consider a charge Q distributed uniformly throughout a nonconducting sphere of radius r_o. In Example 22–4 of the text, it is shown that the electric field created by this sphere is $E = kQ/r^2$ for $r > r_o$ and $E = kQr/r_o^3$ for $r < r_o$.
   (a) Arbitrarily define $V = 0$ at $r = \infty$. Based on this choice for zero potential, derive an expression for the electric potential $V(r)$ for the region within the sphere $(r < r_o)$.
   (b) Use the expression for $V(r)$ from part (a) to find the potential difference ΔV between the sphere's center and its surface [i.e., $\Delta V = V(r = 0) - V(r = r_o)$].
   (c) Now, rework this problem with an alternate definition for "zero potential." Arbitrarily define $V = 0$ at $r = 0$. Based on this choice for zero potential, derive an expression for the electric potential $V(r)$ for the region within the sphere $(r < r_o)$. Note that this expression for $V(r)$ differs from that in part (a) due to the differing choice for zero potential.
   (d) Use the expression for $V(r)$ from part (c) to find the potential difference ΔV between the sphere's center and its surface [i.e., $\Delta V = V(r = 0) - V(r = r_o)$]. Note this potential difference is the same as that found in part (b).

8. (III) The volume charge density ρ_E within a sphere of radius r_o is distributed in accordance with the following spherically symmetric relation

$$\rho_E(r) = \rho_o \left[1 - \frac{r^2}{r_o^2} \right]$$

where r is the radial distance from the center of the sphere and ρ_o is a constant. A point P resides within the sphere at $r < r_o$. Determine the electric potential V at point P, if $V = 0$ at infinity.

Section 23–3

9. (II) A positive point charge $+q$ is situated on the x-axis at the point $x = +a$. At what point on the x-axis should you place a "twice-as-great" negative charge $-2q$, so that it would then require zero net work to bring a third charge $+q$ from infinity to the point $x = 0$?

10. (II) A positive charges $Q = +10$ nC is rigidly anchored at each of the four corners of a square with side length $L = 10$ cm. Point A is at the exact center of the square, while point B is midway along one side of the square as shown in Fig. 23–3.

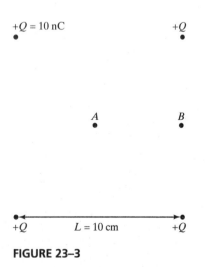

FIGURE 23–3

(a) Find the electric potential at A and at B.
(b) A point charge $q = +0.10$ nC whose mass is $m = 1.0 \times 10^{-11}$ kg travels from A to B. If its velocity at A is $v_o = 50$ m/s, what is its velocity at B?

Section 23–4

11. (II) A rod of length $L = 4.0$ m lies along the x-axis with its left end at the origin. This rod has a nonuniform charge per length $\lambda = Ax$, where $A = 1.0$ nC/m² (see Fig. 23–4).
    (a) Find the electric potential at point P, a distance $y = 3.0$ m above the left end of the rod.
    (b) A proton, initially at rest at point P, is repelled by the rod. What velocity does this proton acquire after being repelled to a distance that is infinitely far away from the rod?

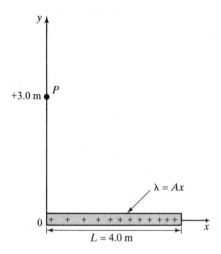

FIGURE 23–4

12. (II) A thin glass rod is bent into a semicircle of radius R. A total charge $+Q$ is uniformly distributed along the rod's upper half, while a total charge $-Q$ is uniformly distributed along the lower half. Point P is at the center of the semicircle (Fig. 23–5).
    (a) Find the electric potential at point P.
    (b) A small positive charge $+q$ of mass m is initially at point P with velocity v_o. If some time later, the charge $+q$ has moved to a position that is infinitely far away from point P, how fast will it be moving?

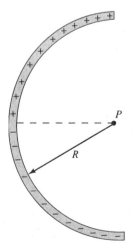

FIGURE 23–5

13. (II) Two lines of charge on the x-axis are symmetrically placed about the origin, separated by a distance $2a$. Each line has charge $+Q$ uniformly distributed over its length L. The y-axis is the perpendicular bisector of this charge distribution; point P is located at $y = +b$ (see Fig. 23–2).
    (a) Using Eq. 23–6b in the text, find the electric potential at point P.
    (b) If a positive point charge $+q$ of mass m is released from rest at point P, what is its velocity after it has been repelled to a distance infinitely far away from the line charges?

14. (III) A thin circular ring of radius R carries a total charge $+Q$ that is nonuniformly distributed along its circumference (Fig. 23–6). The charge dQ per unit arclength dl is given by
$$\lambda = \frac{dQ}{dl} = A \sin^2 \theta, \text{ where } A \text{ is a constant.}$$

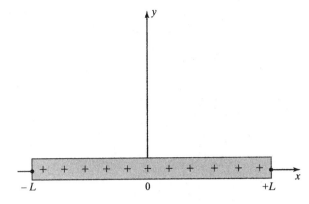

FIGURE 23-7

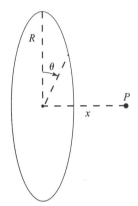

FIGURE 23-6

(a) Determine the constant A in terms of the total charge $+Q$.
(b) Determine the electric potential at a point P on the axis of the ring a distance x from its center.

15. (III) A charged conducting sphere of radius r_0 has a total charge Q uniformly distributed on its surface. In Example 23–4 of the text, Eq. 23–3 is used to determine that the resulting electric potential at a point outside the sphere $(r > r_0)$ is $V = \frac{kQ}{r}$.
Re-derive this result starting with Eq. 23–6b in the text:
$$V(r) = \int_{\text{spherical surface}} \frac{k \, dq}{l}$$
where l and other helpful information are described in Appendix C.
[Hint: Define the uniform charge per unit area $\sigma = Q/4\pi r_0^2$. Then $dq = \sigma \, dA$, where dA is the infinitesimal area of the circular strip shown in Fig. C–1. Finally, depending on how you evaluate your integral, you may wish to let $x = \cos \theta$.]

Section 23-7

16. (III) A thin rod of length $2L$, which carries a uniformly distributed charge Q, lies on the x-axis with its midpoint at the origin (Fig. 23–7).

(a) Show that the potential along the rod's perpendicular bisector (i.e., the y-axis) is given by
$$V(y) = \frac{Q}{8\pi \epsilon_o L} \ln\left(\frac{\sqrt{L^2 + y^2} + L}{\sqrt{L^2 + y^2} - L}\right). \text{ Let } V = 0 \text{ at infinity.}$$
(b) Using $E_y = -\frac{\partial V}{\partial y}$, determine the electric field along the rod's perpendicular bisector. Is your result consistent with the answer given for Problem 47 in Chapter 21 of the text?

Section 23-8

17. (II) In Example 23–12 of the text, the ionization energy of hydrogen is calculated to be 13.6 eV. Assume you have one mole of hydrogen atoms. If you used the energy provided by a 5.0-hp lawn mower engine to ionize all of these atoms, how long would it take?

18. (II) Many chemical reactions are able to release energy because, at the conclusion of the reaction, attracting charges within the involved molecules have a smaller separation than they had at the beginning of the reaction. To demonstrate this concept, let us assume that at the beginning of a reaction, an electron and a proton are separated by 0.11 nm. Further assume that changes due to the chemical reaction leave these charges with a final separation of 0.10 nm. How much electrical potential energy was lost in this reaction (in units of eV)? This lost potential energy is obviously converted into some other form(s) of energy. What are some of these possible forms?

19. (II) A deuterium nucleus consists of a proton and a neutron bound closely together by the strong nuclear force (this force is of significant strength only over very short nucleus-sized distances). Two deuterium nuclei, separated by distances greater than about $d = 3 \times 10^{-15}$ m, will repel each other due to the electrical repulsion of the proton in each nucleus. However, if two such deuterium nuclei come within a separation of d, the strong nuclear attractive force will be able to overpower the electrical repulsive force between the protons and the two nuclei will "fuse" into a single, larger (helium) nucleus with the release of a large amount of energy. This process powers the Sun. Assume two deuterium nuclei, infinitely far apart, are each given a speed v, which causes them to move toward each other. As their separation decreases, the nuclei will slow down due to their protons' electrical repulsion. Find the value of v so that the approaching deuterium

nuclei are not stopped by electrical repulsion until their separation equals d (at this point, the strong nuclear force will take over and fuse the nuclei). From the value of v that you obtain, explain why nuclei only fuse in high-temperature environments.

20. (II) The liquid drop model of the nucleus suggests that high-energy oscillations of certain nuclei can split ("fission") a large nucleus into two unequal fragments plus a few neutrons. Consider the case of a uranium nucleus fissioning into two spherical fragments, one with charge and radius of $q_1 = +38e$ and $r_1 = 5.5 \times 10^{-15}$ m and the other with $q_2 = +54e$ and $r_2 = 6.2 \times 10^{-15}$ m. Assume that, at the moment just after the uranium nucleus fissions, the two fragments are at rest with their spherical surfaces in contact. Further, assume that the charge of each fragment is uniformly distributed throughout the fragment's volume. Calculate the electric potential energy (MeV) of the two fragments in this configuration. Subsequently, this electrical potential energy will be entirely converted to kinetic energy as the fragments repel each other. Does your predicted kinetic energy of the fragments agree with the observed value associated with uranium fission (approximately 200 MeV)? 1 MeV = 10^6 eV.

21. (III) In a region of otherwise empty space, two small metal spheres of mass $m_1 = 5.00$ g and $m_2 = 10.0$ g carry equal positive charges $q = +5.00\,\mu$C. The spheres are connected by a massless, insulating string of length 3.00 m, which is much longer than the radii of the spheres. Assume any energy stored in the string due to its elongation under tension is negligible. The string then breaks. A long time after the string breaks, what is the speed of each sphere?

General Problems

22. (II) A charge $-q_1$ of mass m sits on the y-axis at a distance b above the x-axis. Two positive charges of magnitude $+q_2$ are fixed on the x-axis at $x = +a$ and $x = -a$, respectively (Fig. 23–8). If the $-q_1$ charge is given an initial velocity of v_o in the positive y-direction, what is the minimum value of v_o such that the charge escapes to a point infinitely far away from the two positive charges?

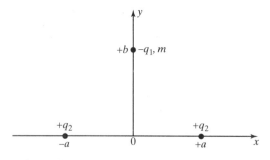

FIGURE 23–8

23. (II) A circular ring of radius R carries an electric charge $+Q$ uniformly distributed around the circumference. A negative point charge $-q$ is situated at the center. How much work would it take to move the point charge out to "infinity" (i.e., far away from the ring)?

24. (II) Consider a circular ring of radius R that carries a uniformly distributed charge $+Q$ as in Example 23–8 of the text. An electron is placed on the axis of this ring at a distance x from its center. If $x \ll R$, show that the electron's electric potential energy, apart from an additive constant, has the same form as that of a mass attached to a spring. Identify the "spring constant" for the electron-plus-ring system and use your knowledge of simple harmonic motion to predict the oscillatory frequency f of the electron.

25. (II) In the Bohr model for the hydrogen atom, the electron ($q = -e$) with speed v is held in a circular orbit of radius r by its electrical attraction to the single proton ($q = +e$) in the atom's nucleus. Apply Newton's second law to the orbiting electron to find its speed v in terms of r. Then go on to show that for this system the ratio of electrical potential energy U to kinetic energy K equals -2. [This result is a consequence of a famous theorem from classical mechanics called the Virial Theorem.]

26. (II) A proton, located between two charged parallel conducting plates separated by distance d, has a velocity $v_o = 2.0 \times 10^4$ m/s when near the 10.0-V plate (Fig. 23–9). The field due to the plates accelerates the proton and it attains a velocity v upon arrival at the 2.0-V plate. Determine v
(a) using conservation of energy.
(b) using the force method (Newton's second law, followed by the kinematic equations). Use $E = V_{ba}/d$ to determine the electric field.

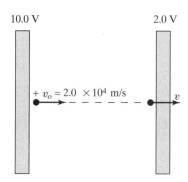

FIGURE 23–9

27. (III) Two identical nonconducting spherical objects of radius R and mass m are in a region of space far removed from any other objects that might influence their motion. Each of the objects carries a uniformly distributed charge $+Q$. The two objects are held at rest with their surfaces touching (i.e., their centers are separated by a distance $d = 2R$), and then released. Noting $v = dx/dt$, so that $dt = dx/v$, evaluate the integral $t = \int_{x_{\text{initial}}}^{x_{\text{final}}} \dfrac{dx}{v}$ to determine how much time elapses until the centers of these two objects are separated by a distance $4R$. You may ignore the influence of gravity.
[*Hint*: In the integral, make the substitution $u = \sqrt{x - d}$. Also, $\int \sqrt{d + u^2}\, du = \dfrac{u\sqrt{d + u^2}}{2} + \dfrac{d}{2}\ln(u + \sqrt{(d + u^2)})$]

28. (III) The work done by a conservative force on an object as it moves from point A to point B is independent of the path taken and depends only on the coordinates of A and B. Consider a point particle with charge $+q$ moving from one

location to another along an arbitrary path within an electric field **E**.

(a) Prove that if **E** is the uniform electric field $\mathbf{E} = E\mathbf{i}$, it produces a conservative force on the particle.

(b) Prove that if **E** is the electric field due to a point charge $+Q$ given by $E = \dfrac{1}{4\pi\epsilon_o}\dfrac{Q}{r^2}\hat{\mathbf{r}}$, it produces a conservative force on the particle.

29. (III) Breakdown of air ("sparking") under standard atmospheric conditions occurs when an electric field with magnitude exceeding $E_B = 2.7 \times 10^6$ V/m exists within the air's volume. It is known that about 15 eV of energy is required to liberate an electron from ("ionize") an air molecule. Consider the following two models for explaining why breakdown of air occurs in the presence of an electric field:

Model A: An electron within each air molecule is accelerated across the molecule's diameter d by the electric field E, gaining enough energy to liberate itself from the molecule. These liberated electrons cause the air to become electrically conductive.

Model B: Free electrons in air (produced by cosmic rays, light, etc.) have a mean free path (average distance traversed between collisions with air molecules) of $l_M = \dfrac{1}{\pi\sqrt{2}r^2\,(N/V)}$

(see Eq. 18–10b in the text; a factor of four is suppressed for collisions between molecules and point-like electrons), where r is the radius of an air molecule and N/V is the number density of molecules in air. The electric field E accelerates each free electron over a distance l_M, so that it gains enough energy to liberate an electron when it collides with an air molecule. As this "chain-reaction" process is repeated, an avalanche of liberated electrons is produced and the air becomes electrically conductive.

Given $r = 1.5 \times 10^{-10}$ m, use each model to predict E_B for air at a pressure and temperature of 1.0 atm and 20°C. Which model offers the most reasonable prediction for E_B?

30. (III) Consider a metal layer in contact with a semiconductor (Fig. 23–10). Let a charge $-Q$ be placed on the surface area A of the metal, producing a voltage $-V_o$ at the metal-semiconductor boundary. If the semiconductor has positively charged impurity atoms embedded in its lattice structure, the movement of free electrons in the semiconductor will produce a positively charged layer of thickness W and a uniform volume charge density $+\rho_E$ (called the depletion region) in the region of the semiconductor nearest the metal.

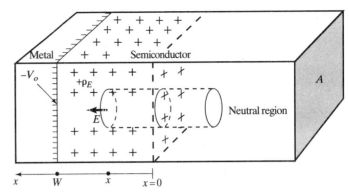

FIGURE 23–10

Beyond the depletion region, the semiconductor is neutral and the effect of $-Q$ is not felt (i.e., the electric field is zero here). This effect is called dielectric screening and W is the screening length.

(a) Using the gaussian surface shown in Fig. 23–10, prove that the electric field at a distance x into the depletion region is given by

$$E = \dfrac{\rho_E}{\epsilon_o} x$$

where $x = 0$ is at the edge of the depletion region.

(b) Defining $V = 0$ at the edge of the depletion region, show that the electric potential as a function of distance x into the depletion region is given by

$$V(x) = -\dfrac{\rho_E}{2\epsilon_o} x^2.$$

(c) Using the fact that $V(x = W) = -V_o$, show that the distance required to screen the applied voltage is

$$W = \sqrt{\dfrac{2\epsilon_o V_o}{\rho_E}}.$$

For a typical semiconductor used in electronic devices, $\rho_E = 1600$ C/m³. Determine W (μm) if $V_o = 10$ V. [In this problem, we ignore the effect of the semiconductor's dielectric constant K (see Section 24–5 in the text). To include this effect, replace ϵ_o by $K\epsilon_o$ in each expression. For the semiconductor silicon, $K = 11.7$.]

CHAPTER 24

Capacitance, Dielectrics, Electric Energy Storage

Section 24–1

1. (II) Compact "ultracapacitors" with capacitance values up to several thousand Farads are now commercially available. One application for ultracapacitors is in providing power for electrical circuits when other sources (e.g., battery) are turned off. To get a feel for how much charge can be stored in such a component, assume a 1000-F ultracapacitor is initially charged to 12.0 V by a battery and then disconnected from the battery. If charge is then drawn off the plates of this capacitor at a rate of 1.0 mC/s, say, to power the backup memory of some electrical gadget, how long (in days) will it take for the potential difference across this capacitor to drop to 6.0 V?

2. (II) In a dynamic random access memory (DRAM) computer chip, each memory cell chiefly consists of a capacitor for charge storage. Each of these cells represents a single binary-bit value of "1" when its 30-fF capacitor is charged at 1.5 V or "0" when uncharged at 0 V.
(a) When it is fully charged, how many excess electrons are on a cell capacitor's negative plate?
(b) After charge has been placed on a cell capacitor's plate, it slowly "leaks" off (through a variety of mechanisms) at a constant rate of 0.3 fC/s. How long does it take for the potential difference across this capacitor to decrease by 1% from its fully charged value? [Because of this leakage effect, the charge on a DRAM capacitor is "refreshed" many times per second.]

Section 24–2

3. (II) A certain soapy water bubble of radius 5 cm and thickness 0.5 µm acquires an excess charge of 1 nC. When this bubble bursts, all of its soapy water, along with the 1 nC excess charge, collapses into a single spherical liquid drop. Assuming that soapy water is electrically conducting and that its density is the same whether in the form of a bubble or a liquid drop, estimate the electric potential of the drop with respect to infinity.

4. (III) Small distances are commonly measured capacitively. Consider an air-filled parallel-plate capacitor with fixed plate-area $A = 25$ mm^2 and a variable plate-separation distance x. Assume this capacitor is attached to a capacitance-measuring instrument that can measure capacitance C in the range 1.0 pF to 1000.0 pF with an accuracy of $\Delta C = 0.1$ pF.
(a) If C is measured while x is varied, over what range $x_{min} \leq x \leq x_{max}$ can the plate-separation distance (μm) be determined by this set-up?
(b) Define Δx to be the accuracy (magnitude) to which x can be determined. Prove that $\Delta x = x^2 \Delta C / \epsilon_o A$.

(c) Determine the percent accuracy to which x_{min} and x_{max} can be measured. [In manufacturing, metal parts can be precisely positioned in this manner. The variable capacitor is realized by a rigidly mounted "sensor" plate of area A, separated by a distance x from the flat surface of a movable conducting part.]

5. (III) In an electrostatic air cleaner ("precipitator"), the strong nonuniform electric field in the central region of a cylindrical capacitor (with outer and inner cylindrical radii R_a and R_b) is used to create ionized air molecules for use in charging dust and soot particles (Fig. 24–1). Under standard atmospheric conditions, if air is subjected to an electric field magnitude that exceeds its dielectric strength $E_S = 2.7 \times 10^6$ N/C, air molecules will be dissociated into positively charged ions and free electrons. In a precipitator, the region within which air is ionized (called the "corona discharge" region) occupies a cylindrical volume of radius r that is typically five times that of the inner cylinder. Assume a particular precipitator is constructed with $R_b = 0.10$ mm and

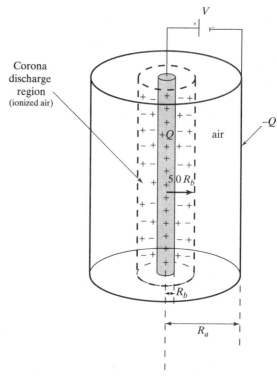

FIGURE 24–1

$R_a = 10.0$ cm. In order to create a corona discharge region with radius $r = 5.0R_b$, what potential difference V should be placed between the precipitator's inner and outer conducting cylinders? [Besides dissociating air, the charged inner cylinder repels the resulting positive ions from the corona discharge region, where they are put to use in charging dust particles, which are then "collected" on the negatively charged outer cylinder.]

6. (III) The controlled placement of excess charge on objects plays a key role in the operation of, e.g., electrostatic air cleaners and photocopiers. Commonly, the strong nonuniform electric field in the central region of a cylindrical capacitor (with outer and inner cylindrical radii R_a and R_b) is used to create ionized air molecules, which in turn attach themselves to, and thereby charge, the desired objects (e.g., dust and soot particles, photocopier drum). See Fig. 24–1. Under standard atmospheric conditions, if air is subjected to an electric field magnitude that exceeds its dielectric strength $E_S = 2.7 \times 10^6$ N/C, air molecules will be dissociated into positively charged ions and free electrons. In an air-ionizing capacitor, the region within which air is ionized (called the "corona discharge" region) occupies a cylindrical volume of radius r that is typically five times that of the inner cylinder. You are designing such a capacitor and decide that, based on available voltage sources, a potential difference of 10,000 V will be placed between its inner and outer conducting cylinders and that the outer cylinder radius will be $R_a = 5.0$ cm. Based on these decisions, what diameter wire (mm) should be used for the inner cylinder to create a corona discharge region with radius $r = 5.0R_b$? Besides dissociating air, the charged wire repels the resulting positive ions from the corona discharge region, where they are put to use in charging objects. [*Hint:* An equation of the form $f(x) = 0$, where $f(x) = x\ln(\alpha/x) - \beta$ with α and β as constants, can be solved for x by, e.g., graphing f vs. x and finding the "zero-crossing" x-value, or by using root-finding software on a calculator or a computer.]

7. (III) A capacitor consists of two equal-length conducting cylinders A and B of radius R_a and R_b, respectively, which have their axes parallel and separated by distance d. The capacitor is charged so that A has a charge $-Q$ and B has a charge $+Q$ (Fig. 24–2). Assume that the cylinder length L is much greater than d, so that end effects can be neglected. Also assume that d is much greater than both R_a and R_b, so that one cylinder does not significantly polarize the surface charge on the other (and thus, to a good approximation, one can assume charge to be uniformly distributed on the surface of each cylinder). Show that, for this capacitor,

$$C \approx \frac{2\pi\epsilon_o L}{\ln(d^2/R_a R_b)}.$$

8. (III) A capacitor consists of two conducting spheres A and B of radius r_a and r_b, respectively, which have their centers separated by distance d. The capacitor is charged so that A has a charge $-Q$ and B has a charge $+Q$ (Fig. 24–3). Assume that d is much greater than both r_a and r_b, so that one sphere does not significantly polarize the surface charge on the other (and thus, to a good approximation, one can assume charge to be uniformly distributed on each spherical surface).
   (a) Determine the capacitance C of this capacitor.
   (b) Show that, if A is much larger than B, your expression for C approaches that of an isolated capacitor.

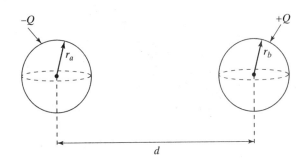

FIGURE 24–3

9. (III) The radii of the concentric outer and inner conductors in a spherical capacitor are r_a and r_b, respectively. The capacitor is designed so that when a potential difference V_S is applied between its two conductors, the electric field magnitude at the surface of the inner conductor exactly equals the dielectric strength E_S of air (the ionized air created here could be put to use in charging objects, e.g., in an air cleaner).
   (a) Show that the required applied potential difference is

$$V_S = r_b\left(1 - \frac{r_b}{r_a}\right)E_S.$$

   (b) Assume that in the design of this capacitor, r_a is constrained to be some given value, but the designer is free to choose the value of r_b. Consider two possible choices for the ratio of radii $\alpha = r_b/r_a$: $\alpha = 1/2$ and $\alpha = 1/10$. Show that, if $\alpha = 1/10$, the required applied voltage can be almost three times smaller than if $\alpha = 1/2$.
   [This problem illustrates that strong electric fields are easier to produce (require less voltage) as the radius of curvature of the field-producing charge distribution becomes smaller (more "point-like").]

Section 24–4

10. (II) For commonly used (CMOS) digital circuits, the charging of their component capacitors C to their working potential difference V accounts for the major contribution of their energy input requirements. Thus, if a given logical operation requires such circuitry to charge its capacitors N times, one can accurately assume that the operation requires an energy of $N\left(\frac{1}{2}CV^2\right)$. In the past 20 years, the capacitance in digital circuits has been reduced by a factor of about 20 and the voltage to which these capacitors are charged has been reduced from 5.0 V to 1.5 V. Also, present-day alkaline batteries hold

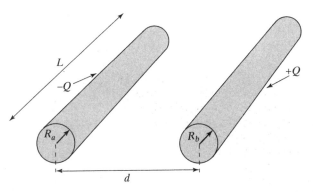

FIGURE 24–2

about five times the energy of older batteries. Two present-day AA alkaline cells, each of which measures 1 cm diameter by 4 cm long, can power the logic circuitry of a hand-held personal digital assistant (PDA) with its display turned off for about two months. If an attempt was made to construct a similar PDA (i.e., same digital capabilities so N remains constant) 20 years ago, how many (older) AA batteries would have been required to power its digital circuitry for two months? Would this PDA fit in a pocket or a purse?

11. (II) (a) A single conducting shell of radius r_b has a charge Q uniformly distributed over its surface. Based on the repulsive interactions between the individual charges from which Q is composed, determine the electrical potential energy associated ("stored") with this configuration.
[Hint: Use the expression for the capacitance of this charge geometry.]
(b) Consider a spherical metal object of radius R, whose center is at point P. Suppose a large number N of excess electrons is placed within the interior of the metal and positioned uniformly about P on a spherical surface whose radius is one-thousandth the size of the metal object radius. If the electrons move en masse outward and spread uniformly on the metal object's outer surface, by what factor will they lower their collective electrical potential energy? Ignore any effects due to other charges within the metal. [It is this desire to lower potential energy that motivates excess charge to be driven to the surface of a conductor.]

12. (III) A Kelvin Balance allows measurement of the permittivity of free space (Fig. 24–4). This experimental set-up consists of a parallel-plate capacitor with two circular conducting plates (area A for each), where the first plate rests on the force-measuring scale and the second plate is rigidly supported at height y above. While a fixed potential difference V is applied between the capacitor plates, the scale measures the force of attraction F that the top plate exerts on the bottom plate.
(a) Using $F = -dU/dy$, prove that ϵ_o can be determined with a Kelvin Balance through the expression $\epsilon_o = \dfrac{2Fy^2}{AV^2}$.
(b) You are an experimental physicist and have constructed a Kelvin Balance to make an accurate measurement of ϵ_o. Assume, from prior research, you know that the order of magnitude of ϵ_o is 1×10^{-11} C²/N·m². In your apparatus, the capacitor plates of area 80 cm² are separated by 0.5 cm. If, to acquire sufficient accuracy from your scale, you need F to be about 0.03 N (equivalent to the weight of a 3-g object), how large does V need to be?

13. (II) The inner and outer cylinders (radius R_b and R_a) of a cylindrical capacitor carry a charge per unit length $+\lambda$ and $-\lambda$, respectively. Determine the electrical energy stored per unit length in this system by two methods:
(a) by using $U = Q^2/2C$.
(b) by determining the total energy per length in the electric field E within the capacitor. That is, integrating $dU = u\, dV$ from $r = R_b$ to $r = R_a$, where u is the energy density of E within the infinitesimal volume dV (contained between radial distances r and $r + dr$).

14. (III) A Kelvin Balance allows measurement of the permittivity of free space (Fig. 24–4). This experimental set-up consists of a parallel-plate capacitor with two circular conducting plates (area A for each), where the first plate rests on the force-measuring scale and the second plate is rigidly supported at height y above. While a potential difference V is applied between the capacitor plates, the scale measures the force of attraction F that the top plate exerts on the bottom plate.
(a) Using $F = -dU/dy$, show theoretically that F and V are related by $F = \dfrac{\epsilon_o A}{2y^2} V^2$.
(b) Based on your result in part (a), explain why graphing F vs. V^2 should yield a straight-line plot. What is the theoretically predicted slope and y-intercept of this straight line?
(c) Using a Kelvin Balance, for which the capacitor plates of area 80 cm² are separated by 0.50 cm, the data in Table 24–1 are obtained.

Table 24–1

V(V)	3000	3500	4000	4500	5000
F(N)	0.013	0.017	0.022	0.029	0.035

Using the data in Table 24–1, graph F vs. V^2. Find the slope of the resulting straight line and use it to determine a value for ϵ_o.

15. (III) A system consists of two charged conducting spheres A and B, with A twice the size of B; i.e., the sphere radii are $r_A = 2R$ and $r_B = R$, respectively (Fig. 24–5). The spheres are far apart so that the electric field created by one does not significantly polarize the surface charge on the other (i.e., each sphere behaves as a single isolated conductor). Answer the following questions using the relations $Q = CV$ and $U = Q^2/2C$.
(a) Initially, A and B each have a charge $+Q$. How much greater is the electric potential V (with respect to infinity) and the stored electric energy U of B compared with that of A? That is, what are the ratios V_B/V_A and U_B/U_A?
(b) The spheres are connected by a conducting wire and a total charge q flows through the wire from B to A until

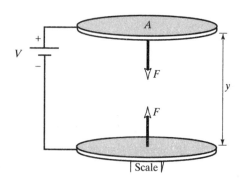

FIGURE 24–4

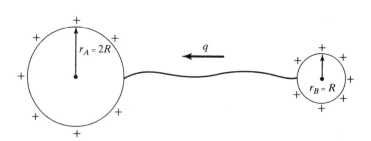

FIGURE 24–5

equilibrium is established. Assume that charge flow ceases when the total electric energy for the system $U = U_A + U_B$ is minimized. Use this "energy minimization" criterion to determine q.

(c) Show that in equilibrium $V_B/V_A = 1$ (or equivalently $V_B = V_A$). Also determine the equilibrium value of U_B/U_A. [Note that the larger sphere must store more energy than the smaller sphere in order to equalize their voltages.]

Section 24–5

16. (II) In a dynamic random access memory (DRAM) computer chip, each memory cell chiefly consists of a parallel-plate 30-fF capacitor for charge storage. Each of these cells represents a single binary-bit value of "1" or "0" when its capacitor is charged or uncharged, respectively. In present-day chips, the cell capacitor's two conducting parallel plates are separated by a 2.0-nm thick insulating material with dielectric constant $K = 25$.
    (a) Determine the area A (μm^2) of the cell capacitor's plates.
    (b) In (older) "planar" designs, the capacitor was mounted on a silicon-wafer surface with its plates parallel to the plane of the wafer. Assuming the plate area A accounts for half of the area of each cell, estimate how many megabytes of memory can be placed on a 3.0-cm^2 silicon wafer with the planar design. 1 byte = 8 bits.
    (c) To pack the cells more densely, trench capacitors in which the capacitor plates are mounted vertically on the four interior walls of a deep hole etched into the silicon wafer are often used. Modeling each wall of a hole to be rectangular in shape, the wall's depth is typically 60 times greater than its width. Assuming that the capacitor (whose cross-sectional area equals its wall-width squared) occupies half the area of a cell, estimate by what factor the number of bits per unit area increases when using the trench (as opposed to planar) design.

17. (III) The quantity of liquid (such as cryogenic liquid nitrogen) available in a storage tank is often monitored by a capacitive level sensor. This sensor is a vertically aligned cylindrical capacitor with outer and inner conductor radii r_a and r_b, whose length L spans the height of the tank. When a nonconducting liquid fills the tank to a height h ($\leq L$) from the tank's bottom, the dielectric in the lower and upper regions between the cylindrical conductors is the liquid (K_l) and its vapor (K_v), respectively (Fig. 24–6).
    (a) Show that the level-sensor capacitance C is related to the fraction F of the tank filled by liquid via
    $$F \equiv \frac{h}{L} = \frac{1}{(K_l - K_v)}\left[\frac{C \ln(r_a/r_b)}{2\pi\epsilon_o L} - K_v\right].$$
    [*Hint*: Consider the sensor as a combination of two capacitors.]
    (b) By connecting a capacitance-measuring instrument to the level sensor, F can be monitored. Assume the sensor dimensions are $L = 2.0$ m, $r_a = 5.0$ mm, and $r_b = 4.5$ mm. For liquid nitrogen ($K_l = 1.4, K_v = 1.0$), what values of C (pF) will correspond to the tank being completely full and completely empty?

18. (III) The quantity of liquid (such as cryogenic liquid nitrogen) available in a storage tank is often monitored by a capacitive lever sensor. This sensor is a vertically aligned cylindrical capacitor with outer and inner conductor radii r_a and r_b, whose length L spans the height of the tank. When a nonconducting liquid fills the tank to a height h ($\leq L$) from the tank's bottom, the dielectric in the lower and upper regions between the cylindrical conductors is the liquid (K_l) and its vapor (K_v), respectively (Fig. 24–6).

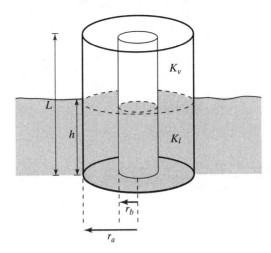

FIGURE 24–6

(a) Show that liquid height h is related to the level-sensor capacitance C by
$$h = \frac{1}{(K_l - K_v)}\left[\frac{C \ln(r_a/r_b)}{2\pi\epsilon_o} - K_v L\right].$$
[*Hint*: Consider the sensor as a combination of two capacitors.]
(b) By connecting a capacitance-measuring instrument to the level sensor, h can be monitored. If this instrument can measure capacitance to an accuracy of ΔC, show that the height can be measured to an accuracy Δh given by
$$\Delta h = \frac{\ln(r_a/r_b)}{2\pi\epsilon_o(K_l - K_v)}\Delta C.$$
[Note that Δh is independent of L].
(c) Assume the sensor dimensions are $r_a = 5.0$ mm and $r_b = 4.5$ mm. If $\Delta C = 1.0$ pF, to what accuracy can the height of liquid nitrogen ($K_l = 1.4, K_v = 1.0$) in its storage tank be determined?

General Problems

19. (II) (a) A rule of thumb for estimating the capacitance C of an isolated conducting sphere with radius r is the following: $C(\text{pF}) \approx r$ (cm); i.e., the numerical value of C in pF is about the same as the numerical value of the sphere's radius in cm. Justify this rule of thumb.
    (b) Modeling the human body as a 1-m radius conducting sphere, use the given rule of thumb to estimate your body's capacitance.
    (c) While walking across a carpet, you acquire an excess "static electricity" charge Q and produce a 0.5 cm spark when reaching out to touch a metallic doorknob. The dielectric strength of air is 30 kV/cm. Use this information to estimate Q (μC).

20. (II) The "induction probe," an instrument used to measure electric field magnitude, consists of a conducting "sensing" plate of area A connected to the upper plate of a capacitor of capacitance C (Fig. 24–7). Assume that both the sensing plate and the capacitor are initially uncharged. If the probe is then positioned so that the field E to be measured impinges

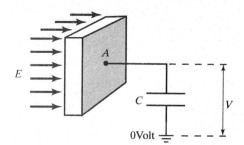

FIGURE 24–7

normally to the sensing plate, induced charge appears on the opposite faces on this plate and in the process, charges the capacitor.
(a) Show that the potential difference V, which develops across the capacitor, is directly proportional to E.
(b) For a particular induction probe, $A = 5.0 \text{ cm}^2$. If, when detecting a field magnitude of 1000 V/m, one wishes V to be an easily measurable value such as, say, 100 mV, what value should be used for C (nF)?

21. (II) To measure the capacitance C of a capacitor, it is first charged to a potential difference of 1350 V (with one plate at 0 V with respect to infinity). It is then connected by a conducting wire to a distant (initially uncharged) metal sphere of radius 3.0 cm. As a result, the capacitor's potential difference drops to 900 V. Determine C.

22. (II) Capacitors can be used as "charge counters." Consider an initially uncharged capacitor of capacitance C with its bottom plate grounded and its top plate connected to a source of electrons.
(a) If N electrons flow onto and are collected by the capacitor's top plate, show that the resulting potential difference V across the capacitor is directly proportional to N.
(b) Assume that the voltage-measuring instrumentation used in a charge-counting set-up can only accurately resolve voltage changes of up to 1 mV. If one wished to detect each new collected electron, what value of C would be necessary?
(c) Using modern semiconductor-device fabrication technology, a micron-sized capacitor can be constructed with parallel conducting plates separated by an insulating oxide of dielectric constant $K = 3$ and thickness $d = 100$ nm. To resolve the arrival of an individual electron on the plate of such a capacitor, assuming square plates of side length L, determine the required value of L (μm).

23. (II) While walking across an insulating floor, what is the maximum "static electricity" charge one might expect to acquire during each step? Assume that the sole of a person's shoe has area $A \approx 150 \text{ cm}^2$ and that while walking, as the foot is lifted from the ground during each step, the sole acquires an excess charge Q due to its rubbing contact with the floor.
(a) Model the sole as a planar conducting surface with Q uniformly distributed across it as the foot is lifted from the ground. If the dielectric strength of the air between the sole and floor as the foot is lifted is $E_S = 3 \times 10^6 \text{ N/C}$, determine Q_{max}, the maximum possible excess charge that can be transferred to the sole during each step.
(b) Modeling a person as an isolated conducting sphere of radius $r \approx 1$ m, estimate the capacitance of a person.
(c) After lifting the foot from the floor, assume the excess charge Q quickly redistributes itself over the entire surface area of the person. Estimate the maximum potential difference that the person can develop with respect to the floor.

24. (II) Consider the following simple model to describe the polarization of charge in a hydrogen atom when it is placed within an electric field. Imagine the atom to be a central positive point charge $+e$ embedded in a uniformly charged sphere of total charge $-e$ and radius $R = 5.3 \times 10^{-11}$ m (Fig. 24–8). When not acted upon by external influences, the positive charge will be located at the center of the negatively charged sphere. Assume that the negative charge remains uniformly distributed over the sphere of radius R even when it is subjected to a field or the positive charge is displaced from its center.

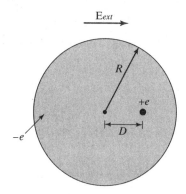

FIGURE 24-8

(a) Focusing on just the uniformly charged sphere, determine the electric field (magnitude and direction) created by this charge distribution at a distance r from its center, where $r \leq R$.
(b) Now assume a hydrogen atom is placed in an external electric field E_{ext}, which causes it to become "polarized;" i.e., the atom's opposite charges are displaced from one another. In the presence of this external field, use the given model to predict D, the equilibrium distance between the positive point charge and the center of the negative sphere.
[*Hint*: In equilibrium, the forces on the proton due to the external field and due to the charged sphere cancel each other out.]
(c) Suppose an experimentalist placed atomic hydrogen between the plates of a parallel plate capacitor with plate separation of 1.0 cm and then applied a potential difference of 100 V between the plates. Show that the given model predicts that, within each hydrogen atom, the displacement of the proton from the electron's spherical center amounts to only about 20 billionths of the atom's radius.
(d) In the given model, $D = R$ can be defined as the ionization of hydrogen. Use this idea to estimate the potential difference necessary to "ionize" a hydrogen atom using the capacitor described in part (c). You will discover that an impossibly large voltage is required.

25. (III) Three capacitors, each with capacitance C, are connected in a triangular-shaped pattern. This network of capacitors is then attached to batteries of potential difference V_1 and V_2 at its left and right ends as shown in Fig. 24–9a, and charges Q_1 and Q_2 are found to flow out of these batteries, respectively, in charging the network's capacitors. Next, three other capacitors, each with capacitance C', are connected in a T-shaped pattern. If this network of capacitors is to be attached to batteries of

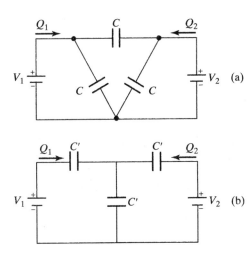

FIGURE 24-9

potential difference V_1 and V_2 at its left and right ends as shown in Fig. 24–9b and one wants the same charges Q_1 and Q_2 to flow out of these batteries as in the case of the triangular network, determine the necessary value for C' (in terms of C).

26. (III) If a volume of matter carries a net charge, the repulsive force between like charges produces an "electrostatic pressure" P at the volume's boundary. For example, consider the following system: A balloon of radius $r_o = 17$ cm contains 1.0 mole of air for which it is assumed that, uniformly throughout its volume, 1 out of every 10^6 air molecules is singly ionized; i.e., there is $+Q = (6.0 \times 10^{23})(10^{-6})$ $(+1.6 \times 10^{-19}\,\text{C}) = 96$ mC distributed uniformly within the balloon's spherical volume.

(a) Using the known spherically symmetric electric field E produced by a solid nonconducting sphere of charge (see Example 22–4 in the text), determine this system's total stored electric potential energy U by summing the electric field's stored energy over all space. That is, evaluate the integral

$$U = \int_{r=0}^{r=\infty} u(r)dv,$$ where $u(r)$ is the field's energy density

at distance r from the sphere's center and $dV = 4\pi r^2 \, dr$ is the infinitesimal volume located between r and $r + dr$.

[*Hint*: The integral breaks into two parts—from 0 to r_o and from r_o to ∞.]

(b) Show, using your expression for U, that when r_o is increased by an amount dr_o, the charged system's potential energy $\Delta U = \dfrac{dU}{dr} dr$ decreases. This loss in potential energy enables the system to do work $W = P\,dV$ on the environment, where P is the system's "electrostatic pressure," which acts at the system's spherical surface as its volume increases by dV. Equating $W = -\Delta U$, find the expression for P.

(c) For the given values of r_o and Q, determine P in atm. Your answer will illustrate why it is so important for matter to be neutral.

27. (III) (a) Consider a single (conducting) liquid drop of radius R, which has a charge Q uniformly distributed over its surface. Based on the repulsive interactions between the individual charges on its surface, determine the electrical potential energy U of this spherical shell of charge.

[*Hint*: Use the expression for the capacitance of an isolated conducting sphere.]

(b) Show, using your expression for U, that when R is increased by an amount dR, the charged drop's potential energy

$$\Delta U = \dfrac{dU}{dR} dR$$ decreases. This loss in potential energy enables

the system to do work $W = P\,dV$ on the environment, where P is the system's "electrostatic pressure" which acts radially outward at all points on the system's spherical surface as its volume increases by dV. Equating $W = -\Delta U$, show that the repulsive electrostatic pressure $P = Q^2/32\pi^2 \epsilon_o R^4$.

(c) Considering just one hemispheric side of the drop, whose axis is defined as the z-axis, show that P exerts a net force along the z-axis on this hemisphere of $F_z = P\pi R^2$ (Fig. 24–10).

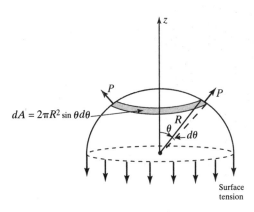

FIGURE 24-10

[*Hint*: On the surface of a sphere, the infinitesimal area located between the angles θ and $\theta + d\theta$ is $dA = (2\pi R \sin\theta)(R\,d\theta) = 2\pi R^2 \sin\theta \, d\theta$ (see Appendix C).]

(d) In order for the drop to remain stable, a surface tension force F_S acting along the length of the circumference C of the hemisphere's base must cancel the electrostatic repulsive force F_z. Show that the drop will break apart if $Q > 8\pi\sqrt{\epsilon_o \gamma R^3}$, where γ is the liquid's surface tension. [*Hint*: $F_S = \gamma C$; see Section 13–12 in the text.]

(e) To break apart a water drop with radius 0.2 mm, what minimum charge must be placed on it? For water, $\gamma = 0.07$ N/m.

[Electrostatic charging is commonly used to "atomize" a liquid such as paint or ink into a known droplet size.]

28. (III) If the charge on a water drop exceeds a threshold value, the drop will become unstable and break apart. Let's investigate this process with the following simple model in which we compare the energy associated with one large drop of charge Q and radius r_1 with that of two smaller drops, each of charge $Q/2$ and radius r_2 (Fig. 24–11). Assume that a drop of radius r has a surface potential energy U_S due to surface tension, which is proportional to its surface area [i.e., $U_S = \gamma 4\pi r^2$, where γ is a constant (called the surface tension)].

(a) In our model, we assume one large drop (radius r_1) breaks into two equal-sized smaller drops (radii r_2). Assuming the liquid's density remains constant, show that $r_2 = r_1/\sqrt[3]{2}$.

(b) Model the large drop as a conducting sphere with charge Q uniformly distributed over its surface and determine an expression for U_1, the sum of its electric and surface energy.

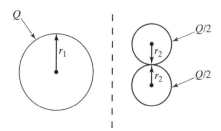

FIGURE 24–11

(c) At the moment of breakup, assume the two equal-size spheres touch and that (to a good approximation) a charge $Q/2$ is uniformly distributed over each surface. Determine an expression for U_2, the sum of its electric and surface energy. [Note: There is stored electric energy on each sphere as well as repulsive electric potential energy between the two spheres].

(d) Assume the criterion for breakup is when the potential energy of the two drops is less than that of the single drop. Show that this criterion predicts a drop of radius r_1 will break apart when its charge exceeds the threshold value $Q_B \approx 100\sqrt{\epsilon_o \gamma r_1^3}$.

(e) Calculate Q_B (nC) for a raindrop with radius of 0.2 mm. For water, $\gamma = 0.07$ N/m.

29. (III) A capacitive displacement sensor consists of a rigidly held inner conducting cylinder of radius R, surrounded by a concentric plastic cylindrical sheath (dielectric constant K) of thickness t, along with a movable outer conducting cylinder that slides along the sheath's outer surface in response to the displacement of an object of interest (Fig. 24–12). Assume that when the object is in its initial position, the overlap between cylinders is (i.e., the length of the resulting cylindrical capacitor) x_o, so that when the object is displaced by Δx, the overlap between the conducting cylinders becomes $x_o + \Delta x$. Prove that, by connecting this sensor to a capacitance-measuring instrument, one obtains a "linear" displacement gauge; i.e., a device that measures displacement using a relation of the form $\Delta x = bC + a$. Determine the constants a and b.

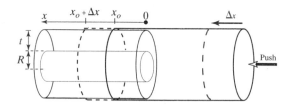

FIGURE 24–12

30. (III) A capacitor can "pump" a fluid uphill. Consider a vertically aligned parallel-plate capacitor whose plates have a width L, height H, and separation d. The initially uncharged capacitor is immersed to a depth h_o in a reservoir of nonconducting fluid (dielectric constant K, mass density ρ). Upon connecting the plates to a battery of voltage V, the fluid between the plates is observed to rise to a height h above its level outside the plates (Fig. 24–13). Assume the reservoir is large so that outside the capacitor the fluid level remains essentially constant at all times.

(a) Show that the capacitor's capacitance C as a function of h is

$$C = \frac{\epsilon_o(K-1)L}{d}(h + h_o) + \frac{\epsilon_o LH}{d}.$$

[*Hint*: The capacitor can be considered to be the combination of two capacitors.]

(b) Assume the battery places a charge of magnitude Q on each capacitor plate. Show that the potential energy of this system is

$$U = \frac{Q^2}{2C} + \frac{1}{2}\rho g d L h^2.$$

(c) Taking Q as fixed (i.e., constant) and the liquid's height h as variable, assume that equilibrium is achieved when $h = h_{eq}$, where h_{eq} is the value of h that causes U to be minimized. Show that this criterion gives

$$h_{eq} = \frac{\epsilon_o(K-1)}{2\rho g d^2}\frac{Q^2}{C^2} = \frac{\epsilon_o(K-1)V^2}{2\rho g d^2}$$

where the last step follows from $Q = CV$.

(d) A capacitor immersed in glycerin ($\rho = 1270$ kg/m^3, $K = 56$) could be used to measure high voltages. If $d = 5$ mm and h_{eq} is observed to be 72 cm, determine V.

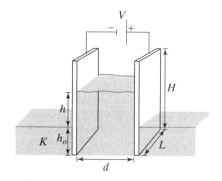

FIGURE 24–13

CHAPTER 25

Electric Currents and Resistance

Sections 25–2 and 25–3

1. (II) A sequence of potential differences V is applied across a wire (diameter 0.32 mm, length 11 cm) and the associated currents I are measured. Table 25–1 summarizes the results of this experiment:

TABLE 25–1

V (V)	0.100	0.200	0.300	0.400	0.500
I (mA)	72	144	216	288	360

(a) If this wire obeys Ohm's law, graphing I vs. V will result in a straight-line plot. Explain why this is so and determine the theoretical predictions for the straight line's slope and y-intercept.
(b) Plot I vs. V. Based on this plot, can you conclude that the wire obeys Ohm's law (i.e., did you obtain a straight line with the expected y-intercept)? If so, determine the wire's resistance R.
(c) Calculate the wire's resistivity and use Table 25–1 in the text to identify the solid material from which it is composed.

2. (II) Suppose that, between times $t = 0$ and $t = 2.0$ s, a pulse of electric current given by

$$I(t) = \begin{cases} (1.0 \text{ A/s})t & 0 \le t \le 1.0 \text{ s} \\ 2.0 \text{ A} - (1.0 \text{ A/s})t & 1.0 \le t \le 2.0 \text{ s} \end{cases}$$

flows onto the plate of a capacitor. During this 2.0-s time interval, how much total charge is collected on the capacitor plate?

3. (II) To eliminate the possible damaging effects of static charge while working with flammable vapors or sensitive electronic components, the potential difference V of a worker with respect to the ground due to "static electricity" must be kept to a low value. Consider the following model for how a "static electricity" voltage can develop on a person while walking across an insulating floor. Assume that the person takes N steps per unit time and that, as a result of rubbing contact between the sole of the person's shoe and the floor, the person acquires an excess charge of q during each step. Further assume that R is the resistance between the body and the ground (mostly due to the person's shoes).
(a) Let Q be the total excess charge on the person. In equilibrium, the rate of increase in Q is balanced by an equal rate of decrease, and thus a (constant) equilibrium charge Q_o is maintained. Show that in equilibrium, if the potential difference of the person with respect to the floor is to be no greater than V_{max}, then $R \le V_{max}/Nq$.
(b) If q (in the worst-case scenario) is estimated to be $0.5 \, \mu C$, V_{max} is chosen as 100 V, and $N = 2$ steps per second, what is the upper limit for an acceptable value of R? [For a person wearing typical shoes, R can be on the order of $10^{11} \, \Omega$. Thus, industrial workers (e.g., in an electronics fabrication clean room) commonly wear conductive footwear to satisfy the restriction on R found in this problem.]

Section 25–4

4. (II) Small changes in the length of an object can be measured using a "strain gauge" sensor, which is simply a wire with undeformed length L_o, cross-sectional area A_o, and resistance R_o. This sensor is rigidly affixed to the object's surface, aligned with its length in the direction in which length changes are to be measured. As the object deforms, the length the wire sensor changes by ΔL and the resulting change ΔR in the sensor's resistance is measured. Assuming that as the solid wire is deformed to a length L, its density, and thus volume, remains constant (this is only approximately correct), prove that the strain ($= \Delta L/L_o$) of the wire sensor, and thus of the object to which it is attached, is $\Delta R/2R_o$.

5. (II) Your research supervisor comes into your lab, hands you a wire coil used to produce magnetic fields, and asks you to make another one exactly like it. Inspecting the coil, you note that it is made from a long length of copper wire that has been wound into a tight spiral of N circular turns, each of radius $r = 5.00$ cm. Using a micrometer, you find that the diameter d of the wire is 1.628 mm, and using an ohmmeter, you find that the total resistance of the coil is $R = 2.45 \, \Omega$. From this information, what number N of 5.00 cm radius turns should the new coil have in order to satisfy your supervisor's request?

6. (II) In contrast with metals, the resistivity of a semiconductor decreases with increasing temperature. This semiconductor property is exploited in a temperature-sensing device called a thermistor whose resistance R as a function of temperature T (on the Kelvin scale) obeys the relation $R = R_o \exp(\beta/T)$ over a temperature range of 270 K $< T <$ 320 K. For a typical thermistor, the constants $R_o = 0.01 \, \Omega$ and $\beta = 4000$ K.
(a) Using the constants given for a typical thermistor, plot R vs. T in the range 270 K $< T <$ 320 K to demonstrate that its semiconductor resistivity does indeed decrease dramatically with increasing temperature over this span of 50 K.
(b) One measure of the usefulness of a temperature sensor at a certain temperature T is its fractional sensitivity

$S \equiv \left(\dfrac{dR}{R}\right)\Big/\left(\dfrac{dT}{T}\right) = \left(\dfrac{T}{R}\right)\dfrac{dR}{dT}$ (i.e., the ratio of its fractional change in resistance per fractional change in temperature). The magnitude of S for many types of temperature sensors (e.g., platinum resistance thermometer) is about one. Show that the magnitude of a thermistor's fractional sensitivity is about a factor of ten greater than these other sensors. [For this reason, thermistors are commonly used to accurately detect and control small changes in temperatures near 300 K.]

7. (II) The temperature dependence of a platinum resistance thermometer very accurately obeys the relation $R = R_o[1 + \alpha(T - T_o)]$ with $R_o = 100.00\ \Omega$, $T_o = 273$ K, and $\alpha = 0.003927$ K^{-1} over the temperature range 70 K $< T <$ 870 K. One measure of the usefulness of a temperature sensor at a certain temperature T (on the Kelvin scale) is its fractional sensitivity $S \equiv \left(\dfrac{dR}{R}\right)\Big/\left(\dfrac{dT}{T}\right) = \left(\dfrac{T}{R}\right)\dfrac{dR}{dT}$ (i.e., the ratio of its fractional change in resistance per fractional change in temperature).
(a) Determine an expression for S as a function of temperature T for the platinum resistance thermometer.
(b) Plot S vs. T over the range 70 K $< T <$ 870 K. For accurate temperature measurement, one typically needs a sensor with $S \geq 1$. Does a platinum resistance thermometer fulfill this criterion over the plotted temperature range?

8. (III) In introductory physics labs, the equipotential lines resulting from various charge geometries are plotted with the aid of conductive paper. Consider the equipotential lines between two electrodes painted on the top side of such paper where the inner electrode at zero electric potential is a small solid conducting disk of radius r_b and the outer electrode at potential $+V_o$ is a solid ring of inner radius r_a (cross-sectional view shown in Fig. 25–1). Define the paper's resistivity and thickness to be ρ and t. Assume that, at a given radius r from the center of the inner electrode, the current that flows between the outer and inner electrode uniformly fills the entire paper thickness and that the entire voltage drop V_o occurs from r_a to r_b (i.e., ignore any small voltage drop that occurs directly under

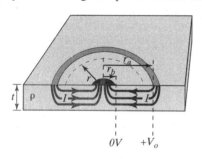

FIGURE 25–1

an electrode). Also note that the total current I must be the same at every r. Show that the resulting equipotential lines between the electrodes are given by the relation $V(r) = \dfrac{V_o}{\ln(r_a/r_b)}\ln\left(\dfrac{r}{r_b}\right)$, which is completely independent of the properties of the paper (as long as it is of uniform construction). [Hint: At a given r, find the potential difference dV across an annulus of radial thickness dr.]

9. (II) The energy contained in a pulse of laser light can be measured using a "rat's nest" calorimeter (RNC). This device consists of a thermally insulated container with a small opening and reflective interior surface (e.g., a thermos flask) inside which a tangled ball of copper wire (length L, radius r) is placed. The copper wire has a thin coating of insulating material so that it is not in electrical contact with itself when tangled. After a light pulse enters the container's opening, assume all of the light energy E is transformed through absorption by the wire into thermal energy, which quickly distributes itself uniformly throughout the wire's volume.
(a) Show that by measuring the resulting change ΔR in the wire's resistance, the light energy can be determined by

$$E = \dfrac{\pi^2 r^4 \rho c\ \Delta R}{\rho_o \alpha}$$

where ρ, c, ρ_o, and α are the wire's mass density, specific heat, resistivity at ambient temperature, and temperature coefficient of resistivity, respectively. Neglect any change in length or radius of the wire.
(b) In a 20°C laboratory, a pulse of laser light enters a RNC constructed using 0.064-mm diameter copper wire (see Tables 13–1, 19–1, and 25–1 in the text for data on copper). If ΔR is measured to be 0.10 Ω, determine the energy contained in this light pulse.

10. (III) Consider a conducting sphere of radius r_o with half of its surface area buried in the ground, which is held at a potential difference $+V_o$ with respect to infinity by a voltage source. A constant current I_o will flow from this sphere into the conducting top soil (resistivity ρ) of the earth. Assume the earth's surface is flat and the radial distance r is measured from the sphere's center.
(a) Determine the total resistance R that the earth presents to the flow of I_o.
(b) Show that the region between the sphere's surface and $r = 2r_o$ accounts for 50% of R (i.e., half of the total resistance is due to the neighboring half-spherical shell of soil with thickness r_o).
[Hint: At a given r, find the resistance dR due to a half-spherical shell of radial thickness dr.]

11. (III) When a typical lightning bolt strikes a point on the earth, it delivers a negative charge of magnitude $Q = 1$ C over a time interval $\Delta t = 40\ \mu$s. Because the earth is a fairly good electrical conductor, over the span Δt, (an assumed constant) current $I = Q/\Delta t$ flows within the earth, bringing a compensating amount of positive charge to the point of impact (Fig. 25–2). Assume the earth has a uniform resistivity ρ.
(a) Let the earth's surface be flat and the radial distance r be measured from the lightning's point of impact. Noting that the total current I must be the same at every r, show that the potential difference V between two radii r_1 and r_2 is given by

$$V = \dfrac{I\rho}{2\pi}\left[\dfrac{1}{r_1} - \dfrac{1}{r_2}\right].$$

[Hint: At a given r, find the potential difference dV across a half-spherical shell of radial thickness dr.]
(b) If a person has the misfortune of standing near the point of a typical lightning-bolt strike with one foot 5.0 m and the other foot 5.5 m from the point of impact, determine V between the person's feet. Assume, for the earth's top soil, $\rho = 10\ \Omega\cdot$m.

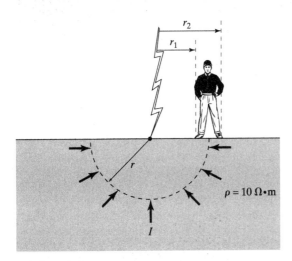

FIGURE 25–2

Sections 25–5 and 25–6

12. (II) An electrical power plant can produce electricity at a fixed power $P = 100$ MW, but the plant operator is free to choose the voltage V at which it is produced. This electricity flows from the plant to the user through a transmission line of resistance $R = 5\,\Omega$ and so arrives to the user at voltage V'. If the reduction in voltage $\Delta V = V - V'$ is to be at most 5% of V, what is the minimum acceptable choice for V?

13. (II) An electrical power plant can produce electricity at a fixed power P, but the plant operator is free to choose the voltage V at which it is produced. This electricity is carried as an electrical current I through a transmission line (resistance R) from the plant to the user, where it provides the user with electrical power P'. Show that the reduction in power $\Delta P = P - P'$ due to transmission losses is given by $\Delta P = P^2 R / V^2$. In order to reduce power losses during transmission, should the operator choose V to be as large or as small as possible?

14. (III) Three identical containers X, Y, and Z are each filled with identical 0°C water-ice mixtures, each of which contains mass m of ice. Identical heating elements (resistance R) are placed inside each container, and the containers are placed in a -10°C room. A voltage difference $V_X = 380$ V and $V_Y = 220$ V is applied to the heating element in X and Y and all of the ice in these containers melts in time $t_X = 4.0$ min and $t_Y = 20.0$ min, respectively. Define P_C to be the (constant) rate at which heat is conducted from the interior of each container to the room while the container's contents are at 0°C. Determine the voltage difference V_Z that should be applied to the heating element in Z in order to maintain a constant mass m of ice in this container indefinitely.

15. (II) A differential calorimeter consists of two isolated metal blocks called the "sample" and the "reference." Each identical block contains an embedded resistor (resistance R), as well as an attached thermometer to monitor its temperature and, outfitted thusly, each has a total mass m and specific heat c. Using this calorimeter, the specific heat c_o of an object (mass m_o) can be measured as follows: Attach the object to the sample block. Then, with both blocks at initial temperature T_o, cause a current I to follow through each embedded resistor for a short time interval Δt, and measure the final temperatures T_S and T_R of the sample and reference blocks, respectively. Show that
$$c_o = \frac{I^2 R\,\Delta t}{m_o} \left[\frac{T_R - T_S}{(T_S - T_o)(T_R - T_o)} \right].$$

Section 25–7

16. (II) For a time-dependent voltage $V(t)$, which is periodic with period T, the root-mean-square voltage is defined to be
$$V_{\rm rms} = \left[\frac{1}{T} \int_0^T V^2\,dt \right]^{1/2}.$$
Use this definition to determine $V_{\rm rms}$ (in terms of the peak voltage V_o) for

(a) a sinusoidal voltage; i.e., $V(t) = V_o \sin\left(\dfrac{2\pi t}{T}\right)$ for $0 \le t \le T$.

(b) a positive square-wave voltage; i.e.,
$$V(t) = \begin{cases} V_o & 0 \le t \le T/2 \\ 0 & T/2 \le t \le T \end{cases}.$$

Section 25–8

17. (II) A resistor of length L, constant cross-sectional area A, and uniform resistivity ρ carries an electric current with uniform current density j. Determine the power transformed into thermal energy per unit volume within this resistor (in terms of ρ and j).

18. (II) A potential difference V is applied across a wire of resistivity ρ, length L, and cross-sectional area A. Assume the wire is made of a metal with n free electrons per unit volume. As each free electron travels the length of the wire in response to the applied voltage, it loses an amount $\Delta U = eV$ of electric potential energy. A portion of this lost potential energy is transformed into the kinetic energy associated with the electron's drift velocity v_d.

(a) Determine v_d in terms of ρ, L, A, n, and/ or V. Then compare, as a fractional ratio, the electron's kinetic energy associated with v_d with its lost potential energy ΔU. That is, determine $F \equiv \dfrac{1}{2} m v_d^2 / \Delta U$.

(b) If 10 V is applied across a 0.5-m long copper wire, calculate F.

(c) You should find that $F \ll 1$. That is, very little of the lost potential energy is transformed into the electron's kinetic energy. What becomes of most of the energy ΔU?

19. (II) A model for electric current in a metallic wire is given in Section 25–8 of the text. In this model, the wire contains n free electrons per unit volume and between successive collisions with the wire's atoms, an applied electric field $\mathbf{E}$ accelerates each free electron with acceleration $\mathbf{a}$, resulting in an average ("drift") velocity $\mathbf{v}_d$ for the electrons taken as a whole.

(a) Define the average time τ between successive electron-atom collisions by $\mathbf{v}_d \equiv \mathbf{a}\tau$; then derive an expression for the wire's resistivity ρ using this model.

(b) Determine τ (s) for an electric current in a copper wire.

General Problems

20. (II) At the International Electrical Congress in Paris in 1889, the "international ohm" was declared to be the electrical resistance of a column of mercury at the temperature of melting ice with a length of 106.3 cm and a uniform cross-sectional

area of one square millimeter. Show that the definition of an "international ohm" is consistent, but not exactly the same, as today's SI definition (i.e., find how many SI ohms are equivalent to one international ohm). For mercury at 20°C, the resistivity and the temperature coefficient of resistivity are 957.83 n$\Omega \cdot$m and 0.00089(C°)$^{-1}$, respectively.

21. (II) When at the same temperature T_o as its surroundings, a resistor has resistance R_o. A large current I is then passed through this resistor over a time period long enough for equilibrium to be achieved and the resistor's temperature and resistance increase to $T_o + \Delta T$ and R, respectively. In this equilibrium situation, the electrical power P dissipated by the resistor is balanced by the rate it transfers heat to its surroundings due to the temperature difference ΔT. Assume that P and ΔT are related by $P = \gamma \Delta T$, where γ is a constant (this relation, called "Newton's law of cooling," often holds true). Show that, if the resistor is made of a metal with temperature coefficient of resistivity α and ΔT is large enough so that $R \gg R_o$ the current is independent of P and given by $I = \sqrt{\gamma/R_o\alpha}$.

22. (II) Capacitors are commonly used as timekeepers. Assume an electronic circuit can provide a constant current I and that this flow of charge is collected on the top plate of a capacitor C, whose bottom plate is connected to the ground. A voltmeter measures the potential difference $V(t)$ across the capacitor as a function of time. Let's demonstrate that this system can be used as a clock.
(a) If the capacitor is uncharged at $t = 0$, prove that, as time moves forward, $V(t)$ is directly proportional to time and determine the constant of proportionality.
(b) Assume $I = 1.0$ mA. If the maximum voltage that the voltmeter can measure is 10 V and the largest capacitance that can be found to use in the clock is 10 mF, what is the longest interval over which one will be able to measure time?

23. What amount of "static electricity" charge can develop on a person while walking across an insulating floor? Assume that the person takes N steps per unit time and that, as a result of rubbing contact between the sole of the person's shoe and the floor, the person acquires an excess charge of q during each step. Further assume that C and R are the capacitance of the person's body with respect to the floor and the resistance between the body and the ground.
(a) Let Q be the total excess charge on the person and V be the person's resulting potential difference with respect to the floor. In equilibrium, the rate of increase in Q is balanced by an equal rate of decrease, and thus a (constant) equilibrium charge Q_o and potential difference V_o is maintained. Show that $Q_o = NqCR$.
(b) For a shoe-wearing person walking across a vinyl floor, $q \approx 30$ nC, $R \approx 1 \times 10^{11}\ \Omega$, and $C \approx 100$ pF. If $N = 2$ steps per second, determine Q_o and V_o.

24. (II) The fairly large (>10 kΩ) resistance R between a person's dry hands is due almost entirely to the outer layers of skin. That is, $R = R_{in} + R_S \approx R_S$, where R_{in} and R_S are the resistances due to the interior of the body and due to the skin, respectively. To demonstrate this fact, model the interior portion of the conducting path between outstretched hands (i.e., the portion that produces R_{in}) as a 2.0-m long cylinder with a 4-cm diameter. Given the ion content of fluids within the body, assume that the resistivity of this cylinder is about the same as that for seawater (0.2 $\Omega \cdot$m).
(a) Use this model to estimate R_{in}. If $R = 10$ kΩ, what is R_S?

(b) After swimming in the ocean, assume a person's skin is saturated with seawater so that R_S becomes almost zero. According to our model, what potential difference applied between these hands will cause the potentially lethal current of 70 mA to flow through the body?

25. (II) The level of liquid helium (temperature ≤ 4 K) in its storage tank is commonly monitored using a vertically aligned niobium-titanium (NbTi) wire, whose length L spans the height of the tank. In this level-sensing set-up, an electronic circuit maintains a constant electrical current I at all times in the NbTi wire and a voltmeter monitors the voltage difference V across this wire. Since the superconducting transition temperature for NbTi is 10 K, the portion of the wire immersed in the liquid helium is in the superconducting state, while the portion above the liquid (in helium vapor with temperature above 10 K) is in the normal state. Define $F \equiv x/L$ to be the fraction of the tank filled with liquid helium (see Fig. 25–3) and V_o to be the value of V when the tank is empty ($F = 0$). By measuring V, F can be found through a linear relation. Determine this linear relation between F and V (in terms of V_o).

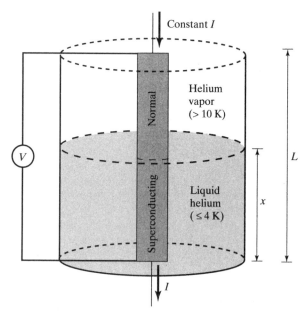

FIGURE 25–3

26. (II) The cost of battery electricity is expensive when compared with that available from a wall receptacle. D-cell (cost US $1.70) and AA-cell (cost US $1.25) alkaline batteries have the capacities to provide a continuous current of 25 mA for 820 h and 117 h, respectively, at 1.5 V (we have assumed the favorable condition of small current for maximum battery life). In the United States, wall receptacle power is provided at 120 V (rms) and typically costs $ 0.10 per kWh. For both D and AA batteries, use this information to find how much more expensive (dc) battery electricity is than wall receptacle (rms) electricity.

27. (III) In introductory physics labs, the equipotential lines resulting from various charge geometries are plotted with the aid of conductive paper. Consider the equipotential lines between two electrodes painted on the top side of such paper where the inner electrode at zero electric potential is a small solid conducting disk of radius r_b and the outer electrode at

potential $+V_o$ is a solid ring of inner radius r_a (cross-sectional view shown in Fig. 25–1). In lab manuals, it is claimed that these electrodes produce the same equipotential pattern as that found between the electrodes of a cylindrical capacitor.
(a) Show that, in the region between two infinitely long concentric, cylindrical inner and outer conducting shells (radii r_b and r_a, uniform charge per length $-\lambda$ and $+\lambda$, respectively; i.e., the cylindrical capacitor geometry), the equipotential lines are given by $V(r) = \dfrac{\lambda}{2\pi\epsilon_o} \ln\left(\dfrac{r}{r_b}\right)$, where r is the radial distance from the center of the inner shell and $V(r_b) \equiv 0$.
(b) Now analyze the equipotential pattern created by the concentric disk and ring electrodes on conductive paper. Define the paper's resistivity and thickness to be ρ and t. Assume that, at a given radius r from the center of the inner electrode, the current that flows between the outer and inner electrode uniformly fills the entire paper thickness and that the entire voltage drop V_o occurs from r_a to r_b (i.e., ignore any small voltage drop that occurs directly under an electrode). Noting that the total current I must be the same at every r, show that the resulting equipotential lines between the electrodes are indeed given by a relation of the form $V(r) = \dfrac{A}{2\pi} \ln\left(\dfrac{r}{r_b}\right)$, where the constant A is analogous to λ/ϵ_o for the charged cylinders. What is A (in terms of ρ, t, and I)? [Hint: At a given r, find the potential difference dV across an annulus of radial thickness dr.]

28. (III) A system consists of two charged conducting spheres A and B, with A twice the size of B (i.e., the sphere radii are related by $r_A = 2r_B$). The spheres are far apart so that the electric field created by one does not significantly polarize the surface charge on the other. Initially, A and B each have a charge $+Q_o$. The spheres are connected by a conducting wire and a current flows through the wire from B to A until equilibrium is established (Fig. 25–4).
(a) Show that a total charge $q_E = Q_o/3$ will flow from B to A in the process of establishing equilibrium.
(b) Using $Q^2/2C$, calculate the system's stored electric energy $U = U_A + U_B$, both initially and in equilibrium. After equilibrium is established, determine ΔU, the amount by which U has decreased (in terms of Q_o and r_B) from its initial value.

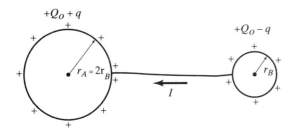

FIGURE 25–4

(c) Determine the potential difference V between the spheres after a charge q has been transferred to A due to the flow of current through the wire from B. Then determine the total electric energy transformed into thermal energy by the wire in the process of establishing equilibrium by evaluating the integral $\int P\,dt$, where P is power [Hint: $I = dq/dt$]

(d) In establishing equilibrium, is the decrease in the system's stored electric energy perfectly accounted for by an increase in thermal energy in the wire? What fraction of the initial U ends up as thermal energy?

29. (III) When high-frequency f ($> 100\,\text{kHz}$) alternating current is passed through a cylindrical conducting wire of radius r_o the current density, rather than being spread uniformly across the wire's cross-section, is distributed according to $J(r) = J_o \exp\left[-(r_o - r)/\delta\right]$, where r is the radial distance from the wire's central axis and J_o is a constant. This relation indicates that the alternating current flows most significantly within a distance δ of the conductor's surface. The constant δ is called the "skin depth."
(a) Let I_o be the total current flowing within the wire. Estimate the thickness t (expressed as a multiple of δ) of the outer portion of the wire which carries 99% of the total current (see Fig. 25–5). [Hint: Define $r_1 = r_o - t$ to be the radius within which 1% of the current is carried. Assume $t \ll r_o$, and so $\delta \ll r_o$. Also, $\int x e^{\alpha x}\,dx = \dfrac{e^{\alpha x}}{\alpha}\left(x - \dfrac{1}{\alpha}\right).$]

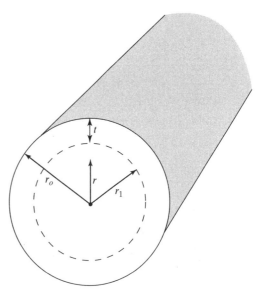

FIGURE 25–5

(b) For copper wire, advanced analysis shows that $\delta = (65\,\text{mm}\cdot\text{Hz}^{1/2})/\sqrt{f}$. If $f = 1\,\text{MHz}$, determine t.

30. Consider the flow of electric current through a cylindrical conducting wire of radius r_o in response to an applied voltage V. Let r be the radial distance from the wire's central axis.
(a) Assuming that a constant current density J_o exists across the wire's cross-section, determine the resistance R_o of the wire.
(b) Next assume that, rather than being distributed uniformly, the current is concentrated near the wire's surface according to $J(r) = J_o\left(\dfrac{r}{r_o}\right)^5$, where J_o is the same constant as in part (a). Determine the resistance R of the wire.
(c) Calculate the ratio R/R_o.
[This problem offers a crude model for the "skin effect"; i.e., the resistance increase observed for high-frequency ($>100\,\text{kHz}$) alternating currents in a conducting wire.]

CHAPTER 26

DC Circuits

Section 26–2

1. (II) A network consists of one resistor of resistance R_0 and three resistors, each of resistance R_1, as shown in Fig. 26–1. Determine the value of R_1 (as a multiple of R_0), if the equivalent resistance of the network (i.e., the resistance between A and B) is R_0.

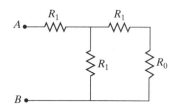

FIGURE 26–1

2. (II) The performance of the starter circuit in an automobile can be significantly degraded by a small amount of corrosion on a battery terminal. Fig. 26–2a depicts a properly functioning circuit with a battery (12.5 V emf, 0.02 Ω internal resistance) attached via corrosion-free cables to a starter motor of resistance $R_s = 0.15\,\Omega$. Assume after some time, corrosion between a battery terminal and a starter cable introduces an extra series resistance of just $R_c = 0.10\,\Omega$ into the circuit as shown in Fig. 26–2b. Let P_o and P be the power delivered to the starter in the circuit free of corrosion and the power delivered to the circuit that includes corrosion, respectively. Determine the ratio P_o/P.

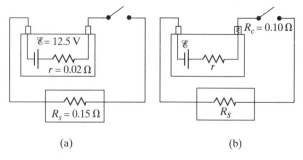

FIGURE 26–2

3. (II) The two terminals of a voltage source with emf $\mathscr{E}$ and internal resistance r are connected to the two sides of a load resistance R. For what value of R will the maximum power be delivered from the source to the load?

4. (III) You are an experimental physicist designing a chamber within which a wire resistance heater provides heat to an enclosed volume of gas. For the apparatus to function properly, this heater must transfer heat to the gas at a constant rate. While in operation, the resistance of the heater will always be close to the value $R = R_0$, but unfortunately, due to changing conditions within the chamber during an experimental run, the heater's temperature is expected to fluctuate slightly causing its resistance to vary a small amount $\Delta R\,(\ll R_0)$. To maintain the heater at constant power, you design the circuit shown in Fig. 26–3, which includes two resistors, each of resistance r. These two resistors are mounted external to the chamber in such a way that their resistance value will always remain fixed. Determine the required value for r so that the heater power will remain constant even if its resistance fluctuates by a small amount.

[*Hint*: If $\Delta R \ll R_0$, then $\Delta P \approx \dfrac{dP}{dR}\bigg|_{R=R_0} \Delta R$.]

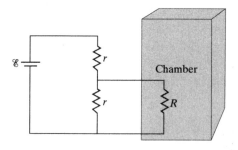

FIGURE 26–3

5. (III) The resistor network shown in Fig. 26–4 is called the R-2R ladder. Prove that the four currents $I_0, I_1, I_2,$ and I_3 are related by $I_n = 2^n I_0$, where $n = 0, 1, 2, 3$. [The R-2R ladder plays a key role in digital-to-analog conversion circuits where its "current-doubling" property is exploited in converting a binary number in a computer's memory to an equivalent real voltage.]

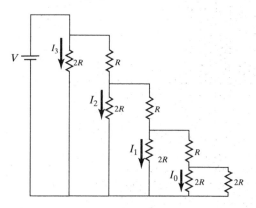

FIGURE 26–4

6. (III) Consider a "load" resistor whose resistance R_L can vary (e.g., a semiconducting thermistor subjected to varying temperature). A "constant current source" circuit will supply the same current I to this load resistor independent of its varying resistance. A crude version of such a circuit is shown in Fig. 26–5 where R is chosen to be much greater than any possible value of R_L.

(a) Define $\Delta I \equiv \dfrac{\mathcal{E}}{R} - I$. Show that, as long as $R \gg R_L$, the fractional deviation of I from the value of $\dfrac{\mathcal{E}}{R}$ is $\dfrac{\Delta I}{(\mathcal{E}/R)} \approx \dfrac{R_L}{R}$.

(b) A particular load resistance is expected to vary within the range of 1.0 kΩ to 5.0 kΩ. At all times, we wish to maintain the current through this resistor at the constant value of 0.10 mA with a precision of 1.0 %. If the circuit in Fig. 26–5 is used to accomplish this feat, what minimum values should be used for R and $\mathcal{E}$?

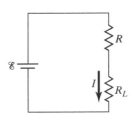

FIGURE 26–5

7. (III) For the circuit shown in Fig. 26–6, the battery voltage $\mathcal{E}$ and the resistances R_o are fixed, but the resistance R can be varied.

(a) Show that the electric power P dissipated in this circuit as a function of R can be written in dimensionless form as
$y = \dfrac{x+1}{2x+1}$, where $y \equiv \dfrac{P}{(\mathcal{E}^2/R_o)}$ and $x \equiv R/R_o$.

(b) Plot (or sketch) y vs. x. Based on this plot, what is $y_{\max}$ and $y_{\min}$, the maximum and minimum value for y, respectively? For what x-values $x_{\max}$ and $x_{\min}$ do these extremal values of y occur?

(c) Translate your results from part (b) to determine $P_{\max}$ and $P_{\min}$, the maximum and minimum power dissipated by the circuit, respectively, and the R-values $R_{\max}$ and $R_{\min}$ for which these extrema in power occur. Offer qualitative explanations for why the power extrema occur for these values of R.

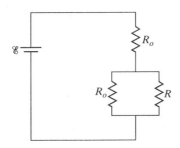

FIGURE 26–6

Section 26–3

8. (III) Three resistors, each with resistance R, are connected in a triangular-shaped pattern. This network of resistors is then attached to batteries $\mathcal{E}_1$ and $\mathcal{E}_2$ at its left and right ends as shown in Fig. 26–7a; the currents I_1 and I_2 are found to flow out of these batteries, respectively. Next, three other resistors, each with resistance R', are connected in a T-shaped pattern. If this network of resistors is to be attached to batteries of potential difference $\mathcal{E}_1$ and $\mathcal{E}_2$ at its left and right ends as shown in Fig. 26–7b and one wants the same currents I_1 and I_2 to flow out of these batteries as in the case of the triangular network, determine the necessary value for R' (in terms of R).

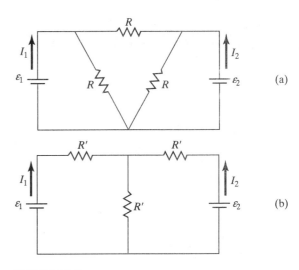

FIGURE 26–7

9. (II) For the circuit shown in Fig. 26–8, find the value of R'' that will make the current I_3 vanish.

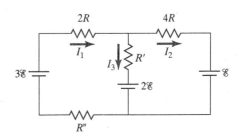

FIGURE 26–8

10. (III) (a) A network of five equal resistors R is connected to a battery $\mathcal{E}$ as shown in Fig. 26–9. Determine the current I that flows out of the battery.

(b) Use the value determined for I to find the single resistor R_{eq} that is equivalent to the five-resistor network.

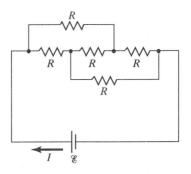

FIGURE 26–9

Section 26-4

11. (II) A parallel-plate capacitor is filled with a dielectric composed of a highly insulating material of resistivity ρ and dielectric constant K. This capacitor can be modeled as a pure capacitance C in parallel with a resistance R. Assume a battery is used to place charges $+Q$ and $-Q$ on the capacitor's opposing plates and is then disconnected. Show that the capacitor discharges over subsequent times with a time constant $\tau = K\epsilon_0\rho$ (known as the "dielectric relaxation time"). Evaluate τ, if the dielectric is glass with $\rho = 1 \times 10^{12}\,\Omega\cdot\text{m}$ and $K = 5$.

12. (II) An RC circuit with time constant $\tau = RC$ is included as the output stage of many laboratory instruments (to eliminate unwanted high-frequency voltages and electronic noise). Such an instrument takes an input (e.g., electrical or optical) signal S_{in} from an experimental set-up, performs a useful process (e.g., amplification), and produces a meaningful output voltage V_{out}. A rule of thumb for researchers is that, when S_{in} changes, one must wait for a time interval equal to at least five time constants in order to obtain an accurate reading of the resulting V_{out}. Let's explore the basis of this rule.
(a) Assume that, in reaction to a change in S_{in}, an instrument's internal processing produces a voltage $\mathscr{E}$, which is then passed through an RC circuit. The instrument's output V_{out} is the potential difference across the RC circuit's capacitor. This situation can be modeled in Fig. 26–10 by closing the switch at time $t = 0$ with C initially uncharged. Define V_n to be the value of V_{out} at a time $t = n\tau$. Prove that if one wishes V_{out} to be within 1% of $\mathscr{E}$, n must be about 5 or greater.
(b) Suppose a researcher, who has been previously waiting 5τ before reading an instrument's output, decides to wait 10τ in an effort to increase the precision of the experimental data. The drawback of this approach is that the experiment now will take twice as long. Determine the percent change in precision that will result from this decision (i.e., find $\dfrac{V_{10} - V_5}{\mathscr{E}} \times 100\%$). [In many cases, this small gain in accuracy is not worth the huge sacrifice in time.]

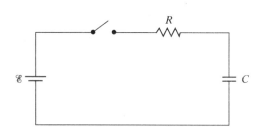

FIGURE 26–10

13. (II) Consider the circuit shown in Fig. 26–11, where all resistors have the same resistance R. At $t = 0$, with the capacitor C uncharged, the switch is closed.
(a) At $t = 0$, the three currents can be determined by analyzing a simpler, but equivalent, circuit. Identify this simpler circuit and use it to find the values of I_1, I_2, and I_3 at $t = 0$.
(b) At $t = \infty$, the currents can be determined by analyzing a simpler, equivalent circuit. Identify this simpler circuit and implement it in finding the values of I_1, I_2, and I_3 at $t = \infty$.
(c) At $t = \infty$, what is the potential difference across the capacitor?

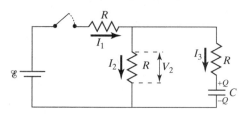

FIGURE 26–11

14. (III) Even if the dielectric that fills the space between a capacitor's plates is a good electrical insulator, it has a finite resistivity. Thus, an isolated, initially charged capacitor will discharge through its dielectric over some (possibly very long) time period. The "leaky capacitor" model, consisting of a capacitance C in parallel with "leakage" resistance R_L, thus provides an accurate picture of a real-life capacitor. Consider a leaky capacitor in series with a resistor R in the RC circuit shown in Fig. 26–12. Define $Q(t)$ to be the charge on C as a function of time t and assume the capacitor is uncharged when the switch is closed at $t = 0$ [i.e., $Q(t = 0) = 0$].
(a) Apply Kirchhoff's rules and go on to show that $Q(t)$ obeys the differential equation

$$\frac{dQ}{dt} + \left(\frac{1}{R_L C} + \frac{1}{RC}\right)Q = \frac{\mathscr{E}}{R}.$$

(b) Demonstrate that $Q(t) = A[1 - \exp(-t/\tau)]$ is the solution to the above differential equation and determine the constants A and τ in terms of $\mathscr{E}$, R, C, and R_L.
(c) Show that the leaky capacitor charges to the same final voltage as an ideal capacitor would if it were placed in this circuit. An ideal capacitor has $R_L = \infty$.
(d) Show that, if $R_L \gg R$, then $\tau \approx RC\left(1 - \dfrac{R}{R_L}\right)$. [For typical capacitors, $R_L > 100\,\text{M}\Omega$, so $\tau = RC$ is an excellent approximation in real circuits where $R \le 1\,\text{M}\Omega$.]

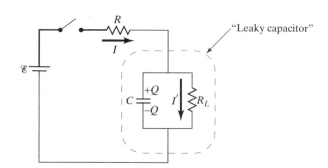

FIGURE 26–12

15. (III) Electronics enthusiasts commonly use the handy "555 timer" to create a simple clock using the charging and discharging of an RC circuit. The 555 timer is an inexpensive integrated circuit that acts as a voltage-controlled switch when connected to the RC network shown in Fig. 26–13. Here is what the 555 does during a single cycle ("tick") of the clock: After an initial start-up period, the 555's switch is open as the capacitor C is charged from 5 V to 10 V by current flowing through the series combination of resistors R_1 and R_2 (Fig. 26–13a). At the moment that the timer's "Sense" input determines C has been charged to 10 V, the 555's internal switch is toggled closed and C is discharged from 10 V to 5 V through resistor R_2 (Fig. 26–13b). When the timer's "Sense" input determines C has been discharged to 5 V, its internal switch is toggled open and a new cycle of the clock begins. The "Sense" input has a large resistance so no current ever flows into or out of it.

(a) Determine T_c, the charge-time portion of the clock cycle (i.e., the time it takes to charge C from 5 V to 10 V via current through R_1 and R_2).

(b) Determine T_d, the discharge-time portion of the clock cycle (i.e., the time it takes to discharge C from 10 V to 5 V via current through R_2).

(c) Show that the clock period $T = (R_1 + 2R_2)C \ln(2)$.

(d) You want to construct a clock that "ticks" every 1.0 ms. If $R_1 = R_2 = 10 \text{ k}\Omega$, what is the required value for C?

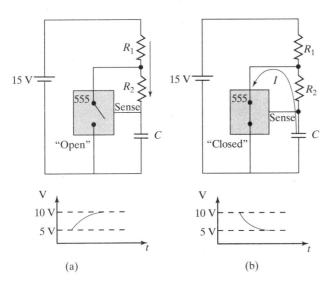

FIGURE 26–13

16. (III) Consider the following model to describe the build-up of a "static electricity" charge on a person while walking across an insulating floor. Assume that the person takes N steps per unit time and that, as a result of rubbing contact between the sole of the person's shoe and the floor, he or she acquires an excess charge of q during each step. Further assume that C and R are the capacitance of the person's body with respect to the ground and the resistance between the body and the ground, respectively.

(a) Let $Q(t)$ be the total excess charge on the person as a function of time t. Using our model, show that Q obeys the differential equation $\dfrac{dQ}{dt} + \dfrac{Q}{CR} = Nq$.

(b) Demonstrate that the solution for this differential equation is $Q(t) = Q_o[1 - \exp(-t/\tau)]$ and determine expressions for the equilibrium total charge Q_o and the time constant for charge build-up τ.

(c) For a shoe-wearing person walking across a vinyl floor, $q \approx 30$ nC, $R \approx 1 \times 10^{11}\,\Omega$, and $C \approx 100$ pF. If $N = 2$ steps per second, determine Q_o (nC) and τ (s). Sketch Q vs. t.

(d) What is the person's potential difference V_o (V) with respect to the floor for times $t \gg \tau$?

17. (III) Suppose that an insulating liquid with volume charge density ρ (C/m³) and flow rate u (m³/s) is flowing into and being collected in an electrically insulated storage tank. Let R be the resistance from the tank to the ground and C be the tank's capacitance with respect to the ground. Further, define the time-dependent quantities $Q(t)$ and $V(t)$ to be the total excess charge within the tank and the tank's potential difference with respect to the ground, respectively.

(a) Prove $V(t) = \rho u R[1 - \exp(-t/RC)]$.

(b) Take $R = 1 \times 10^{10}\,\Omega$ and $\rho = 1$ mC/m³. If, for safety reasons, $V(t)$ is never to exceed 1000 Volts, at what maximum rate (in L/min) can the storage tank be filled?

(c) When the tank is full, the flow of liquid into it is shut off. If $C = 200$ pF, on what time scale does V decay toward zero (i.e., what is the decay's time constant τ)?

18. (III) The capacitor C in the RC circuit shown in Fig. 26–10 is initially uncharged. After the switch is closed at $t = 0$, a time-dependent current I flows out of the battery $\mathscr{E}$ and through the resistor R until C is fully charged at $t = \infty$.

(a) The power delivered by the battery is $\mathscr{E}I$. Determine the total energy that flows out of the battery during the process of fully charging the capacitor.

(b) Show that one-half of the energy that flows out of the battery during the process of fully charging C never makes it to the capacitor, but ends up being converted into thermal energy within the resistor.

19. (III) Consider the circuit shown in Fig. 26–11, where all resistors have the same resistance R. At $t = 0$, with the capacitor C uncharged, the switch is closed.

(a) Define $Q(t)$ to be the magnitude of charge on each capacitor at time t. Apply Kirchhoff's rules, and then go on to show that Q obeys the differential equation

$$\dfrac{dQ}{dt} + \dfrac{2}{3RC}Q = \dfrac{\mathscr{E}}{3R}.$$

(b) Show that $Q(t) = Q_{\max}[1 - e^{-t/\tau}]$ is the solution for the differential equation found in part (a), if the constants $Q_{\max}$ and τ are chosen correctly. What are the correct values for these two constants in terms of $\mathscr{E}$, R, and C?

(c) Show that the time-dependent potential difference across the resistor through which current I_2 flows is given by $V_2(t) = \dfrac{\mathscr{E}}{6}\left[3 - e^{-2t/3RC}\right]$. Sketch V_2 vs. t, indicating significant voltages and times on the axes of your plot.

Section 26–5

20. (II) When measuring the resistance R_S of a highly conductive sample (e.g., a sample near its superconducting transition temperature), one must be aware of the following experimental problem: The resistance R_W of each wire cable connecting the sample to measurement instruments may be on the order of (or even greater than) R_S, thereby introducing an error in its determined value.

(a) In Fig. 26–14a, a digital ohmmeter (whose equivalent circuit is given within the dotted box, where r is the ohmmeter's internal resistance) is connected to a sample via two cables. The voltmeter and ammeter within the ohmmeter measure the potential difference V and current I, respectively; then the meter calculates the ratio $R = \dfrac{V}{I}$ and displays this value as the measured resistance. The internal resistance R_V of the voltmeter is so large in comparison to other resistances in the circuit that it may be assumed infinite. With this assumption, what resistance is actually being displayed by the meter? If the sample resistance happens to equal the cable resistance, (i.e., $R_S = R_W$), and one mistakenly interprets the displayed R as a measure of R_S, what percent error is made?

(b) In Fig. 26–14b, the "four-wire resistance" technique is shown. Here, one instrument creates and measures a current I and a voltmeter measures a potential difference V. The internal resistance R_V of the voltmeter is so large in comparison with R_S that it may be assumed infinite. The resistance R_W of each of the four required connecting cables is shown. If the measured values for I and V are taken from the instruments in this circuit and the ratio $R = \dfrac{V}{I}$ is calculated, what does R equal?

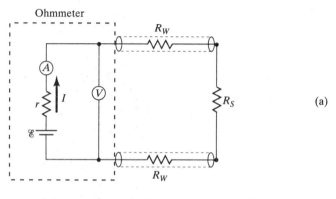

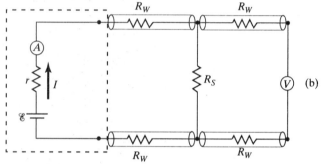

FIGURE 26–14

21. (III) Measurements made on circuits that contain large resistances can be confusing. Consider a circuit powered by a battery $\mathcal{E} = 15.000\text{ V}$ with a 10.00-MΩ resistor in series with an unknown resistor R. As shown in Fig. 26–15, a particular voltmeter reads $V_1 = 366\text{ mV}$ when connected across the 10.00-MΩ resistor and this meter reads $V_2 = 7.317\text{ V}$ when connected across R. Determine the value of R. [*Hint*: Define R_V as the voltmeter's internal resistance.]

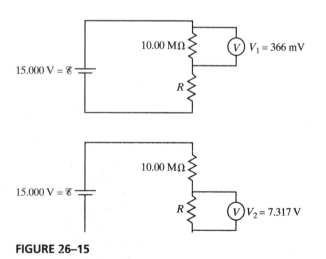

FIGURE 26–15

General Problems

22. (II) Fig. 26–16 shows a portion of an electronic circuit in which a single wire carrying a current I_T connects at a junction with two resistors R_1 and R_2. Let I be the amount of current that flows through R_1 so that $(I_T - I)$ is the current flowing through R_2.
    (a) Write an expression for the power P dissipated in this portion of the circuit in terms of the currents I_T and I.
    (b) Assuming I_T is a fixed quantity, determine I_{min}, the value of I that minimizes P.
    (c) Show that, when $I = I_{min}$, the voltage drop across R_1 is equal to that across R_2. [One familiar case when P is minimum is when R_1 and R_2 are in parallel; i.e., the resistors are also connected at their other ends.]

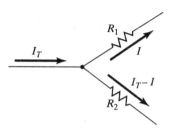

FIGURE 26–16

23. (II) A typical voltmeter has an internal resistance of 10 MΩ and can only measure voltage differences of up to several hundred volts. Fig. 26–17 shows the design of a probe to measure a very large voltage difference V using a voltmeter. If you want the voltmeter to read 50 V when $V = 50\text{ kV}$, what value R should be used in this probe?

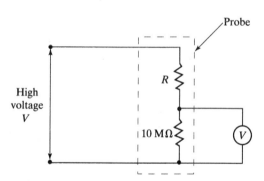

FIGURE 26–17

24. (II) To ensure that a worker, while working with flammable vapors or sensitive electronic components, will not retain static charge nor lose it too quickly, the path from body to ground through the worker's footwear must have a resistance in the range of 1 MΩ to 50 MΩ. To ensure that this condition is met, the testing unit shown in Fig. 26–18 is commonly used. Here, the worker stands on a grounded foot plate while touching a hand to the conducting plate on the wall-mounted test unit. The voltmeter then measures the potential difference V across the "sense" resistor $R_S = 1.0\text{ M}\Omega$, which is then used to find the resistance R from the person's hand to footplate (mostly due to the person's footwear). Assume the voltmeter has an internal resistance $R_V = 10\text{ M}\Omega$.
    (a) Show that $R = \dfrac{\mathcal{E} - V}{V} R_{eq}$, where $R_{eq} = \dfrac{R_S R_V}{R_S + R_V}$.

(b) If $\mathcal{E} = 50\,\text{V}$, what voltages V_{min} and V_{max} will correspond to the minimum and maximum acceptable values for R, respectively.

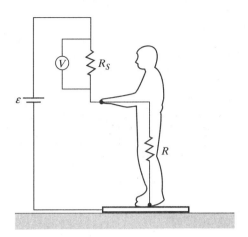

FIGURE 26–18

25. (II) The circuit shown in Fig. 26–19 is a primitive 4-bit digital-to-analog converter (DAC); i.e., it can convert a binary number into a voltage of equivalent decimal value. In this circuit, to make the 2^n place of a binary number equal "1" ("0"), the nth switch is closed (open). For example, 0010_2 is represented by closing switch $n = 1$, while all other switches are open. Show that the voltage V across the $1.0\,\Omega$ resistor for the binary numbers 0001_2, 0010_2, 0100_2, and 1000_2 follows the pattern that you expect for a 4-bit DAC.

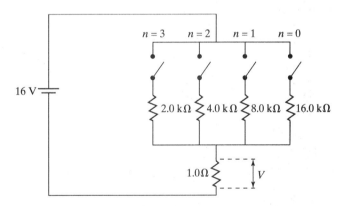

FIGURE 26–19

26. (II) X and Y are two identical cylindrical pieces of material, each possessing the same resistivity ρ_o, length L_o, and cross-sectional area A_o.
(a) A single resistor Z is formed by placing X and Y end to end as shown in Fig. 26–20a. Determine the resistance of Z (in terms of ρ_o, L_o, and A_o) using $R = \rho L/A$.
Compare your result with that obtained from identifying this configuration as a series combination of resistors so that $R_Z = R_X + R_Y$.
(b) A single resistor Z is formed by placing X and Y side by side as shown in Fig. 26–20b. Determine the resistance of Z (in terms of ρ_o, L_o, and A_o) using $R = \rho L/A$. Compare your result with that obtained from identifying this configuration as a parallel combination of resistors so that $R_Z^{-1} = R_X^{-1} + R_Y^{-1}$.

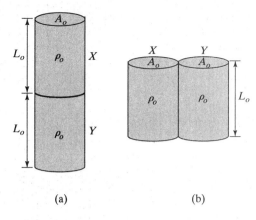

FIGURE 26–20

27. (III) When a typical lightning bolt strikes a point on the earth, it delivers a negative charge of magnitude $Q = 1.0\,\text{C}$ over a time interval $\Delta t = 40\,\mu\text{s}$ (Fig. 26–21). Because the earth is a fairly good electrical conductor, over the span Δt, a current $I = Q/\Delta t$ flows within the earth, bringing a compensating amount of positive charge to the point of impact.
(a) Assume the earth's surface is flat and the radial distance r is measured from the lightning's point of impact. Noting that the total current I must be the same at every r, show that the earth's resistance R_E between two radii r_1 and r_2 is given by

$$R_E = \frac{\rho}{2\pi}\left[\frac{1}{r_1} - \frac{1}{r_2}\right]$$

where ρ is the resistivity of the earth.
[*Hint*: At a given r, find the potential difference dV across a half-spherical shell of radial thickness dr.]
(b) If a person has the misfortune of standing near the point of a typical lightning-bolt strike with one foot 5.0 m and the other foot 5.5 m from the point of impact, the body's resistance R_P (from one foot to the other) is in parallel with R_E. If the total current due to lightning is 25 kA, determine the current I_P that flows through the person, if ρ for the earth's top soil is $10\,\Omega\cdot\text{m}$ and $R_P = 50\,\text{k}\Omega$ (due to dry feet). Is this level of current dangerous? How about if the person has wet feet so that $R_P = 5.0\,\text{k}\Omega$?

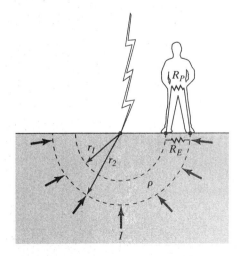

FIGURE 26–21

28. (III) Consider the circuit shown in Fig. 26–22 in which a thermistor is placed in series with a fixed resistance R and voltage source $\mathcal{E}$. The resistance R_T of the semiconducting thermistor decreases with increasing temperature, enabling this circuit to detect small changes in room temperature.

(a) Assume the thermistor's temperature is changed by a small amount ΔT so that its resistance changes by the small amount ΔR_T. Show that the voltage difference across the thermistor then changes by $\Delta V_T \approx \dfrac{\mathcal{E} R\, R_T}{(R + R_T)^2} \dfrac{\Delta R_T}{R_T}$.

(b) Prove that ΔV_T is maximized if $R = R_T$.

(c) At 20°C, the resistance of a particular thermistor is 10 kΩ and this resistance decreases by 5% per 1 C° increase. If this thermistor is used in the given circuit with $R = 10\,\text{k}\Omega$ and $\mathcal{E} = 15\,\text{V}$, determine ΔV_T when room temperature rises from 20°C to 21°C.

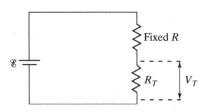

FIGURE 26–22

29. (III) Suppose that an insulating liquid with volume charge density ρ (C/m³) and flow rate u (m³/s) is flowing into and being collected in an electrically insulated storage tank. Let R be the resistance from the tank to the ground and C be the tank's capacitance with respect to the ground. Further, define Q and V to be the total excess charge within the tank and the tank's potential difference with respect to the ground, respectively.

(a) While filling the tank, the inflow of liquid contributes to an increase in Q, while the flow of electrical current from the tank to the ground tends to decrease Q. Explain why dQ/dt equals $+\rho u$ and $-Q/RC$ for the former and latter process, respectively. Also explain qualitatively what characteristic of the electrical current's contribution impedes the accumulated charge from increasing to larger values as time goes on.

(b) In equilibrium, the rate of increase in Q is balanced by an equal rate of decrease and thus, a (constant) equilibrium charge Q_o is maintained. Determine an expression for Q_o as well as the tank's potential difference V_o when in equilibrium.

(c) Take $R = 1 \times 10^{10}\,\Omega$ and $\rho = 1\,\text{mC/m}^3$. If, for safety reasons, V_o is not to exceed 1000 Volts, at what maximum rate (in L/min) can the storage tank be filled?

30. (III) Thevenin's theorem (in part) states that any complicated combination of voltage sources and resistors can be replaced by an equivalent circuit consisting of simply an emf $\mathcal{E}_T$ in series with a resistor R_T. Consider the circuit within the dotted box in Fig. 26–23a, which consists of three equal resistors R and a voltage source $\mathcal{E}$, along with its Thevenin equivalent circuit (Fig. 26–23b). Points A and B are considered the output of each circuit (i.e., the points across which a "load" resistor can be attached). Demonstrate the veracity of Thevenin's theorem as follows:

(a) Connect an infinite "load" resistor $R_L = \infty$ between A and B in each circuit. Determine the value of $\mathcal{E}_T$ (in terms of $\mathcal{E}$) so that the potential difference between these points is the same in each circuit.

(b) Connect a zero-resistance "load" resistor $R_L = 0$ between A and B in each circuit. Determine the value of R_T (in terms of $\mathcal{E}$ and R) so that the current flowing between these points is the same in each circuit.

(c) Now that you know the values of $\mathcal{E}_T$ and R_T, make $R_L = 2R$ for each circuit. Analyze each circuit and determine the current I through R_L in each case. Are the circuits "equivalent"?

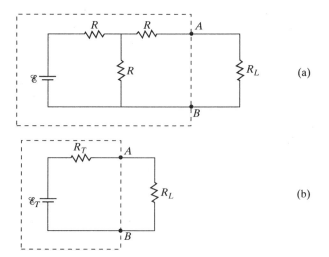

FIGURE 26–23

Magnetism

Section 27–3

1. (II) A current-carrying circular loop of wire (radius r, current I) is only partially immersed in a magnetic field of magnitude B directed out of the page as shown in Fig. 27–1. Determine the net force on the loop due to the field in terms of θ_o.

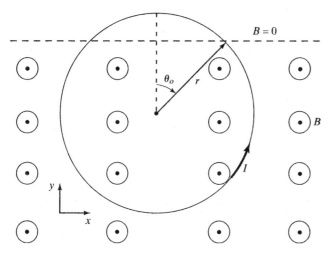

FIGURE 27–1

2. (II) A 2.0-m long wire carries a current of 10 A and is immersed within a uniform magnetic field **B**. When this wire lies along the x-axis with its current flowing in the $+x$-direction, a magnetic force $\mathbf{F} = (-2.5\,\mathbf{j})$ N acts on the wire, and when it lies on the y-axis with its current flowing in the $+y$-direction, the force is $\mathbf{F} = (2.5\,\mathbf{i} - 5.0\,\mathbf{k})$ N. Find **B**.

3. (II) A vertical conducting square loop of side length $L = 10$ cm and weight w_o rests on two pegs (Fig. 27–2). The loop is partially immersed in a magnetic field of magnitude $B = 0.20$ T, which is directed perpendicular to the plane of the loop. It is found that when a current $I_o = 0.20$ A flows through the loop, it levitates slightly off of the pegs where it then remains motionless. A weight w is next attached to the bottom segment of the loop and the current flowing through the loop is adjusted until its value is I, at which point the loop again levitates slightly off of the pegs and there remains motionless. Show that this device is a linear (weight) scale in that an object's weight can be determined from a measured current via a relation of the form $w = aI + b$. Determine the values of the constants a (N/A) and b (N).

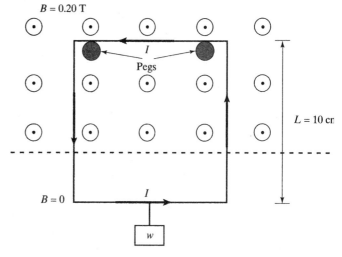

FIGURE 27–2

Section 27–4

4. (II) An electron moving in the magnetic field $\mathbf{B} = (0.25\text{ T})\mathbf{i}$ experiences a force given by $\mathbf{F} = (-4.0\,\mathbf{j} + 8.0\,\mathbf{k}) \times 10^{-15}$ N. Write the electron's velocity as $\mathbf{v} = v_x\mathbf{i} + v_y\mathbf{j} + v_z\mathbf{k}$.
(a) Which component of velocity cannot be determined from the given information? Explain why.
(b) Find the two components of velocity (m/s) that can be determined from the given information.
(c) Explicitly calculate $\mathbf{v} \cdot \mathbf{F}$. Explain why your result equals zero.

5. (II) A particle with charge q moves along a path immersed in a magnetic field **B**. Given that the particle's infinitesimal displacements along this path can be written as $d\mathbf{l} = dx\,\mathbf{i} + dy\,\mathbf{j} + dz\,\mathbf{k}$, prove that the magnetic field does no work on the particle. In light of the work-energy principle, what does this result imply about the motion of a charged particle under the influence of a magnetic field?

6. (II) Magnetic fields are used in high-energy physics experiments to determine the charges and momenta of elementary particles. Consider a particle with charge q and momentum of magnitude p aligned with the x-axis, which enters a region in which a uniform magnetic field $\mathbf{B} = B_o\mathbf{k}$ exists over a spatial extent L in the x direction. As shown in Fig. 27–3, the particle is deflected by a distance d in the $+y$ direction as it traverses the field. Determine (a) whether q is positive or negative, and (b) p.

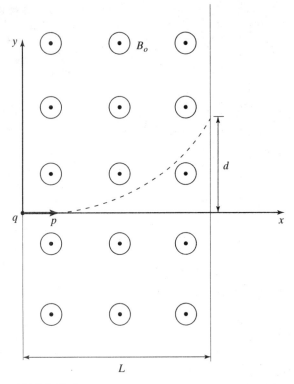

FIGURE 27–3

8. (III) A particle with charge $+q$ and mass m travels in a uniform magnetic field $\mathbf{B} = B_o\mathbf{k}$. At time $t = 0$, the particle's speed is v_o and its velocity vector lies in the x-y plane directed at an angle of 30° with respect to the y-axis as shown in Fig. 27–5. At a later time $t = t_\alpha$, the particle will cross the x-axis at $x = \alpha$. In terms of q, m, v_o, and B_o, determine (a) α, and (b) t_α.

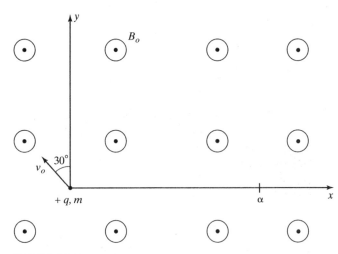

FIGURE 27–5

7. (III) A group of particles, each of mass m and charge $+q$, has a range of speeds v. Consider the following design for a velocity selector, which selects only those particles from the group with speed v_o. In a region of space where uniform magnetic field B (into the page) exists, the particles are injected at the origin with their velocity vectors in the x-y plane, oriented at an angle θ with respect to the y-axis. A metal plate with a small hole drilled in it is placed perpendicular to the y-axis with its hole centered on the point $(x,y) = (0,y_o)$ (Fig. 27–4). For all of the particles entering the field region, only those with speed v_o will pass through the hole. Determine v_o. The influence of gravity can be assumed to be negligible.

Section 27–5

9. (II) You are given a fixed length L of wire with a current I flowing through it. If you fashion this wire into a single, circular loop, let μ_1 be the loop's resulting magnetic dipole moment. If, instead, you fashion the wire into a coil with N loops, find the resulting dipole moment μ_N in terms of μ_1.

10. (II) (a) In Bohr's model for the hydrogen atom, an electron is held in a circular orbit of radius r about a stationary proton by the attractive electric force between these two particles. Based on this model, the electron's orbital motion produces both a magnetic dipole moment and an angular momentum of magnitude μ and L, respectively. Show that these two quantities are related by $\mu = \dfrac{e}{2m}L$.

(b) Bohr's model goes on to postulate that the electron, when orbiting in the smallest allowed orbit (called the "ground state"), has angular momentum $L = \dfrac{h}{2\pi}$, where h is Planck's constant (one of the fundamental constants of nature; see the front cover of the text). Use this idea from Bohr's model to predict the hydrogen atom's magnetic moment $(\text{A}\cdot\text{m}^2)$ when in the ground state. [This moment is called the Bohr magneton μ_B. The magnetic moments of real atoms are typically on the order of μ_B.]

11. (III) Model the needle of a compass as a uniform rod of mass $m = 0.5\,\text{g}$ and length $l = 2\,\text{cm}$, which is free to rotate about an axis normal to its length and through its center. This rod is magnetized, meaning that it has a magnetic dipole moment of magnitude μ directed along its length toward its "North-Pole" indicator tip. Assume that the compass needle lies in a horizontal plane slightly above Earth's surface with its rotational axis perpendicular to Earth's surface and that it is initially in its equilibrium position $\theta = 0$ (i.e., aligned with the horizontal

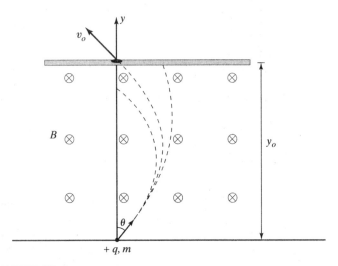

FIGURE 27–4

component of Earth's magnetic field B_x). When the needle is displaced by a small angle θ from equilibrium and released, it performs simple harmonic motion about $\theta = 0$ with a period T.

(a) Prove that $T = 2\pi\sqrt{\dfrac{ml^2}{12\mu B_x}}$.

(b) If, at a certain location on Earth's surface where $B_x = 3 \times 10^{-5}$ T, the compass needle is found to oscillate with period $T = 1$ s, estimate the value of μ.

(c) What current should flow in a 10-turn loop of radius 4 mm in order to produce the same magnetic dipole moment as this needle?

12. (III) A disk-shaped piece of nonconducting material (mass M, radius R) has a uniform mass distribution. A charge $+Q$ is uniformly spread over its top surface and then the disk is spun at constant angular velocity ω about the axis that passes through its center (and is perpendicular to the disk's surface). Determine the relation between the resulting magnetic dipole moment μ for this spinning disk and its angular momentum $\mathbf{L}$.

13. (III) A thin hoop of radius R and mass m has charge q uniformly distributed over it and spins about its central axis with angular velocity ω.

(a) Show that the hoop's magnetic dipole moment μ and angular momentum $\mathbf{L}$ are related by $\boldsymbol{\mu} = \dfrac{q}{2m}\mathbf{L}$.

(b) This spinning, charged hoop is placed in a magnetic field $\mathbf{B}$ with its axis of rotation oriented at an angle ϕ to the direction of the field. Show that a torque given by $\boldsymbol{\tau} = \dfrac{q}{2m}\mathbf{L} \times \mathbf{B}$ acts on the hoop.

(c) Note that $\boldsymbol{\tau}$ is perpendicular to $\mathbf{L}$. Thus, the hoop will precess about the direction of the magnetic field (see Section 11–8 in the text). Prove that the precession (angular) frequency $\Omega = \dfrac{qB}{2m}$. This frequency is called the Larmor frequency.

(d) Take the spinning hoop as a simple model for an orbiting electron in a hydrogen atom. If this atom had its magnetic moment inclined relative to a magnetic field of magnitude $B = 1.0$ T, determine its resulting precession frequency f (Hz).

14. (III) The angular momentum $\mathbf{L}$ of an atomic-level object (e.g., nucleus or electron) produces a magnetic dipole moment $\boldsymbol{\mu}$. These two quantities are related by $\boldsymbol{\mu} = \gamma\mathbf{L}$, where γ is defined as the object's "gyromagnetic ratio." This relation is exploited in the Nuclear Magnetic Resonance (NMR) experimental technique and the medical-related Magnetic Resonance Imaging (MRI) method where the magnetic field of a short pulse of radio waves is used to rotate ("flip") the magnetic moment of protons within a sample.

(a) Assume that, with an object's magnetic moment $\boldsymbol{\mu}$ oriented at angle ϕ with respect to the x-axis, a uniform magnetic field of magnitude B along the x-axis is applied (i.e., $\mathbf{B} = B\mathbf{i}$). Show that a torque given by $\boldsymbol{\tau} = \gamma\mathbf{L} \times \mathbf{B}$ acts on the object.

(b) Note that $\boldsymbol{\tau}$ is perpendicular to $\mathbf{L}$, so the object will precess about the direction of the magnetic field (see Section 11–8 in the text). Show that the object's precession frequency $f = \gamma B/2\pi$. This frequency is called the Larmor frequency (in Hz).

(c) In NMR and MRI, a magnetic field (due to a radio wave) is commonly turned on just long enough to cause a proton's magnetic moment to precess exactly 180° about the x-axis. This process is called a "spin flip" and the short-duration magnetic field is termed a "π-pulse." If $B = 0.15$ T, what is the duration (μs) of the π-pulse for a proton ($\gamma = 43$ MHz/T)?

Section 27–8

15. (II) A Hall probe, consisting of a rectangular slab of current-carrying material, is calibrated by placing it in a known magnetic field of magnitude 0.10 T. When the field is oriented normal to the slab's rectangular face, a Hall emf of 12 mV is measured across the slab's width. If the probe is then placed in a magnetic field of unknown magnitude B, and a Hall emf of 63 mV is measured, determine B assuming that the angle θ with which the unknown field is oriented at relative to the plane of the slab's rectangular face is (a) $\theta = 90°$, or (b) $\theta = 60°$.

16. (II) Let's investigate the best design for a Hall probe, a device used to measure magnetic field strengths. Consider a rectangular slab of material (free-electron density n) with width l and thickness t, carrying a current I along its length L. The slab is immersed in a magnetic field of magnitude B oriented perpendicular to its rectangular face (of area $l\,L$), so that a Hall emf $\mathscr{E}_H$ is produced across its width. The probe's magnetic sensitivity, defined as $K_H \equiv \dfrac{\mathscr{E}_H}{IB}$, indicates the magnitude of Hall emf achieved for a given applied magnetic field and current flow. A slab with a large K_T value is a good candidate for use as a Hall probe.

(a) Show that $K_H = 1/ent$. Thus, a good Hall probe has small values for both n and t.

(b) As possible candidates for the material used in a Hall probe, consider a typical metal and a (doped) semiconductor with n approximately equal to 1×10^{29} m^{-3} and 3×10^{22} m^{-3}, respectively. Assume that a manufacturing process allows production of a semiconductor slab with the thickness 0.15 mm. How thin (nm) would a metal slab have to be to yield a K_H value equal to that of the semiconductor slab? Compare this metal slab thickness with the 0.3-nm size of a typical metal atom.

(c) For the reasons examined above, commercial Hall probes are made of thin semiconductors. For the typical semiconductor slab described in part (b), what is the expected value for $\mathscr{E}_H$ when $I = 100$ mA and $B = 0.1$ T?

17. (III) The Hall effect is commonly used to determine the density n of free electrons in a semiconductor sample. A rectangular slab of (doped) semiconductor material, whose cross-sectional area is defined by its width l and thickness t, has a constant current I flowing along its length L. The slab is immersed in a magnetic field oriented perpendicular to its rectangular face (of area $l\,L$). The field's magnitude B is varied and the associated Hall emfs $\mathscr{E}_H$ are measured.

(a) Show that $\mathscr{E}_H$ and B are related by $\mathscr{E}_H = \dfrac{I}{ent}B$. Explain why a plot of $\mathscr{E}_H$ vs. B should yield a straight line. What are the theoretically expected slope and y-intercept of this line?

(b) The described experiment is carried out on a 0.50-mm thick semiconductor slab with $I = 100$ mA yielding the data shown in Table 27-1 (see the next page). Graph $\mathscr{E}_H$ vs. B and show that you do indeed obtain a straight-line plot. Determine the slope of the line and use it to obtain n (m^{-3}) for the semiconductor sample. Is the y-intercept of your plot the expected value?

TABLE 27-1

B (T)	0.10	0.20	0.30	0.40	0.50
$\mathcal{E}_H$ (mV)	2.3	4.5	6.8	9.1	11.4

Section 27-9

18. (III) (a) An ion with charge q and mass m moves in a circular path of radius r within a perpendicular magnetic field B. Find the ion's kinetic energy K in terms of $q, m, r,$ and B.
(b) Two ions, moving within a perpendicular magnetic field B, have the same charge and the same kinetic energy. However, the ions have slightly different masses m and $m + \Delta m$, and so move in paths of slightly different radii r and $r + \Delta r$, respectively. Show that $\dfrac{\Delta r}{r} = \dfrac{1}{2}\dfrac{\Delta m}{m}$.
(c) For the situation described in part (b), find the fractional difference in radii $\Delta r/r$ if the two ions are the carbon isotopes ^{12}C and ^{14}C. How about if the ions are the uranium isotopes ^{235}U and ^{238}U?

19. (III) Consider the use of magnetic fields for separating ("refining") two isotopes contained in a heavy-element (near bottom of periodic table) sample. Assume that the atomic isotopes α and β have masses $m_\alpha = m$ and $m_\beta = m + \Delta m$, respectively, where $m > 200$ u and $\Delta m \approx 1-5$ u; also assume that these isotopes are mixed together in a sample (e.g., ^{235}U and ^{238}U in a chunk of uranium). Imagine the following experimental procedure to separate the isotopes: First, the sample is vaporized, then all of the atoms are singly ionized and accelerated from rest by a potential difference V. Next, the moving ionized atoms enter from the same location into a uniform perpendicular magnetic field of magnitude B where the two isotopes travel on circles of differing radii r_α and r_β.
(a) Show that, since $m \gg \Delta m$, the difference in radii $\Delta r = r_\beta - r_\alpha$ is $\Delta r \approx \sqrt{\dfrac{mV}{2eB^2}}\dfrac{\Delta m}{m}$.
(b) Assume that to be effective in separating the isotopes (e.g., making them pass through separated small openings) requires that $\Delta r \geq 1$ mm and that the magnetic field strength is $B = 0.5$ T. Taking $\Delta m/m \approx 0.01$ and $m = 220$ u determine the minimum value needed for V.

General Problems

20. (II) (a) Show that, when a particle with charge q and momentum **p** moves within a magnetic field **B** oriented perpendicular to its velocity, $p = qBr$ where r is the radius of curvature of the particle's path.
(b) A proton, which has been accelerated from rest by a potential difference of 5.0 MV, collides with an at-rest nucleus, which is one of the isotopes of hydrogen. The proton and nucleus are immersed in a magnetic field of magnitude $B = 1.5$ T and, after the collision, are observed to travel in the plane perpendicular to the field on curved paths with radii of curvature 7.2 cm and 28.7 cm, respectively. Assuming the collision is elastic, determine whether the nucleus is ^{1}H (hydrogen), ^{2}H (deuterium), or ^{3}H (tritium). [*Hint:* $K = p^2/2m$.]

21. (II) A thin, flexible wire hangs in a uniform magnetic field of magnitude B with a weight w attached to its lower end. When the wire carries a current I, it remains stationary in the shape of a circular arc of radius r as shown in Fig. 27–6, where the

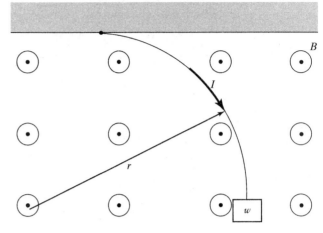

FIGURE 27–6

very bottom portion of the wire is aligned with the vertical. Assuming the wire's weight is negligible, determine r. [*Hint:* The tension force has the same magnitude at opposite ends of the massless wire.]

22. (II) A uniform conducting rod of length L and mass m sits atop a fulcrum, which is placed a distance $L/4$ from the rod's left end and is immersed in a uniform magnetic field of magnitude B directed into the page (Fig. 27–7). If an object whose mass M is ten times greater than the rod's mass is hung from the rod's left end, what current (direction and magnitude) should flow through the rod in order for it to be "balanced" (i.e., be at rest horizontally) on the fulcrum?

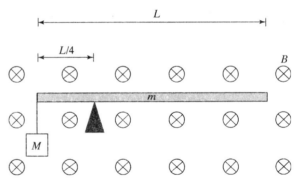

FIGURE 27–7

23. (II) Here is the design for a simple device to measure the magnitude B of a magnetic field. A conducting rod (length $L = 1.0$ m, mass $m = 150$ g) hangs from a friction-free pivot and is oriented so that its axis of rotation is aligned with the direction of the magnetic field to be measured. Through thin, flexible wires (which exert negligible force on the rod) connected to a voltage source, a current $I = 10$ A is produced in the rod, and it deflects to an angle θ with respect to the vertical, as shown in Fig. 27–8, where it remains at rest.
(a) Is the current flowing downward from or upward toward the pivot in Fig. 27–8?
(b) If $\theta = 10°$, determine B.
(c) What is the largest magnetic field magnitude that can be measured using this device?

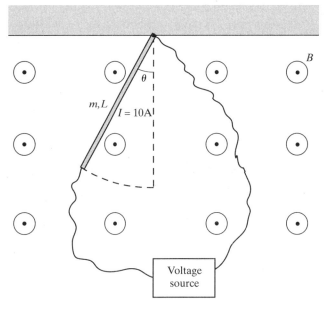

FIGURE 27–8

24. (II) Very hot deuterium (hydrogen-2 with $q = +e, m = 2.0\,\text{u}$) nuclei are confined in a fusion reactor using a "magnetic bottle" as shown in Fig. 27–9a. Near its midpoint M, the bottle's magnetic field can be modeled as uniform and directed completely along the axis so that $\mathbf{B}_M = B_M \mathbf{i}$ (Fig. 27–9b, as viewed from point A). At its neck N, however, the field also includes a radial component so $\mathbf{B}_N = B_M \mathbf{i} - B_R \hat{\mathbf{r}}$ (Fig. 27–9c, as viewed from point A).

(a) Two deuterium nuclei, separated by distances greater than about $d = 3 \times 10^{-15}\,\text{m}$, will repel each other due to the electrical repulsion of the proton in each nucleus. However, if two such deuterium nuclei come within a separation of d, the strong nuclear attractive force will be able to overpower the electrical repulsive force between the protons and the two nuclei will "fuse" into a single, larger (helium) nucleus with the release of a large amount of energy. Assume two deuterium nuclei, infinitely far apart, are each given a speed v, which causes them to move toward each other. Find the value of v so that the approaching deuterium nuclei are not stopped by electrical repulsion until their separation equals d (at this point, the strong nuclear force will take over and fuse the nuclei).

(b) In a fusion reactor, deuterium nuclei of speed v are confined at M by $\mathbf{B}_M$. If $B_M = 0.5\,\text{T}$, determine the radius r of the circular path of the nuclei. The size of this radius informs designers about the required dimensions of the reactor chamber.

(c) If the circulating nuclei wander over to N, they experience the field $\mathbf{B}_N = B_M \mathbf{i} - B_R \hat{\mathbf{r}}$. Explain why this field will push the nuclei back toward M. This effect is called the "magnetic mirror."

25. (II) A Hall probe, along with a small magnet, can be used as a "proximity sensor."

(a) A Hall probe consists of a rectangular slab of semiconductor material (width l, thickness t, and free-electron density n) carrying a current I along its length L. When this slab is immersed in a magnetic field of magnitude B oriented perpendicular to its rectangular face (of area $l\,L$), show that a Hall emf $\mathscr{E}_H = C B$ is produced across its width, where the constant $C = \dfrac{I}{ent}$. If, for a certain Hall probe, $I = 100\,\text{mA}$, $t = 0.15\,\text{mm}$, and $n = 1.6 \times 10^{22}\,\text{m}^{-3}$, determine $C\,(\text{mV/T})$.

(b) Assume that, along its axis, the magnetic field produced by a small cylindrically shaped magnet is known to be directed parallel to the axis with a magnitude that varies with distance as $B(x) = (1.0 \times 10^{-9}\,\text{T} \cdot \text{m}^3)/x^3$. Here, x is the distance along the axis from the cylinder's center (Fig. 27–10a). As shown in Fig. 27–10b, let the axis of this magnet be aligned with the center of the face of the Hall probe described in part (a) and assume the magnet's field is approximately constant over the probe's width. Now imagine that, starting at a large separation, the magnet is moved toward the Hall probe. If the smallest value of $\mathscr{E}_H$ that can be detected is $0.1\,\text{mV}$, at what threshold separation $x_T\,(\text{cm})$ will the probe begin to sense the presence of the magnet? When $x = 5\,\text{mm}$, what is $\mathscr{E}_H$? When determining the magnetic field experienced by the Hall probe, assume its small thickness is negligible in comparison with x.

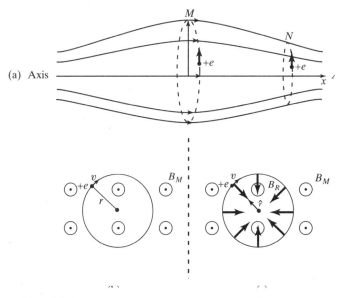

FIGURE 27–9

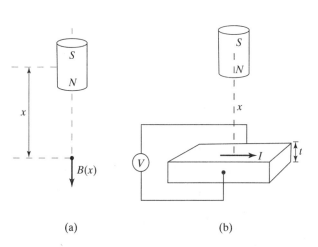

FIGURE 27–10

26. (III) In this problem, we explore the physics underlying the operation of the cyclotron, a device commonly used by hospital staff to accelerate protons (that are then used in the production of radioactive isotopes).
(a) A particle with mass m, charge q, and kinetic energy K enters a region where it travels perpendicularly to a magnetic field of magnitude B. Show that the particle travels on the half-circle of radius $r = \sqrt{2mK}/qB$ for a time $t = \pi m/qB$ before exiting this region.
(b) Assume that a left (L) and a right (R) region of magnetic field exist adjacent to each other, separated by a small "magnetic-field free" gap (Fig. 27–11). Assume further that when a charged particle traverses this gap, going either left or right, its kinetic energy is always increased by a fixed amount E_o (due to an appropriately applied electric field). Consider the following scenario: The particle described in part (a), initially at rest, traverses the gap going left and then travels the small-radius half-circle in L corresponding to $K = 1E_o$. After making this U-turn, it again traverses the gap going right and then travels the larger-radius half-circle in R corresponding to $K = 2E_o$. Next, it traverses the gap going left, entering L with $K = 3E_o$, and so on. After a time T_R, the particle will have kinetic energy K_R and will be traveling in one of the regions on a half-circle of radius R. If the particle is a proton, $B = 1.00\,\text{T}$, and $R = 40.0\,\text{cm}$, what is K_R in MeV? If $E_o = 10.0\,\text{keV}$, how many times has the proton passed through the gap? What is T_R in μs?

27. (III) This problem explores the physics relevant to the operation of a cyclotron, a device commonly used by hospital staff to accelerate protons (that are then used in the production of radioactive isotopes).
(a) A particle with mass m, charge q, and kinetic energy K enters a region where it travels perpendicular to a magnetic field of magnitude B. Show that the particle travels on the half-circle of radius $r = \sqrt{2mK}/qB$ until exiting this region.
(b) Assume that a left (L) and a right (R) region of magnetic field exist adjacent to each other, separated by a small "magnetic-field free" gap (Fig. 27–1). Assume further that when a charged particle traverses this gap, going either left or right, its kinetic energy is always increased by a fixed amount E_o (due to an appropriately applied electric field). Consider the following scenario: The particle described in part (a), initially at rest, traverses the gap going left and then travels on the small-radius half-circle in L corresponding to $K = 1E_o$. After making this U-turn, it again traverses the gap going right and then travels the larger-radius half-circle in R corresponding to $K = 2E_o$. Next, it traverses the gap going left, entering L with $K = 3E_o$, and so on. After traversing the gap N times, the particle will have kinetic energy $K_R = NE_o$ and will be traveling in one of the regions on a half-circle of radius R. If the particle is a proton, $B = 1.00\,\text{T}$, $E_o = 10.0\,\text{keV}$, and $R = 40.0\,\text{cm}$, estimate the total distance the proton has traveled since starting from rest (assuming the gap has negligible spatial extent). [*Hint*: Use a computer (or a calculator) to evaluate the summation numerically.]

28. (III) In a region of space, crossed uniform electric and magnetic fields of magnitudes E and B, respectively, coexist. That is, in this region, $\mathbf{E} = E\mathbf{j}$ and $\mathbf{B} = B\mathbf{k}$. Consider the motion of a particle with mass m and charge q in this region, where $\mathbf{v}(t)$ is defined to be its velocity as a function of time t. Assume the particle is at rest at $t = 0$.
(a) Define a velocity $\mathbf{v}_C$ such that $\mathbf{E} + \mathbf{v}_C \times \mathbf{B} = 0$. Show that the time-independent vector $\mathbf{v}_C = \dfrac{E}{B}\mathbf{i}$ satisfies this definition.
(b) For the reference frame in which the particle's velocity is $\mathbf{v}$, the force on the particle is $\mathbf{F} = q(\mathbf{E} + \mathbf{v} \times \mathbf{B})$. Define a new reference frame that moves with velocity $+\mathbf{v}_C$ relative to this original frame and let the velocity of the particle in this new reference frame be $\mathbf{v}'$. Show that from the vantage point of this new reference frame, the particle moves in a circle of radius $r = mE/qB^2$. [Thus, in the original reference frame, the center of the circular motion moves with constant speed v_C producing a curved trajectory called a cycloid.]

29. (III) A particle with mass m and charge q is released from rest at a location near Earth's surface where Earth's gravitational and magnetic fields are uniform and given by $\mathbf{g} = g\mathbf{j}$ and $\mathbf{B} = B\mathbf{k}$. Let $\mathbf{v}(t)$ be defined to be the particle's velocity with respect to Earth as a function of time t with $\mathbf{v} = 0$ at $t = 0$.
(a) Define a velocity $\mathbf{v}_C$ such that $m\mathbf{g} + q\mathbf{v}_C \times \mathbf{B} = 0$. Show that the time-independent vector $\mathbf{v}_C = \dfrac{mg}{qB}\mathbf{i}$ satisfies this definition.
(b) In Earth's reference frame, the force on the particle is $\mathbf{F} = m\mathbf{g} + q\mathbf{v} \times \mathbf{B}$. Define a new reference frame that moves with velocity $+\mathbf{v}_C$ relative to Earth and let the velocity of the particle in this new reference frame be $\mathbf{v}'$. Show that from the vantage point of this new reference frame, the particle moves in a circle of radius $r = m^2g/q^2B^2$.
(c) Assume the particle is a proton and take $g = 9.8\,\text{m/s}^2$ and $B = 5.0 \times 10^{-5}\,\text{T}$. Describe the (surprising) motion for this proton relative to Earth when it is released from rest.

30. (III) Suppose that within a certain region of space uniform electric and magnetic fields coexist and that both of these fields point in the x-direction (i.e., $\mathbf{E} = E\mathbf{i}$ and $\mathbf{B} = B\mathbf{i}$, where E and B are constants). Suppose further that, at time $t = 0$, a particle with mass m and charge q is injected into this region at the position $x = 0$, $y = \beta$, $z = 0$ with initial velocity $\mathbf{v}_o = v_o\mathbf{k}$. Write the particle's position and velocity at time t as $\mathbf{r}(t) = x\mathbf{i} + y\mathbf{j} + z\mathbf{k}$ and $\mathbf{v}(t) = v_x\mathbf{i} + v_y\mathbf{j} + v_z\mathbf{k}$.

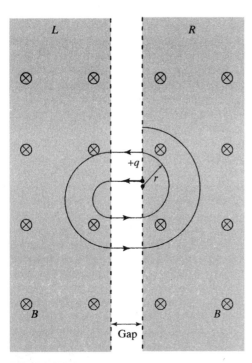

FIGURE 27–11

(a) Apply Newton's second law and show that the particle's motion obeys the three differential equations $\frac{d^2x}{dt^2} = \frac{qE}{m}$, $\frac{d^2y}{dt^2} = \frac{qB}{m}\frac{dz}{dt}$, and $\frac{d^2z}{dt^2} = -\frac{qB}{m}\frac{dy}{dt}$.

(b) Show that the solution to these differential equations (consistent with the initial conditions given at $t = 0$) is $x = \alpha t^2$, $y = \beta\cos(\omega t)$, and $z = \beta\sin(\omega t)$, where α, β, and ω are constants. Determine α, β, and ω in terms of q, m, E, B, and v_o. This solution describes a helical path of radius β whose axis is along the x-axis.

(c) The pitch $p(t)$ of a helix is defined to be the axial distance between points separated by exactly one helical turn (i.e., points at time t and $t+T$, where T is the helical period). Show that p for the helix given in part (b) is a linear function of time. Determine the slope and the y-intercept of this function in terms of q, m, E, B, and v_o.

CHAPTER 28

Sources of Magnetic Field

Sections 28–1 and 28–2

1. (II) Can wires be made to adhere via the magnetic force? Imagine two identical wires of radius r and mass density ρ that are touching while horizontally oriented. The top wire is held in place by rigid supports at both its ends, but the bottom wire can only be supported by its magnetic interaction with the top wire. Now assume that a current I flows in the same direction in each wire and that the wires both have a very thin layer of electrically insulating material on their surfaces so that these currents can flow independently.
(a) Show that the bottom wire will not fall if I equals (or exceeds) the threshold value $I_T = 2\pi\sqrt{\dfrac{\rho g r^3}{\mu_0}}$.
(b) For 14-gauge (diameter 1.63 mm) copper wire, determine I_T. [For this type of wire, even in a well ventilated area, currents above 30 A will cause rapid heating.]
(c) By what factor will I_T decrease if 14-gauge aluminum (rather than copper) wire is used?

2. (II) Two long, insulated wires carrying the same current I cross each other perpendicularly as shown in Figure 28–1. Determine the points in this plane where the net magnetic field is zero (assuming that in the absence of current flow within the wires, there is no external magnetic field in the plane).

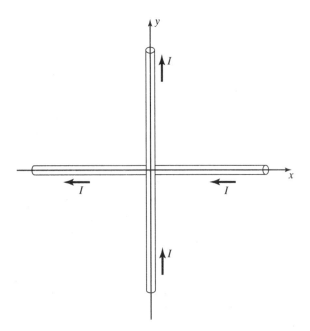

FIGURE 28–1

3. (III) A triangular loop of side length L carries a current I'. If one side of this loop is placed parallel to and a distance d away from a straight wire carrying a current I as shown in Fig. 28–2 determine the force on the loop.

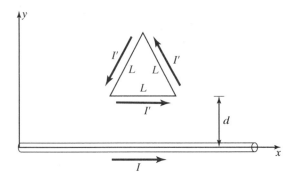

FIGURE 28–2

Sections 28–4 and 28–5

4. (II) A long, hollow metal cylinder of length L with outer and inner radius r_a and r_b, respectively, carries a current that circulates about the cylinder's central axis and has a current density of constant magnitude j at all radii (Fig. 28–3). In its midsection region, find the magnetic field created along the axis of this cylinder. [Hint: The field outside the cylinder is much smaller than the field along its axis.]

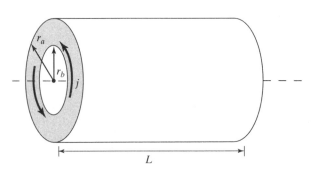

FIGURE 28–3

5. (II) (a) First, consider an isolated very large, flat conducting sheet of thickness t that carries a uniform current density $\mathbf{j} = j\,\mathbf{i}$ throughout its cross-section. Determine the magnetic field (magnitude and direction) created by this sheet at points P and Q, which are perpendicular distances $y > t/2$ above

and below the center of the sheet, respectively. Assume the sheet can be approximated as infinite in extent.

(b) Two identical very large, flat conducting sheets, each of thickness t, are parallel with their centers separated by distance $d > t$. Each sheet carries uniform current density of magnitude j, but these currents are oppositely directed as shown in Fig. 28–4. Use the result of part (a) and the principle of superposition to determine the resulting magnetic field between and outside of the sheets.

(c) Assume the magnetic field outside the two sheets described in part (b) is zero. Use Ampere's law to find the magnetic field between the sheets directly.

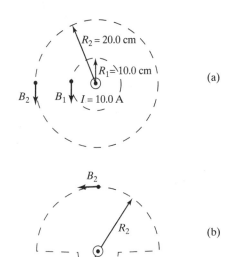

FIGURE 28–5

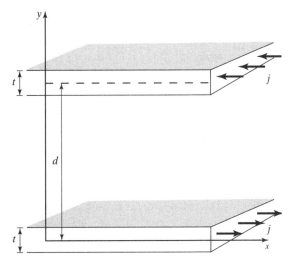

FIGURE 28–4

6. (II) An outer solenoid (radius r_a) and an inner solenoid (radius $r_b < r_a$) are coaxial with n_a and n_b turns per unit length, respectively. The solenoids carry the same current, but in opposite directions. Let r be the radial distance from the common axis of the solenoids. If the magnetic field inside the inner solenoid ($r < r_a$) is to be in the opposite direction as the field between the solenoids ($r_a < r < r_b$), but have one-half the magnitude, determine the required ratio n_b/n_a.

7. (II) Let's demonstrate the veracity of Ampere's law from a numerical standpoint. In Example 28–9 of the text, using only the Biot-Savart law, it is shown that an infinitely long, straight wire carrying a current I produces a magnetic field of magnitude $B = \dfrac{\mu_0 I}{2\pi R}$ at point P, which is a perpendicular distance R from the wire. This magnetic field is directed tangent to the circle of radius R that is centered on the wire.

(a) Assuming $I = 10.0$ A, determine B_1 and B_2 (T) at distances $R_1 = 10.0$ cm and $R_2 = 20.0$ cm, respectively, from the wire.

(b) Imagine two concentric circles with radii $R_1 = 10.0$ cm and $R_2 = 20.0$ cm, respectively (Fig. 28–5a). Find the circumferences C_1 and C_2 (m) of each circle.

(c) For the two circular paths, the integral $\oint \mathbf{B} \cdot d\mathbf{l}$ equals BC. Determine the product BC for each circle. Explain how it is that, even though $B_2 < B_1$, $B_1 C_1 = B_2 C_2$.

(d) Demonstrate numerically that $\mu_0 I$ equals the product BC found in part (c).

(e) Evaluate the integral $\oint \mathbf{B} \cdot d\mathbf{l}$ for the closed path shown in Fig. 28–5b. Do you obtain the same numerical value as that found for the circular paths in part (c)?

8. (II) In a certain atomic physics experiment, the magnetic field along a portion of the axis of a solenoid must change linearly according to $B = 12\text{ mT} + (1.6\text{ mT/cm})x$, where x is the distance in cm along the axis from a point where $B = 12$ mT. If the current flowing through the solenoid's wires is 1.0 A, determine the required "winding profile" $n(x)$ (i.e., the required turns per unit length as a function of x in this portion of the solenoid).

9. (III) A cylindrical conducting wire of radius R carries a total current I. This current is nonuniformly distributed throughout the wire's cross-section such that its radial-dependent current density $j(r) = Cr$, where r is the radial distance from the wire's central axis and C is a constant.

(a) Determine C in terms of I and R.

(b) Determine the magnetic field magnitude B at $r = R/2$ (inside the wire), then find the radial distance $r > R$ (outside the wire), where B has this same value.

Section 28–6

10. (II) A very short wire of length $\Delta l = 0.10$ mm lies on the x-axis, centered on the origin, carrying a current $I = 1.0$ mA directed in the positive x-direction.

(a) Estimate the magnetic field $\Delta \mathbf{B}_P$ (magnitude and direction) created by this wire at P, where P is the point in the x-y plane at $(x\mathbf{i} + y\mathbf{j}) = (1.0\,\mathbf{i} + 1.0\,\mathbf{j})$m.

(b) Estimate the magnetic field $\Delta \mathbf{B}_Q$ (magnitude and direction) created by this wire at Q, where Q is the point on the y-axis at $y = +1.0$ m. How much larger would the magnetic field magnitude be at Q if an infinitely long wire with current $I = 1.0$ mA lay on the x-axis, rather than this very short wire?

11. (II) In Example 28–10 of the text, the actual magnetic field B at a distance x along the axis of circular current-carrying loop (radius R, magnetic dipole μ) is derived and a "far-field" approximation $B_F = \mu_0 \mu / 2\pi x^3$ for when $x \gg R$ is also given. Define α to be the fractional difference between the actual and far-field approximation value for B at a given x; i.e., $\alpha = \dfrac{B_F - B}{B}$. Find the threshold distance x_F (expressed as a multiple of R) beyond which $\alpha < 0.01$; i.e., the error introduced by approximating the loop's magnetic field as B_F rather than using the actual expression for B is less than 1%.

12. (II) You are given a fixed length L of wire with current I flowing through it. If you fashion this wire into a single circular loop, let B_1 be the magnetic field that this loop produces at its center. If, instead, you fashion the wire into a coil with N loops, compare the magnetic field B_N created at this coil's center with B_1 by determining the ratio B_N/B_1.

13. (II) A hairpin turn of radius R redirects the current I in a very long wire as shown in Fig. 28–6. Determine the magnitude and direction of the magnetic field at P, the point at the center of curvature for the turn.

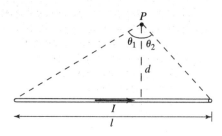

FIGURE 28–7

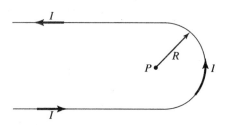

FIGURE 28–6

14. (II) At a particular location on Earth's surface, a horizontal compass is at the center of a vertical circular loop of wire and the compass needle is aligned with the horizontal component B_h of Earth's magnetic field. The radius of the wire loop is $R = 10$ cm and its vertical plane is aligned with the compass needle. When a current $I = 3.7$ A flows in the wire loop, the compass needle realigns itself at an angle of 45° relative to its original ("zero-current") direction. Determine B_h at this location on Earth's surface.

15. (II) The magnetic field at point P, which is perpendicular distance r away from an infinitely long, straight current-carrying wire, is $B = \mu_0 I/2\pi r$. At P, the angle subtended by the entire wire is π. Determine the angle θ subtended at P by the (nearest) portion of the wire that produces one-half of the total field at that location; then compare, as a ratio, θ with π.

16. (II) In Bohr's model for the hydrogen atom, an electron can be held in a variety of circular orbits about a stationary proton by the attractive electric force between these two particles. The smallest such circular orbit (called the ground state) has radius $r = 5.3 \times 10^{-11}$ m.
 (a) Use the Bohr model to find the orbital speed v of the electron when in the ground state.
 (b) From the electron's point of view, the proton orbits it with speed v in a circular orbit of radius r. Use this perspective to predict the magnitude of the magnetic field produced by the proton at the position of the electron.
 (c) An electron has a magnetic dipole moment $\mu = 9.3 \times 10^{-24}$ A·m² that can be parallel or antiparallel to the magnetic field due to the proton. Determine the difference in the electron's magnetic potential energy when its dipole moment is parallel and antiparallel to the proton's field. The differing magnetic energies associated with these two possible orientations of the electron's moment results in two atomic states of slightly different energies. [A more advanced analysis of this problem reduces the answer in part (b) by a factor of two because the electron is not in an inertial reference frame.]

17. (II) Point P is a perpendicular distance d away from a straight segment of wire carrying current I as shown in Fig. 28–7. Find an expression for the magnetic field magnitude B at P in terms of the angles θ_1 and θ_2. If d equals one-half the wire's length l, how much larger is B at the point directly over the wire's midpoint than it is directly over one of its ends?

18. (III) A thin, insulating rod of radius r and length L carries a uniform charge per unit area σ on its outer cylindrical surface. Define the rod's central axis to be the x-axis with $x = 0$ at the midpoint between the rod's two ends. If the rod spins with angular velocity ω about its central axis, find the resulting magnetic field along the x-axis for $x > +L/2$. [Hint: Think of the rotating rod as a collection of current-carrying loops.]

19. (III) A certain experiment, which takes place within a spherical chamber of radius $r = 10.0$ cm, requires that Earth's magnetic field be cancelled at the center of the chamber. At the chamber's center, which is defined as the origin of an x-y-z coordinate system, Earth's field is known to be $\mathbf{B}_E = (3.0\mathbf{i} + 1.0\mathbf{j} - 2.0\mathbf{k}) \times 10^{-5}$ T.
 (a) If 10-turn coils X, Y, and Z, each of radius $R = 5.0$ cm, are placed with their centers at $x = +r$, $y = +r$, and $z = +r$, respectively, and their planes tangent to the chamber at those points, what current (magnitude and direction) should flow in each coil in order to cancel $\mathbf{B}_E$? Indicate the current direction (clockwise or counterclockwise) as seen by an observer at the coil's center and facing the origin.
 (b) Show that just one 10-turn coil, placed with its center tangent to the chamber at point $(x_o\mathbf{i} + y_o\mathbf{j} + z_o\mathbf{k})$, can be used to cancel $\mathbf{B}_E$. That is, find the proper $(x_o\mathbf{i} + y_o\mathbf{j} + z_o\mathbf{k})$ and determine the current (magnitude and direction) that should flow through this single coil.

20. (III) (a) Assign the axis of a circular loop to be the x-axis with the origin at the loop's center. If this loop has radius R and carries current I, it creates a magnetic field of magnitude $B_o = \mu_0 I/2R$ at $x = 0$. Let's arbitrarily define the range λ of the magnetic field created by this loop along its axis to be the value of x at which the field has dropped to 1.0% of B_o. Using this definition, find λ in terms of R (Fig. 28–8a).
 (b) Consider the magnetic field produced by two nested circular loops of radii r and $3r$ with oppositely directed currents I as shown in Fig. 28–8b. Show that the magnetic field created along the common x-axis of these loops can be written in dimensionless form as

$$y = (1 + z^2)^{-3/2} - 9(9 + z^2)^{-3/2}$$

where $y = B / \left(\dfrac{\mu_0 I}{2r}\right)$ and $z = x/r$.

 (c) Using a computer or a graphing calculator, plot y vs. z in the range $0 \le z \le 15$. Determine B_o and λ for each loop individually; then use these values to offer a qualitative explanation for the shape of your y vs. z plot. [The current configuration in

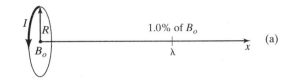

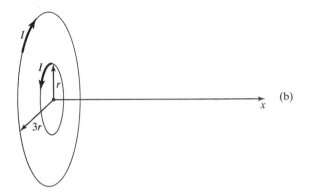

FIGURE 28–8

Fig. 28–8b can be used to model the magnetic field produced by a ring-shaped permanent magnet with inner and outer radii r and $3r$, respectively, which is magnetized in the direction along its axis.]

21. (III) Consider the magnetic field magnitude B at distance x along the axis of a circular current-carrying loop (radius R, current I). Let's investigate the conditions under which an approximate (and thus simpler) representation for B can be used with only a small compromise in precision. Define $z = R^2/x^2$.

(a) Show that B can be written as $B = \dfrac{\mu_0 \mu}{2\pi x^3}\left[1 + T_1 + T_2 + T_3 + \cdots\right]$, where $T_1 = -\dfrac{3}{2}z$, $T_2 = +\dfrac{15}{8}z^2$, and $T_3 = -\dfrac{35}{16}z^3$. $T_1, T_2,$ and T_3 are the first-, second- and third-order "correction terms in z" when $x > R$.

(b) When x is sufficiently large, one expects that z will become small enough for the T-terms to be negligible compared with 1. Then the magnetic field will be accurately described by the simplified relation $B \approx \dfrac{\mu_0 \mu}{2\pi x^3}$. One plausible definition for "sufficient largeness" of x is that the first-order term provides less than a 1.0% correction to the "1" within the square brackets of the expression for B (i.e., $|T_1| < 0.010$). Find the threshold value x_T such that for $x > x_T$ this criterion will be fulfilled.

(c) When the correction due to T_1 is small, the corrections due to the higher-order terms are guaranteed to be even smaller. Demonstrate this fact by calculating the ratios $|T_2/T_1|$ and $|T_3/T_1|$ when $x > x_T$.

Section 28–9

22. (III) A solenoid with length l, cross-sectional area A, and N loops carries a current I.
(a) Defining the magnetic dipole moment of this solenoid as $\mu = NIA$, show that the magnetic field within it is given by $B = \mu_0 \mu / V$, where V is the volume enclosed by the solenoid.

(b) Assume a solenoid with volume $V = 1000 \text{ cm}^3$ is filled with iron and that this solenoid carries the minimum current necessary to saturate the iron. From Fig. 28–26 in the text, determine the "saturation" magnetic field B_S that exists within the solenoid; then use your expression from part (a) to find the solenoid's "saturation" magnetic moment $\mu_S (\text{A} \cdot \text{m}^2)$. [Note: Here μ is magnetic dipole moment, not magnetic permeability.]

(c) Each iron atom has a magnetic dipole moment μ_{Fe}. Assume that the μ_S you found in part (b) is due to the alignment of all of these μ_{Fe} within the solenoid and that the solenoid's current flow makes a negligible contribution to μ_S. Use iron's density (Table 13–1 in the text) and atomic mass (back cover of the text) to estimate μ_{Fe}.

(d) Precision experiments yield $\mu_{Fe} = 2.1 \times 10^{-23} \text{ A} \cdot \text{m}^2$. Why is your answer in part (c) a bit smaller than this value?

General Problems

23. (II) (a) A very large, flat conducting sheet of thickness t carries a uniform current density $\mathbf{j} = j\mathbf{i}$ throughout its cross-section (see lower sheet in Fig. 28–9). Determine the magnetic field (magnitude and direction) created by this sheet at point P, a perpendicular distance $y > t/2$ above the center of the sheet. Assume the sheet can be approximated as infinite in extent.

(b) A second, identical sheet (see upper sheet in Fig. 28–9) also carrying a uniform current density $\mathbf{j} = j\mathbf{i}$ throughout its cross-section is placed parallel to the first sheet at a (center-to-center) separation distance of $y > t$. Determine the force per unit area that the first sheet produces on the second. Does the first sheet attract or repel the second sheet?

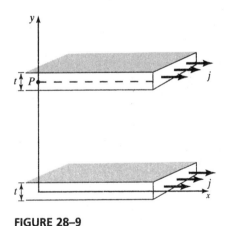

FIGURE 28–9

24. (II) A long solenoid of length L consists of N tightly packed loops, each of radius $R = L/20$. Define the x-axis to be along the solenoid's axis with $x = 0$ at one of the solenoid's ends. Compare, as a ratio, the magnetic field magnitude at $x = 0$ (the solenoid's end) with that at $x = +L/2$ (the solenoid's center). [Hint: Think of the solenoid as a collection of current loops.]

25. (II) Consider the circular region enclosed by a single wire loop of radius R carrying a current I. Let's estimate the variation of the magnetic field within this region. Define B_C and B_W to be the magnetic field magnitude at the loop's center and next to the loop's wire, respectively, as shown

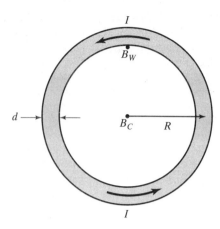

FIGURE 28–10

in Fig. 28–10. Assuming that the wire diameter d is much less than the loop radius R, estimate the ratio B_W/B_C. If $R = 25d$, evaluate the ratio. [*Hint*: At locations next to a thin wire with a large radius of curvature, the field can be estimated easily.]

26. (II) Earth's magnetic field is caused by the motion of charged particles in its liquid core. The liquid core occupies the region from $r = 0.20R_E$ to $r = 0.55R_E$, where r is the radial distance from Earth's center and R_E is the radius of Earth. Let's assume that Earth's rotation is the source of the relevant motion and model the liquid core currents as roughly equivalent to a rotating ring of charge, centered on $r = 0$, with a radius $R = R_E/3$ and period $T = 1\,\text{d}$. Use this "rotating ring" model to estimate
    (a) Earth's magnetic dipole moment μ, given that at the magnetic poles on Earth's surface, $B = 7 \times 10^{-5}\,\text{T}$.
    (b) the amount of net charge Q in Earth's liquid core, which produces its magnetic field.
    (c) the fraction of atoms within the liquid core that have to be singly ionized to produce the value of Q found in part (b). Assume the liquid core is composed mostly of iron under high pressure with density of about $10\,\text{g/cm}^3$. [The rotating ring model cannot explain some of the known features of Earth's field such as its pole reversals throughout history. A more sophisticated model called the geodynamo takes into account both rotational and convective motion to provide an accurate picture of the source of the field.]

27. (III) Here is the design for an electromagnetic clock. An thin aluminum hoop of radius $r = 5.0\,\text{cm}$ and mass $m = 15\,\text{g}$ (and negligible width) is free to rotate about an axis along its diameter and carries a current $I_H = 1.5\,\text{A}$. This hoop is placed inside a long solenoid with its axis of rotation perpendicular to the solenoid's axis. The solenoid has n turns per unit length and carries current $I_S = 2.0\,\text{A}$. In equilibrium, the hoop is at $\theta = 0°$, where θ is the angle between the solenoid's axis and the normal to the hoop's plane. When the hoop is rotationally displaced to a small angle θ and released, prove that it will subsequently perform simple harmonic oscillation and so can be used as a timekeeper. If one wishes the period of this oscillation to be $T = 1.0\,\text{s}$, what is the required value for n?

28. (III) A pair of identical circular wire coils, each with radius R and N turns, which carry current I in the same direction, is often used to create a uniform magnetic field for an experiment. Assuming that the coils share a common axis (defined as the x-axis) with their centers separated by a distance d, let's investigate the magnetic field created along this axis between these coils. Define $x = 0$ to be at the midpoint between the coils (Fig. 28–11).
    (a) Show that the magnetic field at position x along the x-axis is given by
    $$B(x) = \frac{\mu_o NIR^2}{2}\left\{\left[R^2 + \left(\frac{d}{2} - x\right)^2\right]^{-3/2} + \left[R^2 + \left(\frac{d}{2} + x\right)^2\right]^{-3/2}\right\}.$$
    (b) For small excursions Δx from the origin, the variation in B is approximately $\Delta B \approx \left.\frac{dB}{dx}\right|_{x=0} \Delta x$. Determine dB/dx and evaluate this derivative at $x = 0$. What does your result indicate about the uniformity of the field near the origin?
    (c) A result from advanced calculus (called Taylor's Theorem) reveals that a better approximation for the variation in B about the origin is
    $$\Delta B \approx \left.\frac{dB}{dx}\right|_{x=0} \Delta x + \frac{1}{2}\left.\frac{d^2B}{dx^2}\right|_{x=0}(\Delta x)^2.$$
    Show that, for the coil pair,
    $$\frac{1}{2}\left.\frac{d^2B}{dx^2}\right|_{x=0} = 3\mu_o NIR^2 \frac{(d^2 - R^2)}{(R^2 + d^2/4)^{7/2}}.$$
    Thus, by choosing $d = R$, the second derivative at the origin will vanish, as well as the first derivative.
    (d) When constructed with $d = R$, the coils are called Helmholtz coils. What is the magnitude of the very uniform field near the origin for Helmholtz coils; i.e., what is $B(0)$?

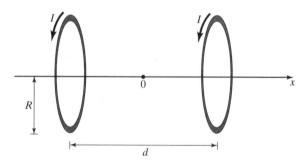

FIGURE 28–11

29. (III) The design of a magneto-optical atom trap requires a magnetic field B that is directly proportional to position x along an axis. Such a field perturbs the absorption of laser light by atoms in the manner needed to spatially confine atoms in the trap. Let's demonstrate that anti-Helmholtz coils will provide the required field $B = Cx$, where C is a constant.

(a) Anti-Helmholtz coils consist of two identical circular wire coils, each with radius R and N turns, carrying current I in opposite directions (Fig. 28–12). The coils share a common axis (defined as the x-axis with $x = 0$ at the midpoint between the coils). Assume that the centers of the coils are separated by a distance equal to the coil radius R.

(a) Show that the magnetic field at position x along the x-axis is given by

$$B(x) = \frac{4\mu_0 NI}{R}\left\{\left[4 + \left(1 - \frac{2x}{R}\right)^2\right]^{-3/2} - \left[4 + \left(1 + \frac{2x}{R}\right)^2\right]^{-3/2}\right\}.$$

(b) For small excursions from the origin where $|x| \ll R$, prove that the magnetic field is given by $B \approx Cx$, where the constant $C = \dfrac{48\mu_0 NI}{25\sqrt{5}R^2}$.

(c) For optimal atom trapping, dB/dx should be about 0.15 T/m. Assume an atom trap uses anti-Helmholtz coils with $R = 4.0$ cm and $N = 150$. What current should flow through the coils? [Coil separation equal to coil radius, as assumed in this problem, is not a strict requirement for anti-Helmholtz coils.]

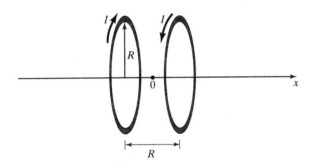

FIGURE 28–12

30. (III) Model the needle of a compass as a uniform rod of mass $m = 0.5$ g and length $l = 2$ cm, which is free to rotate about an axis normal to its length and through its center. This rod is magnetized iron with its magnetic dipole moment of magnitude μ directed along its length toward its "North-Pole" indicator tip. Assume that the compass needle lies in a horizontal plane slightly above (and its axis perpendicular to) Earth's surface and that it is initially in its equilibrium position $\theta = 0$ (i.e., aligned with the horizontal component of Earth's magnetic field B_h). When the needle is displaced by a small angle θ from equilibrium and released, it performs simple harmonic motion about $\theta = 0$ with a period T.

(a) Prove that $T = 2\pi\sqrt{\dfrac{ml^2}{12\mu B_h}}$.

(b) If, at a certain location on Earth's surface where $B_h = 3 \times 10^{-5}$ T, the compass needle is found to oscillate with period $T = 1.0$ s, estimate the value of μ.

(c) The magnetic dipole moment of an individual iron atom is $\mu_{Fe} = 2.1 \times 10^{-23}$ A·m². Use iron's density (Table 13–1 in the text) and atomic mass (back cover of the text) to estimate the fraction of the needle's iron atoms that are cooperating to produce its ferromagnetic dipole moment.

Electromagnetic Induction and Faraday's Law

Sections 29–1 and 29–2

1. (II) A 10-turn coil of cross-sectional area $A = 80.0\ \text{cm}^2$ is immersed in a uniform magnetic field of magnitude $B = 0.100\ \text{T}$. At a certain instant in time, the coil's area vector $\mathbf{A}$ is oriented at angle $\theta = \pi/3$ rad relative to the field and A, B, and θ are all changing simultaneously at the rates $+1.0\ \text{cm}^2/\text{s}$, $B = -0.010\ \text{T/s}$, and $-\pi$ rad/s, respectively. Find the induced emf in the coil at that moment.

2. (II) A circular wire loop of radius $r = 10\ \text{cm}$ is immersed in a uniform magnetic field $B = 0.500\ \text{T}$ with its plane normal to the direction of the field. If the field magnitude then decreases at a constant rate of $-0.010\ \text{T/s}$, at what rate should r increase so that the induced emf within the loop is zero?

3. (II) While demonstrating Faraday's law to her class, a physics professor inadvertently moves the gold ring on her finger from a location where a 0.80-T magnetic field points along her finger to a zero-field location in 50 ms. The 1.5-cm diameter ring has a resistance and mass of 50 $\mu\Omega$ and 15 g, respectively.
(a) Estimate the thermal energy produced in the ring due to the flow of induced current.
(b) Find the temperature rise of the ring, assuming all of the thermal energy produced goes into increasing the ring's temperature. The specific heat of gold is 129 J/kg·C°.

4. (II) A power line carrying a sinusoidally varying current with frequency $f = 60\ \text{Hz}$ and peak value $I_o = 10\ \text{kA}$ runs at a height of 7.0 m across a farmer's land (Fig. 29–1). The farmer constructs a vertically oriented 2.0-m high 10-turn rectangular wire coil below the power line. The farmer wants to use the induced voltage in this coil to power "120-Volt" electrical equipment, which requires a sinusoidally varying voltage with frequency $f = 60\ \text{Hz}$ and peak value $V_o = 170\ \text{V}$. What should the length l of the coil be?

5. (II) An elastic conducting loop with resistance $R = 2.0\ \Omega$ is immersed in a magnetic field, which is directed out of the page. The field's magnitude is uniform spatially, but varies with time t according to $B(t) = \alpha t$, where $\alpha = 0.60\ \text{T/s}$. Additionally, the area A of the loop increases at a constant rate according to $A(t) = A_o + \beta t$, where $A_o = 0.50\ \text{m}^2$ and $\beta = 0.70\ \text{m}^2/\text{s}$. Find the magnitude and direction (clockwise or counterclockwise, when viewed from above the page) of the induced current within the loop at time $t = 2.0\ \text{s}$ the magnetic field
(a) is parallel to the plane of the loop.
(b) is perpendicular to the plane of the loop.

6. (III) The bottom and left sides of a 2.0 cm × 2.0 cm wire loop lie on the x- and y- axis, respectively, with the square's bottom left corner at the origin. The loop is immersed in a magnetic field (oriented in the $+\mathbf{k}$ direction) whose magnitude in the x-y plane is a function of both the y-coordinate and time t according to the following relation: $B = \alpha y t^2$, where the constant $\alpha = 4.0\ \text{T/m}\cdot\text{s}^2$. At $t = 2.5\ \text{s}$, determine the (a) emf induced in the loop and (b) direction of the induced current in the loop (indicate whether the current in the loop's bottom side flows in the $+\mathbf{i}$ or $-\mathbf{i}$ direction).

7. (III) A long, straight wire carries constant current I as shown in Fig. 29–2. A square wire loop of side length l and total resistance R is pulled away from the long wire at a constant

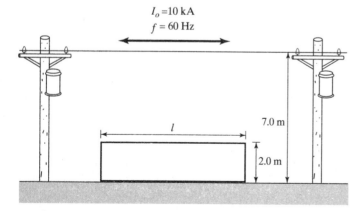

FIGURE 29–1

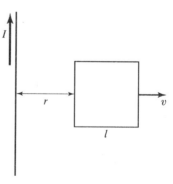

FIGURE 29–2

velocity v. Define r to be the time-dependent distance from the straight wire to the near side of the loop.

(a) Determine the flux through the loop. Then, using Faraday's law, find the magnitude of the induced current in the loop as a function of $I, l, v, R,$ and r. [*Hint*: $v = dr/dt$.]

(b) Does the induced current flow clockwise or counterclockwise?

(c) Electric power is dissipated in the loop via the induced current. This energy is supplied by the agent that is pulling the loop away from the straight wire. Determine the force F due to this agent on the loop.

8. (III) A rectangular wire loop has dimensions $l \times w$ and resistance R. A long, straight wire carrying right-directed current I lays across the loop directed parallel to, and a distance $l/4$ from, the loop's bottom side (Fig. 29–3). The square loop and straight wire each have an insulating coating so that they are not in direct electrical contact. If, during the time interval T, the current in the wire suddenly drops (at a constant rate) from I to zero, find the induced current (magnitude and direction) that flows within the loop. [*Hint*: Use symmetry to simplify the flux calculation.]

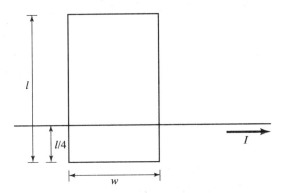

FIGURE 29–3

9. (III) In a popular instructional laboratory experiment, Faraday's law is investigated by monitoring the induced voltage $\mathscr{E}$ in a coil as a small permanent magnet with magnetic dipole moment μ is dropped though it (Fig. 29–4). What variation in $\mathscr{E}$ is expected theoretically in this situation? Model the magnet as a current-carrying loop of radius R and magnetic dipole moment μ. Let the shared axis of this current loop and the coil of N turns

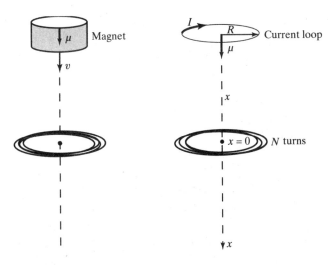

FIGURE 29–4

and radius (slightly greater than) R be defined as the x-axis with $x = 0$ at the coil's center. Assume the current loop travels along this axis at constant speed v. Also assume that when the current loop is at position x its magnetic field across the entire area of the coil can be approximated as directed normal to the plane of the coil with magnitude equal to the value on the axis.

(a) Use this model to show $\mathscr{E} = \dfrac{3\mu_o \mu N R^2 v}{2} \dfrac{x}{(R^2 + x^2)^{5/2}}$.

(b) At what value of x is (the absolute value of) $\mathscr{E}$ maximum? What is this maximum value for $\mathscr{E}$?

(c) Sketch $\mathscr{E}$ vs. x.

Section 29–3

10. (III) A conducting rod (mass m, length l, resistance R) sits on a U-shaped resistanceless conductor that lies on a horizontal tabletop. The spacing between the U-shaped conductor's parallel rails is l. This apparatus is immersed in a vertically directed uniform magnetic field of magnitude B and the rod is connected via a string and pulley to a vertically hanging object of mass M as shown in Fig. 29–5. Let $v(t)$ be the shared speed of the rod and the hanging object as a function of time t and assume $v = 0$ at $t = 0$. Apply Newton's second law to find the differential equation that v obeys. Show that the solution to this equation is of the form $v(t) = v_T(1 - e^{-t/\tau})$ and find the terminal velocity v_T and time constant τ in terms of the given parameters.

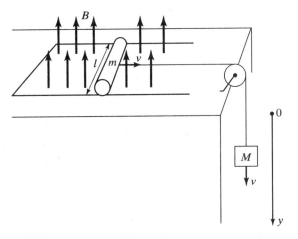

FIGURE 29–5

11. (III) A conducting rod (mass m, length l, resistance R) sits on a U-shaped resistanceless conductor that lies on a horizontal tabletop. The spacing between the U-shaped conductor's parallel rails is l. This apparatus is immersed in a vertically directed uniform magnetic field of magnitude B and the rod is connected via a string and pulley to a vertically hanging object of mass M as shown in Fig. 29–5. Let $v(t)$ be the shared speed of the rod and the hanging object as a function of time t and assume $v = 0$ at $t = 0$. Also, define $y(t)$ to be the (positive) position of the hanging mass at time t with $y = 0$ at $t = 0$. The rate at which thermal energy is gained in the rod is equal to the rate at which the total mechanical (kinetic plus potential) energy is lost by this system. Apply this principle to obtain the differential equation that v obeys. Show that the solution to this equation is of the form $v(t) = v_T(1 - e^{-t/\tau})$ and find the terminal velocity v_T and the time constant τ in terms of the given parameters.

Section 29-4

12. (II) You are stranded on a desert isle near the equator where Earth's magnetic field runs totally parallel to the ground and has a strength of 5.0×10^{-5} T. To power your 12-Volt rms guava juice blender, you wind a wire coil with N circular turns of diameter 25 cm around a coconut. Your monkey can rotate this coil once every 0.30 s. If this coil is to power your blender, calculate what N should be. About what axis should your monkey rotate the coil?

Section 29-6

13. (II) Assume a voltage source supplies an ac voltage of amplitude V between its output terminals. If the output terminals of this source are connected to a circuit and an ac current of amplitude I flows through the terminals, then the equivalent resistance of the circuit is $R_{eq} = V/I$.
(a) If a resistor R is connected directly to this source, what is R_{eq}?
(b) If a transformer with N_P and N_S turns in its primary and secondary, respectively, is placed between the source and the resistor as shown in Fig. 29–6, what is R_{eq}?
[Transformers are used in ac circuits to alter the apparent resistance of circuit elements, such as speakers, in order to maximize transfer of power from a source to the element.]

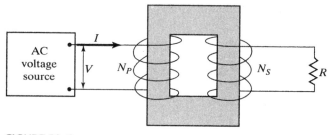

FIGURE 29–6

Section 29-7

14. (II) In a circular region, there is a uniform magnetic field **B** pointing into the page (Fig. 29–7). An x-y coordinate system is defined with the origin at the circular region's center. A free positive point charge $+Q = 1.0\,\mu\text{C}$ is initially at rest at a position $x = +10$ cm on the x-axis. If the magnitude of the magnetic field is now decreased at a rate of -0.10 T/s, what force (magnitude and direction) will act on $+Q$?

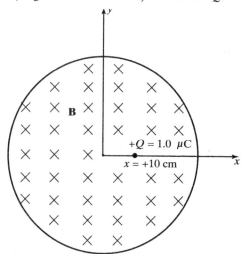

FIGURE 29–7

15. (III) Initially, a particle with charge q, mass m, and linear speed v moves in a circular orbit of radius $r = mv/qB$ in a plane perpendicular to the spatially uniform magnetic field **B** within which it is immersed (see Example 27–4 in the text). Next, assume that the field's magnitude is continuously changed at rate dB/dt, but that this change takes place slowly enough so that at any time t the particle can be considered to be traveling in a circular orbit of radius $r = mv/qB$, where v and B are the instantaneous values of the particle's linear speed and the field's magnitude at time t, respectively.
(a) Use Newton's second law and Faraday's law to analyze how the particle's linear speed changes in response to the changing magnetic field. Prove that its orbital radius changes at a rate $\dfrac{dr}{dt} = -\dfrac{r}{2B}\dfrac{dB}{dt}$, and then integrate this relation to show that $r^2 B = C$, where C is a time-independent constant.
(b) If, over time, B becomes twice as large, by what factors do the particle's orbital radius and linear speed change?

16. (III) The magnetic field in a certain region of space is directed perpendicular to the x-y plane and has a magnitude $B(r)$ that is a function of the radial distance r from the origin. In this region, a particle with mass m and charge q travels in the x-y plane in a circular path of radius R about the origin.
(a) If B then changes with time, use Newton's second law and the general form of Faraday's law to show that the particle will continue in the same circular path if, at all times, $\Phi_B = 2\pi R^2 B(R)$, where Φ is the magnetic flux through the particle's orbit. To satisfy this criterion, B must decrease with increasing r. [Hint: Since the particle remains on the same path as B varies in time, the particle's velocity as a function of time is given by $v = qRB(R)/m$, where $B(R)$ is the magnetic field at radius R at time t.]
(b) Show that the spatially uniform field $B(r) = B_o$, where B_o is a constant, does not satisfy the given criterion.
(c) If $B = B_o(1 - \alpha r)$, where B_o and α are constants, find α (in terms of R) so that the given criterion is satisfied. [The betatron, a particle-accelerating device, works on the physics principles outlined in this problem.]

General Problems

17. (II) A very long, straight wire carries current I. A rod of length l oriented parallel to the wire, starting at distance r_o at time $t = 0$, moves toward the wire at constant speed v. Find the motional emf induced between the ends of the rod as a function of time (until it collides with the wire). In the reference frame of the rod, what is the induced electric field in the rod as a function of time?

18. (II) Two concentric circular coils are coplanar and centered on point P. The outer coil (radius r_a, N_a turns) is connected to an ac voltage source so that current $I(t) = I_o \sin(\omega t)$ flows in it, where I_o and ω are the peak current and angular frequency, respectively. The amplitude $\mathcal{E}_o$ of the sinusoidal induced emf in the inner coil (radius r_b, N_b turns) is monitored by a voltmeter (Fig. 29–8).
(a) Determine the time-varying magnetic field magnitude $B_P(t)$ produced by the outer coil at P.
(b) Assuming that $B_P(t)$ accurately approximates the magnetic field across the entire area of the inner loop, find an expression for $\mathcal{E}_o$.
(c) For accurate measurement, a designer wants $\mathcal{E}_o = 250$ mV. If $I_o = 500$ mA, $\omega/2\pi = 5.0$ kHz, $r_a = 5.0$ cm, $r_b = 2.5$ cm, and $N_a = 20$, how many turns should be placed

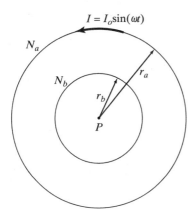

FIGURE 29–8

on the inner coil? [This device can be used as a crude metal detector. If, say, a piece of iron is placed near the center of the inner coil, the voltmeter reading $\mathcal{E}_o$ will increase.]

19. (II) A circular-shaped circuit of radius r, containing a resistance R and capacitance C, is situated with its plane perpendicular to the spatially uniform magnetic field **B** within which it is immersed. The magnetic field is directed into the page (Fig. 29–9). Starting at time $t = 0$, the voltage difference $V_{ba} = V_b - V_a$ across the capacitor plates is observed to increase with time t according to $V_{ba} = V_o(1 - e^{-t/\tau})$, where V_o and τ are positive constants. Determine dB/dt, the rate at which the magnetic field magnitude changes with time. Is B becoming larger or smaller as time increases?

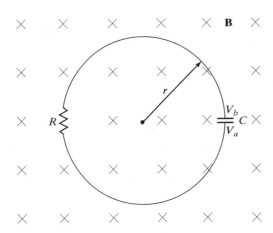

FIGURE 29–9

20. (II) The plane of a square loop with side length l and resistance R is perpendicular to a magnetic field B. The field only extends over a finite area.
    (a) With the magnetic field stationary, the loop is pulled out of the field region at a constant speed v to the right as shown in Fig. 20–10a. Using Faraday's and Lenz's laws, determine the induced current (magnitude and direction) that flows within the loop. Also use $\mathbf{F} = q\mathbf{v} \times \mathbf{B}$ to offer a qualitative explanation for why a current flows within the loop.
    (b) With the loop stationary, the field region is moved to the left at a constant speed v as shown in Fig. 20–10b. In this situation, the magnetic force $\mathbf{F} = q\mathbf{v} \times \mathbf{B}$ is zero on free charges within the loop. Does an induced current flow within the loop? If so, determine its magnitude and direction and offer a qualitative explanation for its cause.

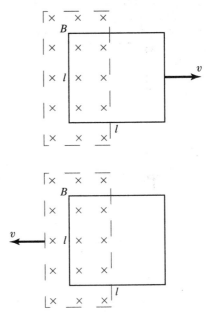

FIGURE 29–10

21. (II) In a certain region of space near Earth's surface, a uniform horizontal magnetic field of magnitude B exists above a level defined to be $y = 0$. Below $y = 0$, the field abruptly ends and so can be assumed to be zero everywhere in that region (Fig. 29–11). A vertical square wire loop (resistivity ρ, mass density ρ_m, diameter d, side length l), initially at rest with its lower horizontal side at $y = 0$, is released with its plane normal to the direction of the field.
    (a) While the loop is still partially immersed in the magnetic field as it falls into the zero-field region, determine the magnetic "drag" force that acts on it at the moment when its speed is v.
    (b) Assume that the loop achieves a terminal velocity v_T before its upper horizontal side exits the field. Show that $v_T = 16\rho\rho_m g/B^2$.
    (c) If the loop is made of copper and $B = 0.80\,\text{T}$, find v_T (cm/s).

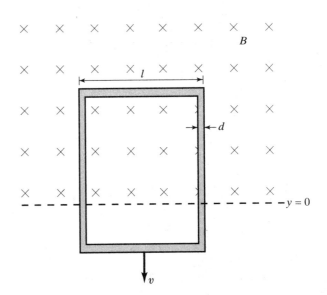

FIGURE 29–11

22. (II) Inductive battery chargers, which allow transfer of electrical power without the need for exposed electrical contacts, are commonly used in appliances that need to be safely immersed in water, such as electric toothbrushes. Consider the following simple model for the power transfer in an inductive charger (Fig. 29–12): Within the charger's plastic base, a primary solenoid of diameter d with n_p turns per unit length is connected to a home's ac wall receptacle so that a time-dependent current $I = I_o \sin(2\pi ft)$ flows within it. When the toothbrush is seated on the base, an N-turn secondary coil encased within the toothbrush's plastic case of diameter only slightly greater than d is centered on the primary. Find an expression for the emf induced in the secondary coil. [This induced emf powers circuitry within the toothbrush that recharges its battery.]

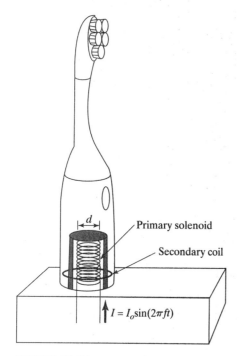

FIGURE 29–12

23. (III) Two concentric circular coils are coplanar and centered on point P. The outer coil (radius $r_a = 5.0$ cm, N_a turns) is connected to an ac voltage source so that current $I(t) = I_o \sin(\omega t)$ flows in it, where $I_o = 500$ mA and $\omega/2\pi = 5.0$ kHz are the peak current and frequency, respectively. The amplitude $\mathcal{E}_o$ of the sinusoidal induced emf in the inner coil (radius $r_b = 2.5$ cm, N_b turns) is monitored by a voltmeter (Fig. 29–8).
(a) Assume that the time-varying magnetic field magnitude $B_P(t)$ produced by the outer coil at P accurately approximates the magnetic field across the entire area of the inner loop. Using this approximation, show that the product of the number of turns in each coil is $N_a N_b = 2r_a \mathcal{E}_o / \mu_o \pi r_b^2 I_o \omega$.
(b) Let l be the total length of wire needed to construct both the inner and outer coils. If, for accurate measurement, a designer wants $\mathcal{E}_o = 250$ mV, what choices for N_a and N_b will require the least use of wire (i.e., minimize l)? [This device can be used as a crude metal detector. If, say, a piece of iron is placed near the center of the inner coil, the voltmeter reading $\mathcal{E}_o$ will increase.]

24. (III) A circular-shaped circuit of radius r, containing a resistance R and capacitance C, is situated with its plane perpendicular to the uniform magnetic field $\mathbf{B}$ within which it is immersed. The magnetic field is directed into the page (Fig. 29–9). Starting at time $t = 0$, when the magnetic field magnitude has the value B_o, the voltage difference $V_{ba} = V_b - V_a$ across the capacitor plates is observed to increase with time t according to $V_{ba} = \alpha t$, where α is a positive constant. Determine an expression for $B(t)$, the magnetic field magnitude as a function of time. Is B becoming larger or smaller as time increases?

25. (III) Near Earth's surface, a circular metal ring of mass m, radius r, and resistance R is falling in a magnetic field. The vertical component of this field changes with height y above Earth's surface as $B = B_o(1 + ky)$, where B_o and k are positive constants. Assume that the plane of the ring remains horizontal as it falls and that the ring is not subjected to air resistance.
(a) Show that, as the ring falls, it has an induced magnetic dipole moment $\boldsymbol{\mu} = \dfrac{B_o k \pi^2 r^4 v}{R} \mathbf{j}$, where v is the ring's speed along the y-axis.
(b) Show that, when it is at a height y with speed v, the ring has magnetic potential energy $U = -\dfrac{B_o^2 k \pi^2 r^4}{R}(v + kvy)$.
(c) If the ring falls over a long enough distance, it will achieve a (constant) terminal velocity v_T. Use $F = -dU/dy$ to determine the magnetic "drag" force that acts on the ring when $v = v_T$, then show $v_T = \dfrac{mgR}{B_o^2 k^2 \pi^2 r^4}$.

26. (III) In the Nuclear Magnetic Resonance (NMR) experimental technique called "free induction decay," a sample's net magnetic dipole moment $\boldsymbol{\mu}$ is set rotating at frequency ω in the x-y plane and subsequently "decays" with a time constant T_2 according to the following relation:

$$\boldsymbol{\mu} = \mu_o [e^{-t/T_2} \cos(\omega t) \mathbf{i} + e^{-t/T_2} \sin(\omega t) \mathbf{j}].$$

In such experiments, a "pickup coil" is placed with its axis along the x-axis so the sample produces a flux Φ through this coil at time t proportional to the instantaneous value of the x-component of $\boldsymbol{\mu}$ (i.e., $\Phi = C\mu_x$), where C is a constant. Show that the amplitude of the ac emf induced in the pickup coil is approximately $C\mu_o \omega e^{-t/T_2}$, if $\omega \gg 1/T_2$. [The value of T_2 gives information about the magnetic interactions between the sample's atoms. By monitoring the decay of the pickup coil's ac emf, T_2 can be measured.]

27. (III) The coil of a generator is constructed from a single wire wound into N turns, each of cross-sectional area A. The two ends of this wire are connected to an external circuit so that as an applied external torque τ rotates the coil at constant angular velocity ω within the generator's magnetic field B, a current I flows in the circuit. Let the total resistance of the circuit (including the coil's resistance) and the generator's output emf as a function of time be R and $NBA\omega \sin(\omega t)$, respectively. Show that, at any time t, the power delivered by the torque to the rotating coil equals the power dissipated in the circuit's resistance. [*Hint*: See Sections 10-10 and 27-5 in the text.]

28. (III) In a certain region of space near Earth's surface, a uniform horizontal magnetic field of magnitude B exists above a level defined to be $y = 0$. Below $y = 0$, the field abruptly ends and so can be assumed to be zero everywhere in that region (Fig. 29–11). A vertical square wire loop (resistivity ρ, mass density ρ_m, diameter d, side length l), initially at rest with its lower horizontal side at $y = 0$, is released with its plane normal to the direction of the field. Let $v(t)$ be the loop's velocity as time t increases.
(a) While the loop is still partially immersed in the magnetic field as it falls into the zero-field region, show that a magnetic "drag" force acts on it that is directly proportional to its instantaneous velocity (i.e., $F_D = -bv$) and determine the constant b in terms of ρ, d, l, and B.
(b) Apply Newton's second law to the loop as it falls from the field region (but while its upper horizontal side is still in the field). Show that the solution to the resulting differential equation for $v(t)$ is of the form $v(t) = v_T[1 - \exp(-t/\tau)]$. Determine the constants v_T (terminal velocity) and τ (time constant) in terms of ρ, ρ_m, d, l, and B.
(c) If the loop is made of copper and $B = 0.80\,\text{T}$, find v_T (cm/s) and τ (ms).

29. (III) Assume that in a certain region of space, the electric field is parallel to the x-axis, (i.e., $E_x \neq 0$, $E_y = 0$, $E_z = 0$). Further, assume an electrostatic situation so that there are no time-dependent fields in this region.
(a) Prove that E_x is independent of y and z [i.e., $E_x = E_x(x)$].
(b) If, in addition, there is no charge in this region, prove $E_x(x) = $ constant.

30. (III) A square wire loop of side length l is pulled away from a long, straight wire carrying current I at a constant speed v (Fig. 29–2). Find the emf induced in the loop at the moment when its left-hand side is a distance r away from the straight wire by using the following two methods:
(a) direct application of Faraday's law to the loop.
(b) summing the motional emf contributed by each of the loop's four moving sides.

CHAPTER 30

Inductance; and Electromagnetic Oscillations

Section 30–2

1. (II) To demonstrate the enormity of the Henry unit, a physics professor wants to wind an air-filled solenoid with self-inductance of 1.0 H on the outside of a 10-cm diameter plastic hollow tube using copper wire with a 0.81 mm diameter. The solenoid is to be tightly wound with each turn touching its neighbor (the wire has a thin insulating layer on its surface so the neighboring turns are not in electrical contact). How long will the plastic tube need to be and how many kilometers of copper wire will be required? What will be the resistance of this solenoid?

2. (II) What is the self-inductance L of a circular wire loop of radius R? When the loop carries a current I, the magnetic field at its center is $B_C = \mu_0 I/2R$. However, the magnetic field is larger at all other positions within the loop, with its largest magnitude at locations nearest the wire.
   (a) Carry out a "back-of-the-envelope" calculation for L, assuming that the magnetic field at every point within the loop is approximately equal to the value B_C at its center. Call this the estimate for the self-inductance L_e.
   (b) An advanced-level calculation, which accounts for the variation of B within the loop's interior, determines that the self-inductance of the loop is $L = \mu_0 R \left[\ln\left(\frac{8R}{r}\right) - \frac{7}{4} \right]$, where r is the radius of wire used in constructing the loop. Let the loop radius be measured as a multiple of the wire radius (i.e., $R = \alpha r$, where α is the multiplicative factor). Find the threshold value α_T such that, if $\alpha > \alpha_T$ then L_e will be within a factor of two of L.

Section 30–3

3. (II) A long, straight wire of radius R carries current I uniformly distributed across its cross-sectional area. Find the magnetic energy stored per unit length in the interior of this wire.

4. (II) Assume an infinitely long, straight wire of radius R initially carries no current, but then a power source causes current I to flow in it. By means of the following calculation, show that a problem exists in this seemly benign scenario. Consider a portion of the wire of length l. This portion defines a hollow cylindrical-shaped volume outside of the wire whose inner and outer radii are $r = R$ and $r = \infty$, respectively. Calculate how much energy the power source must supply to create the requisite magnetic field within this volume when it establishes the current in the wire. Qualitatively, what does the result of this calculation mean? [The infinite extent of the wire is the problem here. According to the Biot-Savart law, when far enough away from a finite length wire, the field will fall off as $1/r^2$, rather than $1/r$.]

5. (III) An LC circuit consists of an inductor with self-inductance L in parallel with a capacitor with capacitance C. Assume that the current in this circuit is given by $I = I_o \sin \omega t$, where the circuit's oscillatory period is $T = 2\pi/\omega = 2\pi\sqrt{LC}$.
   (a) Taking L and C as fixed values, show that the average energy stored by the inductor during one oscillation cycle is
   $\overline{U} = \frac{1}{4} L I_o^2$. [Hint: $\overline{U} = \int_0^T U \, dt/T$.]
   (b) Now assume that the inductor's self-inductance L' is changed slightly over the course of an oscillatory cycle (e.g., by partially inserting or removing an iron core within the inductor). In particular, assume L' oscillates about the value L with small amplitude α at a frequency that is twice that of the LC circuit's oscillation frequency; i.e., $L' = L[1 - \alpha \cos(2\omega t)]$. Assuming that α is small enough that the current within the circuit is still given by $I = I_o \sin \omega t$, show that the average power stored in the inductor during one oscillation cycle is
   $\overline{U} = \frac{1}{4} L I_o^2 \left(1 + \frac{\alpha}{2}\right)$. [This is an example of a parametric oscillator in which a variation in a system parameter at the correct frequency adds energy to the oscillator.]

Section 30–4

6. (II) You want to turn on the current through a coil of self-inductance L in a controlled manner so you place it in series with a resistor $R = 2200 \, \Omega$, a switch, and a dc voltage source $V_o = 100$ V. After closing the switch, you find that the current through the coil builds up to its steady-state value with a time constant τ. You are pleased with the current's steady-state value, but want τ to be half as long. What new values should you use for R and V_o?

7. (II) A 10-V battery has been connected to an LR circuit for sufficient time so that a steady current flows through the resistor $R = 10 \, \Omega$ and inductor $L = 10$ mH. At $t = 0$, a switch is used to remove the battery from the circuit and the current decays exponentially through R. Determine the emf $\mathscr{E}$ across the inductor as time t increases. At what time is $\mathscr{E}$ greatest and what is this maximum value (V)?

8. (II) Although not easy to prove, the self-inductance L of a straight wire of diameter d (m) and length l (m) is given by
$$L = 2l\left[\ln\left(\frac{4l}{d}\right) - 0.75\right] \text{ nH}.$$ Consider a 1.0-mm diameter copper wire that is 0.30 m long.
   (a) Determine the wire's resistance and self-inductance.
   (b) If, at time $t = 0$, the wire is attached between the positive and negative terminals of a 0.10-V voltage source, about how long will it take for the wire's current to build up to the value dictated by Ohm's law? Assume, after being powered for a time interval equal to five time constants, a more-or-less steady current flows in an LR circuit.

9. (II) (a) In Fig. 30–1a, assume that the switch has been in position a for sufficient time so that a steady current $I_o = V_o/R$ flows through the resistor R. At time $t = 0$, the switch is quickly switched to position b and the current through R decays according to $I = I_o e^{-t/\tau}$. Show that the maximum emf $\mathcal{E}_{max}$ induced in the inductor during this time period equals the battery voltage V_o.
   (b) In Fig. 30–1b, assume that the switch has been in position a for sufficient time so that a steady current $I_o = V_o/R$ flows through the resistor R. At time $t = 0$, the switch is quickly switched to position b and the current decays through resistor R' (which is much greater than R) according to $I = I_o e^{-t/\tau'}$. Show that the maximum emf $\mathcal{E}_{max}$ induced in the inductor during this time period is $\left(\frac{R'}{R}\right)V_o$. If $R' = 50R$ and $V_o = 100$ V, determine $\mathcal{E}_{max}$. [When a mechanical switch is opened, a high-resistance air gap is created, which is modeled as R' here. This problem illustrates why high-voltage sparking can occur if a current-carrying inductor is suddenly cut off from its power source.]

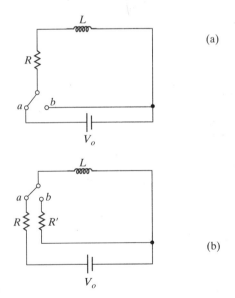

FIGURE 30–1

10. (II) At $t = 0$, a switch is closed so that a battery $+V_o$ is connected to an LR circuit. Due to its self-inductance, the inductor resists the influence of this battery by inducing an opposing emf $-\mathcal{E}$. Define $F = \dfrac{V_o - \mathcal{E}}{V_o}$ as the inductor's fractional effectiveness at canceling out the applied voltage of the battery. Determine F at (a) $t = 0$, (b) $t = \tau$, and (c) $t = \infty$.

11. (II) At $t = 0$, a switch is closed so that a battery V_o is connected to an LR circuit. At what time is the voltage across the resistor R equal to that across the inductor L?

12. (II) To detect vehicles at traffic stops, a wire loop with self-inductance L is often buried horizontally oriented in roadways. This loop is part of an LC circuit whose oscillation frequency is f. When a vehicle is located on the roadway above the buried loop, the loop's self-inductance changes by an amount ΔL (due to induced eddy currents in the car's metal parts) and consequently the LC circuit's oscillation frequency changes by Δf.
   (a) Show that $\dfrac{\Delta f}{f} \approx -\dfrac{1}{2}\dfrac{\Delta L}{L}$.
   (b) When a car is above the loop, the loop's inductance will typically be decreased by 10%. As a result, by what percentage will f change? Will this cause f to increase or decrease?
   (c) A bicycle will commonly only decrease the loop's inductance by 0.1%. If $f = 20$ kHz, what will be the resulting Δf? [By noting changes in f, electronic circuitry detects the presence of vehicles.]

Section 30–5

13. (II) To detect vehicles at traffic stops, a wire loop with self-inductance L is often buried horizontally oriented in roadways. This loop is part of an LC circuit whose oscillation frequency is f.
   (a) Taking typical values of $L = 500$ μH and $f = 20$ kHz, what is the capacitance C in this circuit?
   (b) When a car is located above the buried loop, the loop's self-inductance decreases by 10% (due to induced eddy currents in the car's metal parts). Determine the new resonant frequency of the LC circuit. [By noting changes in f, electronic circuitry detects the presence of vehicles.]

14. (II) A solenoid with self-inductance $L = 2.0$ H is being used in an experiment to produce a uniform magnetic field. In normal operation, the solenoid carries a steady current of 1.0 A. In the event that its source of current is suddenly disconnected (e.g., in a power outage), a researcher places a capacitor C between the two terminals of the solenoid. When the power is disconnected, an LC circuit is formed with its capacitor initially uncharged and an initial 1.0-A current flowing in its inductor. If the researcher wants to prevent the potential difference generated by the collapse of the solenoid's magnetic field from rising above 300 V, what value should be chosen for C? [The small resistance in its wires, which we have ignored in this problem, will eventually dissipate the energy in this circuit.]

15. (III) Distances are commonly measured capacitively. Consider forming an LC circuit using a parallel-plate capacitor with plate area A and a known inductance L.
   (a) If charge is found to oscillate in this circuit at frequency $f = \omega/2\pi$ when the capacitor plates are separated by distance x, show that $x = 4\pi^2 A\epsilon_o f^2 L$.
   (b) When the plate separation is changed by Δx, the circuit's oscillation frequency will change by Δf. Show that $\dfrac{\Delta x}{x} \approx 2\dfrac{\Delta f}{f}$. If f is on the order of 1 MHz and can be measured to a precision of $\Delta f = 1$ Hz, with what percent accuracy can x be determined? Assume fringing effects at the capacitor's edges can be neglected.

16. (III) The circuit shown in Fig. 30–2 includes a battery V_o, an inductor L, and three identical resistors R. The switch is closed at time $t = 0$.
(a) After $t = 0$, apply Kirchhoff's law to obtain three equations involving the currents I_1, I_2, and I_3.
(b) Find the differential equation obeyed by I_3. Then use the results of Section 30–4 in the text to write down the solution for I_3 as a function of time. What is the time constant for the "build-up" of this current?
(c) From your expression for $I_3(t)$, what is the value of this current at $t = \infty$? Offer an explanation for the veracity of this value by considering the simplified equivalent circuit for Fig. 30–2 at $t = \infty$.

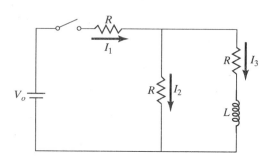

FIGURE 30–2

17. (III) An LC circuit consists of an inductor L and two capacitors C and C', all attached in parallel. With L and C fixed, the circuit's oscillation frequency f is measured as a function of C'. The resulting data are shown in Table 30–1.
(a) Explain (theoretically) why graphing $1/f^2$ vs. C' should yield a straight-line plot. What are the line's slope and y-intercept expected to be theoretically (in terms of L and C)?
(b) Using the data in Table 30–1, graph $1/f^2$ vs. C' and show that a straight-line plot does indeed result. Determine L and C from the slope and the y-intercept of this line.

TABLE 30–1

$C'(\mu F)$	1.00	2.00	3.00	4.00	5.00
f (Hz)	570	505	440	415	378

Section 30–6

18. (II) A solenoid with self-inductance $L = 2.0$ H is being used in an experiment to produce a uniform magnetic field. In normal operation, the solenoid carries a steady current of 1.0 A and there is a negligible voltage difference between the terminals at its two ends (i.e., the solenoid has negligible resistance). In the event that its source of current is suddenly disconnected (e.g., in a power outage), a researcher places a resistor R in series with a capacitor C between the two terminals of the solenoid. When the power is disconnected, an LRC circuit is formed with its capacitor initially uncharged and an initial 1.0-A current flowing in its inductor. If the researcher wants the charge flow through the solenoid to be critically damped such that 3.0 s after the power has been disconnected, the current has fallen to 37% of its initial value, what values should be chosen for R and C?

19. (II) Consider the following method for converting the dc voltage from a battery into a high-frequency ac voltage. In Fig. 30–3, initially switches X and Y are closed and opened, respectively, so that the battery V_o charges the capacitor C fully. At time $t = 0$, switches X and Y are opened and closed, respectively. The LRC circuit then oscillates and a high-frequency ac voltage is available for use across C. Assume the resistance R in this circuit has the small value of 50 Ω (e.g., due to the inductor's wire) and that $R \ll \sqrt{4L/C}$.
(a) The designer of this system wishes to produce a pulse of 1.0-MHz ac voltage that lasts for 1.0 ms. Over the course of this 1.0-ms time interval, the amplitude of 1.0-MHz voltage will be allowed to decay to 50% of V_o. At that point, the switching process is to be repeated so that the capacitor can be recharged by the battery and another 1.0-ms pulse can be produced. Determine the required values for L and C.
(b) Using the numerical values obtained for L and C, show that the assumption $R \ll \sqrt{4L/C}$ is indeed valid.

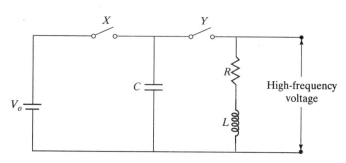

FIGURE 30–3

20. (III) Use the trigonometric identity $\cos(A + B) = \cos A \cos B - \sin A \sin B$ to write Eq. 30–19 in the text as $Q = Q_o e^{-\alpha t}[\cos \phi \cos(\omega' t) - \sin \phi \sin(\omega' t)]$. Take the first and second derivatives of this expression with respect to time; then show that it is the solution to Eq. 30–17 if

$$\omega' = \sqrt{\frac{1}{LC} - \frac{R^2}{4L^2}} \quad \text{and} \quad \alpha = \frac{R}{2L}.$$

General Problems

21. (II) At time $t = 0$, the switch in the circuit shown in Fig. 30–4 is closed. After a sufficiently long time, steady currents I_1, I_2, and I_3 flow through resistors R_1, R_2, and R_3, respectively. Determine these three currents.

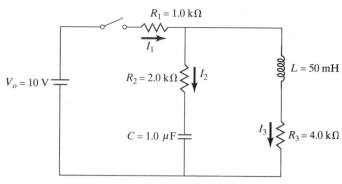

FIGURE 30–4

22. (II) A battery has been connected to an LR circuit for sufficient time so that a steady current I_o flows through the resistor R and inductor L. At time $t = 0$, a switch is used to remove the battery from the circuit and the current through R decays according to $I = I_o e^{-t/\tau}$. Show that the magnetic energy stored in L at $t = 0$ is completely dissipated as thermal energy in R as time progresses.

23. (II) A ring-shaped sample composed of a high-temperature superconductor has self-inductance $L = 6$ nH. At time $t = 0$, with the sample cooled so that it is in its superconducting state, a current I_o is established in this ring. The time-dependent current $I(t)$ is then measured by monitoring the magnetic field produced at a point along the ring's axis. In such an experiment, an effect called "flux creep" will cause the superconducting sample to exhibit a small resistance R. Over the course of 20 days, the current is found to decrease by about 5%.
    (a) Determine R for this superconducting sample due to flux creep.
    (b) If a similarly shaped sample (so that its self-inductance is the same) is instead made of copper, its resistance will be 40 $\mu\Omega$. How long will it take an initial current I_o to drop by 5% in this copper sample?

24. (II) You have borrowed a homemade solenoid from the neighboring lab to produce the uniform magnetic field needed in your newest experiment. You measure its diameter and length to be 5.0 cm and 30.0 cm, respectively, but to calculate its interior field you need to know its total number of turns N. Unfortunately, the person who loaned you the solenoid cannot find any written record of the value for N. So you place the solenoid in series with resistance $R = 1000 \, \Omega$, a switch, and a dc voltage source V_o. With an oscilloscope connected across R, you find that 40 μs after the switch is closed, the voltage across this circuit element has risen to 63% of V_o. Assuming that the solenoid has self-inductance L and negligible resistance (compared with 1000 Ω), use the given information to determine N.

25. (III) A transformer consists of a primary coil with N_P turns connected to an alternating voltage source V_P and a secondary coil with N_S turns to which a load resistor R is attached (Fig. 30–5). Define M to be the mutual inductance of the transformer's two coils and let L_P and L_S be the self-inductance of the primary and secondary coils, respectively.
    (a) In an ideal transformer, the magnetic flux through each coil is exactly the same at all times. With this assumption, show that $M^2 = L_P L_S$ and that $\dfrac{L_S}{L_P} = \left(\dfrac{N_S}{N_P}\right)^2$.
    (b) Let's show that, depending on the ratio of N_S and N_P, the voltage V_S across R may be larger or smaller than V_P. Applying Kirchhoff's loop rule gives $V_P - L_P \dfrac{dI_P}{dt} - M \dfrac{dI_S}{dt} = 0$ and $V_S - L_S \dfrac{dI_S}{dt} - M \dfrac{dI_P}{dt} = 0$. Eliminate $\dfrac{dI_P}{dt}$ from these two equations and use the identities from part (a) to show $\dfrac{V_S}{V_P} = \dfrac{N_S}{N_P}$.

26. (III) A large-diameter outer solenoid has $n_1 = 200$ turns/cm. Inside this outer solenoid, a coaxial inner solenoid of cross-sectional area $A_2 = 3.0 \text{ cm}^2$, length $l_2 = 5.0$ cm, and $N_2 = 5000$ total turns is placed with its two ends connected to a resistor $R = 10 \, \Omega$ (Fig. 30–6).
    (a) Assume that between times $0 \leq t \leq 1.0$ s a voltage source causes the current in the outer coil to change at rate $dI_1/dt = 15$ A/s. Determine the emf $\mathcal{E}$ induced in the inner solenoid during the time interval $0 \leq t \leq 1.0$ s.
    (b) If the self-inductance of the inner solenoid is ignored, one would conclude that a constant induced current $I_2 = \mathcal{E}/R$ flows through R over the entire time interval $0 \leq t \leq 1.0$ s. In fact, this induced current will be time-dependent according to $I_2 = \dfrac{\mathcal{E}}{R}(1 - e^{-t/\tau})$. The time constant τ indicates the time scale over which the current I_2 builds up from zero to its final value of $\mathcal{E}/R$. Determine τ.
    (c) How long will it take the induced current to build up from 0 to 99% of its final value? Is this time less than the 1.0-s time interval over which the induced current exists?
    (d) After $t = 1.0$ s, the current in the outer coil does not change any longer. If the self-inductance of the inner coil is ignored, one would conclude that no induced current flows through R for times $t > 1.0$ s. In fact, the initial current I_o existent in the inner solenoid at $t = 1.0$ s will exponentially decay to zero with time constant τ. Determine I_o and τ.

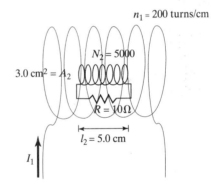

FIGURE 30–6

27. (III) A cylindrical wire of finite length l and radius R carries current I. In the region near the midpoint of the wire, let r be the radial distance from the wire's axis. For positions where $R < r < l/2$, the wire will appear approximately infinite in extent and so we might expect the magnetic field $B \approx \mu_o I/2\pi r$ in this region. Conversely, when $r \gg l$ the wire will appear infinitesimal in extent and so, from the Biot-Savart law with $\theta = 90°$, we expect $B \approx \mu_o I l / 4\pi r^2$. Let's (very) crudely model the magnetic field outside a current-carrying finite wire as $B = \begin{cases} \mu_o I/2\pi r & R \leq r \leq l/2 \\ \mu_o I l / 4\pi r^2 & r \geq l/2 \end{cases}$. Consider a central portion of the wire of length Δl. This portion defines a hollow cylindrical-shaped volume outside of the wire whose inner and outer radii are $r = R$ and $r = \infty$, respectively, and whose length is Δl. Use the given model to calculate the magnetic field energy in this volume. Show that this energy is finite unless the length of the wire is infinite.

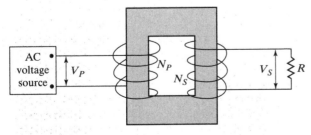

FIGURE 30–5

28. (III) A rectangular loop of dimension $l \times w$, resistance R, self-inductance L, and mass m moves into the region $x \geq 0$ where a uniform magnetic field B directed into the page exists (Fig. 30–7). The loop's position is specified by the x-coordinate of its right-hand side and its velocity along the x-direction is v. Assume that the force due to the magnetic field is the only force that acts on the loop. Also assume that, at time $t = 0$, the loop is located at $x = 0$ and has the velocity $v = v_o$.

(a) As it moves into the field, a motional emf as well as an emf due to its (nonzero) self-inductance will be induced in the loop. Use this idea, along with Newton's second law, to show that the induced current in the loop obeys the following differential equation:

$$L \frac{d^2 I}{dt^2} + R \frac{dI}{dt} + \frac{l^2 B^2}{m} I = 0.$$

(b) Use the results of Section 30–6 in the text, along with the fact that $I = 0$ at $t = 0$, to write the solution for $I(t)$, the induced current as a function of time.

(c) Below a certain threshold value for R, the induced current will oscillate. Determine this threshold resistance value R_T in terms of $B, l, L,$ and m.

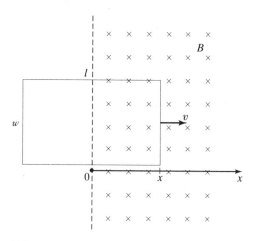

FIGURE 30–7

29. (III) A solenoid with N turns, cross-sectional area A, and length l has its internal volume ($= Al$) filled with air. When this solenoid is used as the inductor in an LC circuit, charge is found to oscillate at angular frequency ω. Suppose that, with no other changes to the LC circuit, the solenoid's internal volume is filled with a diamagnetic material (such as copper), so that the magnetic field within that volume changes by a factor of $K_m = 1 + \chi_m$, where χ_m is the diamagnetic material's magnetic susceptibility, and that, consequently, the circuit oscillates at a new angular frequency ω'. Using the fact that $|\chi_m| \ll 1$ for diamagnetic materials, show that the material's magnetic susceptibility can be found from $\chi_m \approx -\frac{2\Delta\omega}{\omega}$, where $\Delta\omega = \omega' - \omega$. If $f = \omega/2\pi = 1$ MHz, determine the change in frequency Δf (Hz) that would have to be detected in order to measure $\chi_m = -1 \times 10^{-5}$, a value typical of diamagnetic materials.

30. (III) A 30-cm long cylindrical copper wire of diameter 1 mm can be modeled electrically as a 6 mΩ resistance in series with a 13 nH inductance. If, at time $t = 0$, a wire of such construction is attached between the positive and negative terminals of a 0.1-V voltage source, over the ensuing time interval of duration 5τ, the current flow within the wire will be established. Here, τ is the time constant for the LR series circuit. To a good approximation, for times $t > 5\tau$, a steady current I_o flows in the wire. Over the time interval $0 \leq t \leq 5\tau$, the energy supplied to the wire by the voltage source is used to establish the wire's magnetic field and to create thermal energy.

(a) Determine the energy (J) required to establish the wire's magnetic field given that its current changes from zero to I_o.

(b) Over the time interval $0 \leq t \leq 5\tau$, determine the energy (J) from the battery that is transformed into thermal energy in the wire.

(c) During the 5τ time interval, compare (as a ratio) the thermal energy created with the energy stored in the wire's magnetic field.

CHAPTER 31

AC Circuits

Sections 31–1 to 31–4

1. (II) A resistor R is in parallel with a capacitor C, and this parallel combination is in series with a resistor R'. If connected to an ac voltage source of frequency ω, what is the equivalent impedance of this circuit at the two extremes in frequency (a) $\omega = 0$, and (b) $\omega = \infty$?

2. (II) A resistor $R = 1.0\,\text{M}\Omega$ is in parallel with a capacitor $C = 25\,\text{pF}$ in a circuit carrying ac current of frequency f.
   (a) In a certain range of frequencies, this parallel combination will have an effective resistance of approximately $1.0\,\text{M}\Omega$. Define this range to be when the capacitive reactance X_C is at least 100 times greater than the resistor's resistance; then determine the frequency range.
   (b) In a certain range of frequencies, this parallel combination will have an effective resistance of approximately the capacitance's reactance. Define this range to be when the capacitive reactance X_C is at most 100 times smaller than the resistor's resistance; then determine the frequency range. What are the maximum and minimum values of X_C in this frequency range?

Section 31–5

3. (II) An ac voltage source is connected in series with a $1.0\text{-}\mu\text{F}$ capacitor and a $1000\text{-}\Omega$ resistor. Using a digital ac voltmeter, the amplitude of the voltage source is measured to be $4.0\,\text{V}$ rms, while the voltages across the resistor and across the capacitor are found to be $3.0\,\text{V}$ rms and $2.7\,\text{V}$ rms, respectively. Determine the frequency of the ac voltage source. Why is the voltage measured across the voltage source not equal to the sum of the voltages measured across the resistor and across the capacitor?

4. (II) A 60-W light bulb is designed to operate with an applied ac voltage of $120\,\text{V}$ rms. This bulb is placed in series with an inductor L; this series combination is then connected to a 60-Hz 240-V rms voltage source. For the bulb to operate properly, determine the required value for L. Assume the bulb has resistance R and negligible inductance.

5. (II) An LRC series circuit with $R = 150\,\Omega$, $L = 20\,\text{mH}$, and $C = 2.0\,\mu\text{F}$ is powered by an ac voltage source of peak voltage $V_o = 100\,\text{V}$ and frequency $f = 600\,\text{Hz}$.
   (a) Determine the peak current that flows in this circuit.
   (b) Determine the phase angle of the source voltage relative to the current.
   (c) Determine the peak voltage across R and its phase angle relative to the source voltage.
   (d) Determine the peak voltage across L and its phase angle relative to the source voltage.
   (e) Determine the peak voltage across C and its phase angle relative to the source voltage.

6. (II) An LR circuit can be used as a "phase shifter." Assume that an "input" source voltage $V = V_o \sin(2\pi ft + \phi)$ is connected across a series combination of an inductor $L = 50\,\text{mH}$ and resistor R. The "output" of this circuit is taken across the resistor. If $V_o = 10\,\text{V}$ and $f = 200\,\text{Hz}$, determine the value of R so that the output voltage V_R lags the input voltage V by $30°$. Compare (as a ratio) the peak output voltage with V_o.

7. (II) Calculate the average power (W) dissipated as thermal energy in a device carrying current $I = (4.0\,\text{A})\sin(\omega t)$, if the voltage across the device is
   (a) $V = (10\,\text{V})\sin(\omega t + 30°)$.
   (b) $V = (10\,\text{V})\cos(\omega t)$.

8. (II) An LRC circuit with $L = 12\,\text{mH}$, $R = 250\,\Omega$, and $C = 670\,\text{nF}$ is powered by a 1.0-kHz ac voltage source. Determine the time difference Δt (μs) between when the voltage source and the current within the circuit reach their peak values. Does the voltage peak occur before or after the current peak?

9. (II) You have constructed a 5.0-cm diameter solenoid of length 30.0 cm to produce the uniform magnetic field needed in your newest experiment. Unfortunately, while winding on its coils you got distracted talking to your lab mate and lost count of the solenoid's total number of turns N. So, you place the solenoid in an LR series circuit with resistance $R = 1000\,\Omega$ and ac voltage source of frequency $f = 1.0\,\text{kHz}$. You then use a two-channel oscilloscope to simultaneously measure the voltage across R and across the voltage source and find that the peak voltage across the source occurs $140\,\mu\text{s}$ before the peak voltage across R. Assuming that the solenoid has self-inductance L and negligible resistance (compared with $1000\,\Omega$), use the given information to determine N.

10. (II) Consider the following method to measure an unknown capacitance C. Assume the capacitor is placed in series with a known resistance R and that this series combination is connected to an ac voltage source of peak voltage V_o and frequency f. In this circuit, an ac voltmeter (or oscilloscope) is connected across the resistor and the peak voltage V_{Ro} is measured. Prove that the capacitance value can be found from

$$C = \frac{1}{2\pi f R}\frac{V_{Ro}}{\sqrt{V_o^2 - V_{Ro}^2}}.$$

11. (III) An unknown capacitance C is placed in series with a resistance that is known to be $R = 10.0$ kHz. This series combination is connected to an ac voltage source of fixed peak voltage $V_o = 5.00$ V and variable frequency f, and an oscilloscope is connected across the capacitor so that the peak voltage V_{Co} across this circuit element can be measured. With this set-up, the frequency f is varied and the associated values of V_{Co} are recorded in Table 31–1.

TABLE 31–1

f (kHz)	2.00	4.00	6.00	8.00	10.00
V_{Co} (V)	4.68	3.99	3.31	2.76	2.34

(a) Explain why graphing $\left(\dfrac{V_o}{V_{Co}}\right)^2$ vs. f^2 should yield a straight-line plot. What are the line's slope and y-intercept expected to be theoretically (in terms of $V_o, f, R,$ and C)?

(b) Using the data in Table 31–1, graph $\left(\dfrac{V_o}{V_{Co}}\right)^2$ vs. f^2 and show that a straight-line plot does indeed result. Determine C from the slope of this line. Does the y-intercept have the expected value?

12. (III) Assume that a stereo speaker is electrically equivalent to a resistor $R = 8\,\Omega$. Consider placing a speaker in the following ac circuits, where the ac voltage source $V = V_o \sin(2\pi f + \phi)$ represents the audio signal (peak voltage V_o, frequency f) delivered from the stereo amplifier to the speaker. One of these circuits is a "low-pass filter" in which, for small f and large f, the peak voltage V_{Ro} across the speaker approximately equals V_o and zero, respectively. Thus, this circuit "passes" low frequency audio signals to the speaker and suppresses high frequencies. The "crossover frequency" f_{co} is at the boundary between these two regimes and is defined as the frequency at which $V_{Ro} = V_o/\sqrt{2}$. The other circuit is a "high-pass filter" that performs conversely to the low-pass circuit.

(a) If a speaker is placed in the RL circuit shown in Fig. 31–1a, show that the peak voltage across the speaker is given by $V_{Ro} = V_o/\sqrt{1 + (2\pi f L/R)^2}$. Should the speaker in this circuit be a (low-frequency) woofer or a (high-frequency) tweeter? Determine an expression for this circuit's crossover frequency.

(b) If a speaker is placed in the RC circuit shown in Fig. 31–1b, show that the peak voltage across the speaker is given by $V_{Ro} = V_o/\sqrt{1 + (2\pi f RC)^{-2}}$. Should the speaker in this circuit be a (low-frequency) woofer or a (high-frequency) tweeter? Determine an expression for this circuit's crossover frequency.

(c) In the crossover network described in Section 31–4 of the text, these two circuits are used to direct an audio-frequency signal to the appropriate woofer or tweeter in a speaker system. Assume that one wants the same crossover frequency for both circuits. Show then that $L/C = 64\,\Omega^2$. [For a two-speaker system, the crossover frequency is typically chosen near 1.5 kHz.]

13. (III) A certain coil behaves electrically as resistance $R = 30\,\Omega$ in series with inductance $L = 10$ mH. What resistor R' should be placed in series with the coil if one wants a current of 1.5 A rms to flow in this circuit when powered by a 60-Hz 120-V rms voltage source? When the circuit carries 1.5 A rms, what is the rms voltage across the coil?

14. (III) An electromagnet driven by an alternating current supply produces a spatially uniform time-dependent magnetic field $B = B_o \cos(\omega t)$ in the region between its pole faces. A circular wire loop of radius r and resistance R is placed in this region with its plane oriented perpendicular to the field.
(a) If presented with this problem in Chapter 29, we would have (naively) solved it under the assumption that the loop has negligible self-inductance L. Assuming $L = 0$, determine the amplitude I_o of the current induced in the loop.
(b) We now know that a loop does in fact possess nonzero self-inductance L. Include this effect in your analysis and determine the amplitude I'_o of the current induced in the loop.
(c) Show that in the low-frequency regime $\omega \ll R/L$, the presence of self-inductance has no noticeable effect on the induced current (i.e., $I'_o \approx I_o$).
(d) Show that self-inductance makes a dramatic impact at high frequencies (as ω approaches infinity) by finding the limiting values for I_o and I'_o in this regime. At what frequency ω is I'_o half of the value of I_o?

15. (III) Investigate the properties of the circuit shown in Fig. 31–2 where $I = I_o \sin(\omega t)$ is the current flowing from the ac voltage source of frequency $f = \omega/2\pi$ and V_C is the voltage across the capacitor. For this circuit, the peak source voltage V_o is considered to be the "input" and the peak voltage across the capacitor V_{Co} is considered to be the "output."

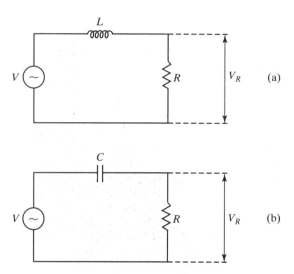

FIGURE 31–1

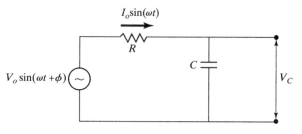

FIGURE 31–2

(a) Define the circuit's frequency-dependent "gain" as $G = V_{Co}/V_o$. Prove that $G = [1 + (f/f_o)^2]^{-1/2}$, where $f_o = 1/2\pi RC$.

(b) Rewrite the expression for G in terms of the dimensionless frequency $f' = f/f_o$; then plot G vs. f' in the range $0 \le f' \le 5$.
(c) This circuit is known as a "low-pass filter" because low-frequency (small f') input voltages V_o are "passed" unattenuated to the output (i.e., $G \approx 1$), while for high frequencies (large f'), the output voltage is much less than the input voltage. The frequency $f' = 1$ (i.e., $f = f_o$) is usually taken as the boundary between these two regimes. What is the value for V_{Co} (in terms of V_o) when $f = f_o$? If $C = 1.0 \, \mu\text{F}$ and you want the circuit to only pass frequencies within the audio range (<20 kHz), what should R be?

Section 31–6

16. (II) You have constructed a 5.0-cm diameter solenoid of length 30.0 cm to produce the uniform magnetic field needed in your newest experiment. Unfortunately, while winding on its coils you got distracted talking to your lab mate and lost count of the solenoid's total number of turns N. So, you place the solenoid in series with a capacitance $C = 250$ nF, a resistance R, an ac ammeter, and an ac voltage source whose frequency f can be varied. With the source's peak voltage kept fixed, you then vary its frequency until you find that the ammeter gives its maximum reading when $f = 725$ Hz. Use this information to determine N.

17. (II) The frequency of the ac voltage source (peak voltage V_o) in an LRC circuit is tuned to the circuit's resonant frequency $f_o = 1/2\pi\sqrt{LC}$.
(a) Show that the peak voltage across the capacitor is $V_{Co} = \dfrac{V_o}{2\pi} \dfrac{T_o}{\tau}$, where T_o is the period of the resonant frequency and $\tau = RC$ is the time constant for charging the capacitor C through the resistor R.
(b) Define $\beta = \dfrac{T_o}{2\pi\tau}$ so that $V_{Co} = \beta V_o$. Then β is the "amplification" of the source voltage across the capacitor. If a particular LRC circuit contains a 2.0-nF capacitor and has a resonant frequency of 5000 Hz, what value of R will yield $\beta = 100$?

18. (III) Capacitors made from piezoelectric materials are commonly used as sound transducers ("speakers"). A disadvantage of certain such speakers is that they require large operating voltages. One method for providing the required voltage is to include the speaker as part of an LRC circuit as shown in Fig. 31–3, where the speaker is modeled electrically as the capacitance $C = 1.0$ nF. Take $R = 35 \, \Omega$ and $L = 50$ mH.

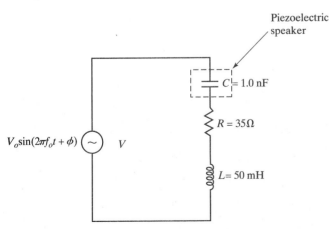

FIGURE 31–3

(a) What is the resonant frequency f_o for this circuit?
(b) If the voltage source has peak amplitude $V_o = 2.0$ V at frequency $f = f_o$, find the peak voltage across the speaker (i.e., the capacitor C). How many times larger is V_{Co} than V_o?

General Problems

19. (II) A function generator is a laboratory instrument that can produce a sinusoidal voltage. The equivalent circuit for a function generator consists of an ac voltage source (peak voltage V_o and frequency f) in series with a resistor $R_G = 50 \, \Omega$ and capacitor $C_G = 50$ pF. The generator's output is across C_G. Assume that an external "load" resistor $R_L = 1000 \, \Omega$ is connected across the generator's output as shown in Fig. 31–4.
(a) Because it is in parallel with R_L, if the reactance of C_G is much greater than R_L (say, at least 100 times greater), then the effect of C_G within this circuit can be ignored. Over what frequency range will this criterion be fulfilled? Explain why (to a good approximation) the voltage across R_L will equal V_o in this frequency range.
(b) If the reactance of C_G is less than ten times R_L, then a significant ac current will flow through C_G and the voltage across R_L will be less than V_o. Within what frequency range will this criterion be fulfilled? [This effect limits the "bandwidth" of the function generator.]

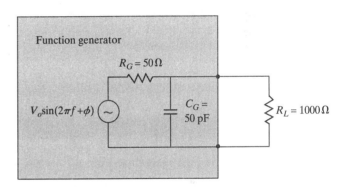

FIGURE 31–4

20. (II) An ac voltage source $V = V_o \sin(\omega t + 90°)$ is connected across an inductor L and current $I = I_o \sin(\omega t)$ flows in this circuit.
(a) Directly calculate the average power delivered by the source over one period T of its sinusoidal cycle via the integral $\overline{P} = \int_0^T VI \, dt/T$.
(b) Apply the relation $\overline{P} = I_{rms} V_{rms} \cos\phi$ to this circuit and show that the answer you obtain is consistent with that found in part (a).

21. (II) In an LRC series circuit, let's investigate the difference between the inductive reactance X_L and the capacitive reactance X_C when the frequency ω of an alternating voltage source is below and above that of the circuit's resonant frequency ω_o.
(a) Prove that $X_L - X_C = \sqrt{\dfrac{L}{C}}\left[\dfrac{\omega}{\omega_o} - \dfrac{\omega_o}{\omega}\right]$.
(b) Define the dimensionless frequency $\omega' = \omega/\omega_o$; then sketch the plot of $(X_L - X_C)$ vs. ω' over the entire possible range of source frequencies (i.e., $0 \le f' \le \infty$).
(c) Based on your sketch, what can you conclude about ϕ, the phase angle between the source voltage and the current in this circuit (i.e., the value of ϕ at a particular ω, the sign of ϕ over a certain range of ω)?

22. (II) In three-phase power, three 120° out-of-phase sinusoidal ac voltages oscillate at frequency ω and peak voltage V_o. Show that the potential difference between any two of the three-phase power lines [i.e., $(V_2 - V_1), (V_3 - V_2),$ and $(V_3 - V_1)$] is itself a sinusoidal ac voltage. What are the frequency and the peak voltage of each of these ac potential differences in terms of ω and V_o?

23. (II) The ac voltage source in an LRC circuit oscillates at the circuit's resonant frequency ω_o so that the peak voltages across the inductor and capacitor are V_{Lo} and V_{Co}, respectively. Imagine then that the value of inductance L in this circuit is increased by a factor α, while at the same time the value of capacitance C is decreased by the same factor α.
(a) What is the resonant frequency of the new circuit?
(b) How does the peak voltage across the inductor and across the capacitor in the new circuit compare with the values in the original circuit?

24. (II) In a certain LRC series circuit, when the ac voltage source has a particular frequency f, the peak voltage across the inductor is ten times greater than the peak voltage across the capacitor. Determine f in terms of resonant frequency f_o of this circuit.

25. (II) In a plasma globe, a hollow glass sphere is filled with low-pressure gas and a small spherical metal electrode is located at its center. Assume an ac voltage source of peak voltage V_o and frequency f is applied between the metal sphere and the ground, and that a person is touching the outer surface of the globe with a fingertip, whose approximate area is 1 cm². The equivalent circuit for this situation is shown in Fig. 31–5, where R_G and R_P are the resistances of the gas and the person, respectively, and C is the capacitance formed by the gas, glass, and finger.

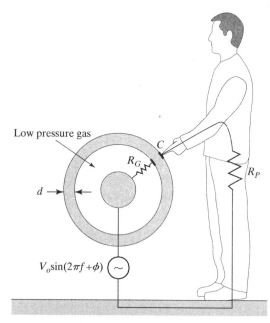

FIGURE 31–5

(a) Model C as a parallel-plate capacitor. The conductive gas and the person's moderately conductive fingertip form the opposing plates of area $A = 1$ cm². The plates are separated by glass (dielectric constant $K = 5$) of thickness $d = 2$ mm. Determine C in pF.
(b) In a typical plasma globe, $f = 10$ kHz. Determine the reactance X_C of C at this frequency in MΩ.

(c) In a typical plasma globe, $V_o = 3000$ V. With this high voltage, the dielectric strength of the gas is exceeded and the gas becomes ionized. In this "plasma" state, the gas emits light ("sparks") and is highly conductive so that $R_G \ll X_C$. Assuming also that $R_P \ll X_C$, estimate the peak current that flows in the given circuit. Is this level of current dangerous?
(d) If the plasma globe operated at $f = 1$ MHz, estimate the peak current that would flow in the given circuit. Is this level of current dangerous?

26. (III) An ac "adapter" transformer, whose "primary-side" plugs into an ac wall receptacle, is commonly used to power and recharge electronic devices (e.g., laptops). Suppose the primary-side of such a transformer is plugged into an ac wall receptacle, but its secondary-side is not attached to any device. Advanced circuit theory determines that the equivalent circuit for this "open-circuit" situation is as shown in Fig. 31–6, where R and L are due to magnetic effects within the transformer's magnetic core material. Let the ac source voltage be $V = V_o \sin(\omega t)$.

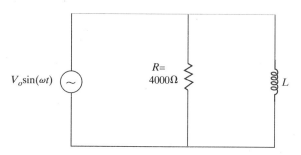

FIGURE 31–6

(a) Show that the current flow from the source is
$$I = \frac{V_o}{R}\left[\sin(\omega t) - \frac{R}{\omega L}\cos(\omega t)\right].$$
(b) Directly calculate the average power delivered by the source over one period T of its sinusoidal cycle via the integral $\overline{P} = \int_0^T VI\, dt/T$. Show that $\overline{P} = V_{rms}^2/R$.
(c) For a typical adapter transformer, $R = 4000\ \Omega$. If its primary-side is plugged into a 60-Hz 120-V rms wall receptacle, but its secondary-side is left unattached, how much power is consumed by the transformer?
(d) If each of the 100 million households in the United States has one adapter transformer plugged in, but not being used to power a device, how much energy is being wasted per year? If electrical energy cost $ 0.10 per kWh, how much money is being spent to power these nonproductive transformers per year?

27. (III) Even if the dielectric that fills the space between a capacitor's plates is a good electrical insulator, it has a finite resistivity. Thus, the "leaky capacitor" model, consisting of a capacitance C in parallel with "leakage" resistance R_L, provides an accurate picture of a real-life capacitor. Consider a leaky capacitor connected to the ac voltage source $V = V_o \sin(\omega t)$ as in Fig. 31–7. The current that flows in this circuit will be "phase-shifted" in comparison with the source voltage and can be written as the sum $I = I_s \sin(\omega t) + I_c \cos(\omega t)$. A special instrument called a lock-in amplifier, if placed in this circuit, can measure both I_s and I_c. Show that C and R_L can be determined from these two current measurements.

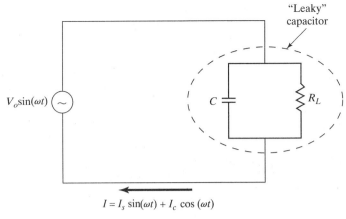

FIGURE 31–7

28. (III) A "notch" filter strongly attenuates frequencies near a particular value f_o, but "passes" all other frequencies from its input to its output unaltered. Let's explore how an LRC series circuit can function in such a manner. In Fig. 31–8, take the ac voltage source of peak voltage V_o and angular frequency ω as the "input" and the voltage V' across both the inductor and the capacitor as the "output."
    (a) Determine an expression for V' as a function of V_o and resistance R as well as X_L and X_C, the inductive and capacitive reactances, respectively.
    (b) Use your expression for V' to show that low-frequency (near $\omega = 0$) and high-frequency (as ω approaches infinity) inputs are not attenuated by this circuit (i.e., $|V'| \approx V_o$).
    (c) Determine the "notch" frequency f_o (i.e., the frequency for which $V' = 0$). If one wishes $f_o = 60\,\text{Hz}$ (a common choice to eliminate power-line noise from an experimental signal), what should the product LC equal?

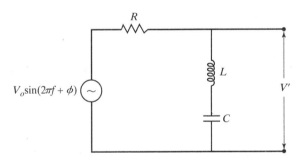

FIGURE 31–8

29. (III) To detect vehicles at traffic stops, wire loops with dimensions on the order of 2 m are often buried horizontally oriented in roadways. Assume the self-inductance of such a loop is $L = 5.0\,\text{mH}$ and that it is part of an LRC circuit as shown in Fig. 31–9 with $C = 0.10\,\mu\text{F}$ and $R = 50\,\Omega$. The ac voltage has frequency f and rms voltage V_{rms}.

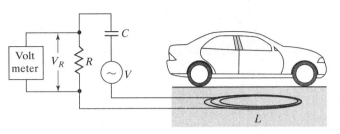

FIGURE 31–9

(a) The frequency f is chosen to match the resonant frequency f_o of the circuit. Find f_o and determine what the rms voltage $(V_R)_{rms}$ across the resistor will be when $f = f_o$.
(b) Assume that f, C, and R never change, but that, when a car is located above the buried loop, the loop's self-inductance decreases by 10% (due to induced eddy currents in the car's metal parts). Determine by what factor the voltage $(V_R)_{rms}$ decreases in this situation in comparison with when there is no car above the loop. [By monitoring $(V_R)_{rms}$ with a voltmeter, the presence of a car can be detected.]

30. (III) In an LRC circuit, the ac source voltage's frequency $f = 60\,\text{Hz}$ is "detuned far off-resonance." That is, $f \ll f_o$, where $f_o = 1/2\pi\sqrt{LC}$. Consider two possible choices for the inductor and the capacitor in this circuit: choice 1 in which $L_1 = 1.0\,\mu\text{H}$ and $C_1 = 10\,\text{nF}$; choice 2 in which $L_2 = 1.0\,\text{mH}$ and $C_2 = 10\,\text{pF}$. Take the peak source voltage to be $V_o = 10\,\text{kV}$ and the resistance to be $R = 100\,\Omega$.
(a) Show that f_o is the same value for both choice 1 and choice 2. Determine f_o in MHz.
(b) Determine the peak current that flows in the circuit for both choice 1 and choice 2.
(c) If one's goal is to create the greatest current flow in this off-resonant circuit, which choice (1 or 2) is best? Offer a qualitative explanation for why this is so. [A circuit similar to this is used as the primary in a Tesla coil.]

Maxwell's Equations and Electromagnetic Waves

Section 32–1

1. (II) Suppose that a circular parallel plate capacitor has radius $R = 3.0$ cm and plate separation $d = 5.0$ mm. A sinusoidal potential difference $V = V_o \sin(2\pi f t)$ is applied across the plates, where $V_o = 150$ V and $f = 60$ Hz. In the region between the plates, let r be the radial distance from the capacitor's central axis. For $r = R$, show that the induced magnetic field is given by $B = B_o(R)\cos(2\pi f t)$ and determine the value of $B_o(R)$ (i.e., the magnitude of the induced magnetic field at the radial distance $r = R$.) Compare (as a ratio) this field strength with that of Earth's field ($50\ \mu$T). Ignore fringing.

2. (III) Within a circle of radius r_o on a page, a time- and space-dependent electric field $E(r, t) = E_o(r_o^2 - r^2)t$ is directed out of the plane of the page, where r is the radial distance measured from the circle's center and E_o is a constant. Outside the circle, $E = 0$ everywhere. Find the magnitude of the induced magnetic field $B(r, t)$ at locations (a) $r < r_o$ and (b) $r > r_o$.

3. (III) Suppose that a circular parallel-plate capacitor has radius $R = 3.0$ cm and plate separation $d = 5.0$ mm. A sinusoidal potential difference $V = V_o \sin(2\pi f t)$ is applied across the plates, where $V_o = 150$ V and $f = 60$ Hz. In the region between the plates, show that the magnitude of the induced magnetic field is given by $B = B_o(r)\cos(2\pi f t)$, where r is the radial distance from the capacitor's central axis, and determine the expression for the amplitude $B_o(r)$ of this time-dependent (sinusoidal) field when $r \leq R$ and $r > R$. Plot $B_o(r)$ in Tesla for the range $0 \leq r \leq 10$ cm.

Section 32–3

4. (II) At sufficiently large distances x along the axis of a current-carrying loop, the magnetic field $\mathbf{B} \approx \dfrac{\mu_o \mu}{2\pi}\dfrac{1}{x^3}\mathbf{i}$, where μ is is the loop's (constant) magnetic dipole moment and the loop's axis defines the x-axis. Let's postulate that, in some region of space, the magnetic field $\mathbf{B} = \dfrac{\mu_o \mu}{2\pi}\dfrac{1}{x^3}\mathbf{i}$ everywhere (not just along a certain line). Prove that this magnetic field cannot exist in nature. What differs between the postulated field and the actual field due to a current-carrying loop?

Section 32–5

5. (II) In the modern SI system of units, the definition of the meter follows from defining the speed of light in free space to be 299,792,458 m/s exactly. The permittivity of free space also is theoretically defined as $\epsilon_o = 1/c^2 \mu_o$.
(a) What is the exact numerical value of ϵ_o?
(b) Determine the numerical value of ϵ_o to six significant figures and show that the value you obtain is consistent with that given at the front of the text.

6. (III) Consider two possible candidates $E(x,t)$ as solutions of the wave equation for an EM wave's electric field. Let A and α be constants. Show that
(a) $E(x, t) = Ae^{-\alpha(x-vt)^2}$ satisfies the wave equation.
(b) $E(x, t) = Ae^{-(\alpha x^2 - vt)}$ does not satisfy the wave equation.
[Any function of the form $f(x \pm vt)$ is a solution for the wave equation.]

7. (III) In three dimensions, the wave equation for the electric-field portion of an EM wave is $\dfrac{\partial^2 \mathbf{E}}{\partial x^2} + \dfrac{\partial^2 \mathbf{E}}{\partial y^2} + \dfrac{\partial^2 \mathbf{E}}{\partial z^2} = \dfrac{1}{v^2}\dfrac{\partial^2 \mathbf{E}}{\partial t^2}.$ Demonstrate that $\mathbf{E} = E_o \sin(k_x x + k_y y + k_z z - \omega t)$ is a solution to the three-dimensional wave equation, as long as the constants $k_x, k_y, k_z,$ and ω are related properly. What is the required relation between these constants?

8. (III) When an EM wave travels in a good conductor (electrical conductivity σ), rather than free space, the right side of Ampere's law will be dominated by the conduction current I and the displacement current will be negligible. In the derivation of Section 32–5 in the text for the EM wave's electric field E and magnetic field B magnitudes, Eq. 32–9 will remain unchanged; however, Eq. 32–10 will be modified. Assume the conductor obeys Ohm's law, which can be written in the form $J = \sigma E$, where J is the current density.
(a) Using $I = J\,dA = (\sigma E)(dx\,\Delta z)$, show that the "modified" Eq. 32–10 in a good conductor is $\dfrac{\partial B}{\partial x} = -\mu_o \sigma E.$
(b) Using the derivation of Eq. 32–13a as a guide, show that in a good conductor, $\dfrac{\partial^2 E}{\partial x^2} = \mu_o \sigma \dfrac{\partial E}{\partial t}.$

9. (III) When an EM wave travels in a good conductor (electrical conductivity σ), rather than free space, the EM wave's electric field E obeys the following differential equation $\dfrac{\partial^2 E}{\partial x^2} = \mu_o \sigma \dfrac{\partial E}{\partial t},$ rather than Eq. 32–13a in the text.
(a) Verify that the solution for E in a good conductor is the damped sinusoidal wave $E = E_o e^{-kx}\sin(kx - \omega t)$ and find the expression for k in terms of σ and ω.

(b) The amplitude of E will be reduced to 37% of its initial value as the EM wave travel distance $d = 1/k$ through the conductor. Thus, when an EM wave enters a metal from air, it will only be able to penetrate the metal to a depth of a few times d. If a 900-MHz EM wave being used in cell-phone communication is incident on a steel plate with $\sigma = 1 \times 10^7 \, (\Omega \cdot m)^{-1}$, determine d (μm). [This problem illustrates a common way in which conducting building materials block cell-phone communications.]

Section 32–6

10. (II) The global positioning system (GPS) functions by determining the travel times of EM waves from various satellites to a land-based GPS receiver. When the receiver is moved, the distance traveled by these EM waves changes and the change in travel times must be sensed. If changes in travel distances on the order of 3 m are to be detected, how accurately must the travel times be measured?

11. (II) When an EM wave of wavelength 1064 nm is input into a "frequency-doubling" crystal, an EM wave with twice the input frequency is output. What is the wavelength of the output EM wave? In which region (radio, microwave, etc.) of the EM spectrum are the input and the output EM waves?

Section 32–7

12. (II) The Arecibo radio telescope in Puerto Rico can detect a radio wave with an intensity as low as 1×10^{-23} W/m². As a "best-case" scenario for communication with extraterrestrials, consider the following: An advanced civilization located at point A, a distance x away from Earth, is somehow able to harness the entire power output of a Sun-like star, converting that power completely into a radio-wave signal, which is transmitted uniformly in all directions from A. In order for Arecibo to detect this radio signal, what is the maximum value for x (ly)? How does this maximum value compare with the 100,000-ly size of our Milky Way galaxy? The intensity of sunlight at Earth's orbital distance from the Sun is 1400 W/m².

13. (II) How practical is solar power for various devices? Assume that on a sunny day, sunlight has an intensity of 1000 W/m² at the surface of Earth and that, when illuminated by that sunlight, a solar-cell panel can convert 10% of the sunlight's energy into electric power. For each device given below, calculate the area A of solar panel needed to power it.
(a) A calculator consumes 50 mW. Find A in cm². Is A small enough so that the solar panel can be mounted directly on the calculator that it is powering?
(b) A hair dryer consumes 1500 W. Find A in m². Assuming no other electronic devices are operating within a house at the same time, is A small enough so that the hair dryer can be powered by a solar panel mounted on the house's roof?
(c) A car requires 20 hp for highway driving at constant velocity (this car would perform poorly in situations requiring acceleration). Find A in m². Is A small enough so that this solar panel can be mounted directly on the car and power it in "real time"?

14. (II) At a certain point in space and at a particular moment in time, the electric and magnetic fields of an EM wave are given by $\mathbf{E} = [3.0\,\mathbf{i} + 6.0\,\mathbf{j}]E_o$ and $\mathbf{B} = B_o\mathbf{k}$, where E_o and B_o are the maximum values of the wave's electric and magnetic fields, respectively. In what direction is the EM wave moving?

15. (II) Earth's magnetic field, with a magnitude on the order of 50 μT, has sufficient strength to deflect a compass. What intensity $\overline{S}$ would visible light have, if the maximum value of its magnetic field were equal to that of Earth's field? If light of this intensity were shone on a compass such that its magnetic field were aligned perpendicular to the compass needle's magnetic dipole moment, would the needle deflect?

Section 32–8

16. (II) Laser light can be focused to a spot with a radius r equal to its wavelength λ. Suppose that a 1.0-W beam of green laser light ($\lambda = 5 \times 10^{-7}$ m) is used to form such a spot and that a cylindrical particle of the same size (i.e., radius and height equal to r) is illuminated by the spot as shown in Fig. 32–1. Determine the acceleration of the particle, if its density equals that of water. [This order-of-magnitude calculation convinced researchers of the feasibility of "optical tweezers."]

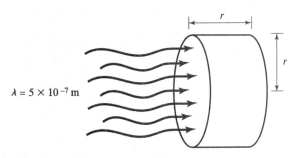

FIGURE 32–1

17. (II) The powerful laser used in a laser light show provides a 3-mm diameter beam of green light with a power of 3 W. While a space-walking astronaut is outside the Space Shuttle, her colleague inside the Shuttle playfully aims such a laser beam at the astronaut's space suit. The masses of the suited astronaut and the Space Shuttle are 120 kg and 103,000 kg.
(a) Assuming the suit is perfectly reflecting, determine the "radiation-pressure" force exerted on the astronaut by the laser beam.
(b) Assuming the astronaut is separated from the Shuttle's center of mass by 20 m, model the Shuttle as a sphere in order to estimate the gravitation force it exerts on the astronaut.
(c) Which of the two forces is larger and by what factor?

Section 32–9

18. (II) Consider a rectangular-shaped tunnel of width w, height h, and length l cut through a conducting material. Advanced analysis shows that the minimum-frequency EM wave that can pass through this tunnel forms a standing wave across the tunnel's width with a node at each wall and an antinode at the center as shown in Fig. 32–2. For a 15-m wide tunnel, show that AM radio waves will not be able to propagate through this tunnel, while FM radio waves will.

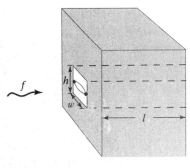

FIGURE 32–2

General Problems

19. (II) (a) The electric field at every point along the length l of a conducting wire varies with time according to $E = E_o \sin \omega t$. Assuming the wire's resistance is R, show that $\dfrac{\bar{P}}{l^2} = \dfrac{E_{rms}^2}{R}$, where $\bar{P}$ is the average power supplied to the wire by the electric field.
(b) An EM wave in free space transports average power $\bar{P}$ through a square area of dimension $l \times l$ whose plane is perpendicular to the direction in which the wave is moving. In analogy with a conducting wire, define the "impedance of free space" Z_o via $\dfrac{\bar{P}}{l^2} = \dfrac{E_{rms}^2}{Z_o}$, where E_{rms} is the rms value of the EM wave's electric field. Prove that $Z_o = \sqrt{\mu_o/\epsilon_o}$ and determine its value in ohms. [The impedance of free space is important in antenna design.]

20. (II) Radio-controlled clocks throughout the United States receive a radio signal from a transmitter in Fort Collins, Colorado that accurately (within a microsecond) marks the beginning of each minute. A slight delay, however, is introduced because this signal must travel from the transmitter to the clocks. Assuming Fort Collins is approximately in the middle of the United States and that the country fits within a 5000-km diameter circle, what is the longest travel-time delay for a clock located within the United States?

21. (II) The metal walls of a microwave oven form a cavity of dimensions $37 \text{ cm} \times 37 \text{ cm} \times 20 \text{ cm}$. When 2.45-GHz microwaves are continuously introduced into this cavity, reflection of incident waves from the walls set up standing waves with nodes at the walls. Along the 37-cm dimension of the oven, how many nodes exist (excluding the nodes at the wall) and what is the distance between adjacent nodes? [Because no heating occurs at these nodes, most microwaves rotate food while operating.]

22. (II) The EM wave used in cell-phone communication commonly has a frequency near 900 MHz. The optimal length of the antenna used to transmit and receive an EM wave is equal to one-fourth of its wavelength. What is the optimal length of a cell phone's antenna?

23. (II) In free space ("vacuum"), where the net charge and current flow is zero, the speed of an EM wave is given by $v = 1/\sqrt{\epsilon_o \mu_o}$. If, instead, an EM wave travels in a nonconducting ("dielectric") material with dielectric constant K, then $v = 1/\sqrt{\epsilon \mu_o}$. For frequencies corresponding to the visible spectrum (near 5×10^{14} Hz), the dielectric constant of water is 1.77. Predict the speed of light in water and compare this value (as a percentage) with the speed of light in a vacuum.

24. (II) The electric field of an EM wave traveling along the x-axis in free space is given by $E_y = E_o \exp[-\alpha^2 x^2 - \beta^2 t^2 + 2\alpha\beta xt]$, where E_o, α, and β are constants.
(a) Is the EM moving in the $+x$ or $-x$ direction?
(b) Express β in terms of α and c (speed of light in free space).
(c) Determine the expression for the magnetic field of this EM wave.

25. (III) A hemisphere of radius R is placed in a region of space within which exists a uniform magnetic field of magnitude B directed perpendicular to the hemisphere's circular base (Fig. 32–3).
(a) Using the definition of Φ_B through an "open" surface, calculate (via explicit integration) the magnetic flux through the hemisphere. [Hint: In Appendix C of the text, it is shown that, on the surface of a sphere, the infinitesimal area located between the angles θ and $\theta + d\theta$ is $dA = (2\pi R \sin \theta)(R d\theta) = 2\pi R^2 \sin \theta \, d\theta$.]
(b) Choose an appropriate gaussian surface, then use Gauss's law for magnetism to obtain the same result as found in part (a) for the magnetic flux through the hemisphere much more easily.

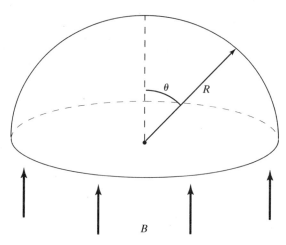

FIGURE 32–3

26. (III) (a) In Section 32–1 of the text, starting with $Q = CV$ for an air-filled capacitor, the expression for I_D is derived. Assuming, instead, that the capacitor is filled with a material of dielectric constant K, identify the (minor) change necessary in this derivation and show that it yields $I_D = K\epsilon_o \dfrac{d\Phi_E}{dt}$.
(b) A capacitor with circular plates of radius r_o and plate separation d has the volume between its plates filled with a material of dielectric constant $K = 5$ and resistivity $\rho = 1 \times 10^4 \, \Omega \cdot \text{m}$. An ac voltage $V = V_o \sin(\omega t)$ of frequency $f = \omega/2\pi$ is applied across the plates. Between the plates, at radial distance $r = r_o$ from the capacitor's central axis, show that the ratio of the maximum values of the conduction current I_C and displacement current I_D is given by $1/K\epsilon_o \rho \omega$.
(c) Evaluate I_C/I_D for $f = 1$ kHz, 1 MHz, and 1 GHz. At which frequency can the material between the plates be considered a conductor and at which frequency can it be considered an insulator?

27. (III) Suppose that a right-moving EM wave overlaps with a left-moving EM wave so that, in a certain region of space, the total electric field in the y-direction and magnetic field in the z-direction are given by $E_y = E_o \sin(kx - \omega t) + E_o \sin(kx + \omega t)$ and $B_z = B_o \sin(kx - \omega t) - B_o \sin(kx + \omega t)$.
(a) Find the mathematical expression that represents the standing electric and magnetic waves in the y- and z-directions, respectively.
(b) Determine the Poynting vector and find the x locations at which it is zero at all times.

28. (III) The electric and magnetic fields of a certain EM wave in free space are given by $\mathbf{E} = E_o \sin(kx - \omega t)\mathbf{j} + E_o \cos(kx - \omega t)\mathbf{k}$ and $\mathbf{B} = B_o \cos(kx - \omega t)\mathbf{j} - B_o \sin(kx - \omega t)\mathbf{k}$.
   (a) Prove that $\mathbf{E}$ and $\mathbf{B}$ are perpendicular to each other at all times.
   (b) For this wave, $\mathbf{E}$ and $\mathbf{B}$ are in a plane parallel to the y-z plane. Show that the wave moves in a direction perpendicular to both $\mathbf{E}$ and $\mathbf{B}$.
   (c) At any arbitrary choice of position x and time t, prove that the magnitudes of $\mathbf{E}$ and $\mathbf{B}$ always equal E_o and B_o, respectively.
   (d) At $x = 0$, draw the orientation of $\mathbf{E}$ and $\mathbf{B}$ in the y-z plane at $t = 0$. Then qualitatively describe the motion of these vectors in the y-z plane as time increases.
   [The EM wave in this problem is "circularly polarized."]

29. (III) Imagine that a steady current I flows in a cylindrical wire of radius r_o and resistivity ρ.
   (a) If the current is then changed at a rate dI/dt, show that a displacement current I_D exists in the wire of magnitude $\epsilon_o \rho \dfrac{dI}{dt}$.
   (b) If the current in a copper wire were changed at the rate of 1.0 A per microsecond, determine the magnitude of I_D.
   (c) Determine the magnitude of the magnetic field B_D (T) created by I_D at the surface of a copper wire with $r_o = 1.0$ mm. Compare (as a ratio) B_D with the field created at the surface of the wire by a steady current of 1.0 A.

30. (III) The current I in a very long solenoid shown in Fig. 32–4 (N turns, length l, radius R) increases with time t according to $I = \alpha t$, where α is a constant. Let $\hat{\mathbf{r}}$ be the unit vector directed radially away from the central axis of the solenoid.

   (a) Determine the magnitude of electric field E induced inside the solenoid at $r = R$.
   (b) Determine the Poynting vector $\mathbf{S}$ (magnitude and direction) inside the solenoid at $r = R$.
   (c) The rate at which energy enters the solenoid in the form of electromagnetic fields can be found through the integral $\int \mathbf{S} \cdot d\mathbf{A}$, where $d\mathbf{A}$ is an inward directed infinitesimal area vector on the solenoid's surface. Compute this integral over the solenoid's cylindrical surface area. Go on to show that your result equals the rate at which energy stored in the solenoid's magnetic field is increasing.

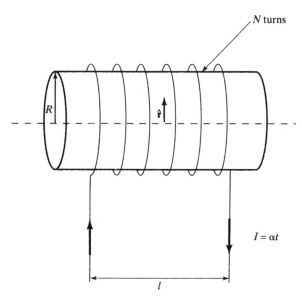

FIGURE 32–4

CHAPTER 33

Light: Reflection and Refraction

Section 33–2

1. (II) The 10-cm height of beaker A is filled with ethyl alcohol ($n_A = 1.36$), while the 10-cm height of identical beaker B contains a layer of water ($n_W = 1.33$) upon which floats a layer of mineral oil ($n_O = 1.47$). If vertically directed light is to take the same time in traversing the 10-cm height of each beaker, what is the required thickness of the mineral oil layer?

2. (II) (a) In Michelson's rotating-mirror experiment, determine the roundtrip time T for the light traveling between Mt. Wilson and Mt. San Antonio.
   (b) Imagine that Michelson wanted to use his experimental apparatus to demonstrate that light travels slower in water than in air. This demonstration could be carried out by causing the light to pass through a water-filled tube of length l during (a small portion of) one leg of its roundtrip between Mt. Wilson and Mt. San Antonio. Assuming that Michelson's set-up was accurate enough to detect a 0.01% increase in T, determine the necessary value for l in order to observe the slowing of light in water.

3. (III) You, the renowned experimental physicist, are setting up a "time-of-flight" experiment to measure the speed of light accurately. In this experiment, a light pulse produced by a laser-diode source will travel to a mirror a distance x away, reflect, and then return to a photodetector mounted next to the laser diode. By connecting the photodetector to an expensive oscilloscope, you will be able to measure the pulse's roundtrip travel time T to an accuracy of $\Delta t = 1$ ns. Compared to the measurement of T, there is negligible error in determining x. If your goal is to measure the speed of light to an accuracy of 0.1%, at what minimum distance from the laser diode should the mirror be placed? Assume that you know the ballpark value for light's speed is 3×10^8 m/s.

Section 33–3

4. (II) A concave mirror has focal length f. When an object is placed a distance $d_o > f$ from this mirror, a real image with magnification m is formed.
   (a) Prove that $m = f/(f - d_o)$.
   (b) Sketch m vs. d_o over the range $f < d_o \leq +\infty$.
   (c) For what value of d_o will the real image have the same (lateral) size as the object?
   (d) To obtain a real image that is much larger than the object, in what general region should the object be placed relative to the mirror?

5. (II) A point source of light emits two rays that diverge from each other at angle θ (Fig. 33–1). If these two rays reflect from a flat mirror, at what angle ϕ do the two reflected rays diverge from each other? Justify your answer with a

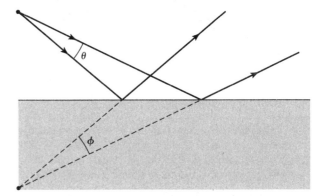

FIGURE 33–1

geometric proof based on applying the law of reflection to each reflected ray.

Section 33–4

6. (II) In devices such as laser printers and barcode readers it is necessary to scan a laser beam along a line. Imagine you are given the task of designing a laser scanner that functions (as is common) by means of a rotating mirror. An input laser beam, whose incoming direction is fixed, reflects from the center of a mirror, then travels a distance of 25 cm to the object that is to be scanned (Fig. 33–2). If the laser beam must scan at 100 cm/s across the object, at what (constant) frequency must the mirror be rotated about the axis through its center? At the position of the mirror, assume the object subtends a small angle.

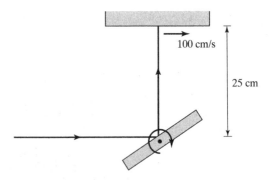

FIGURE 33–2

7. (II) Let the focal length of a convex mirror be written as $f = -|f|$. Show that the magnification m of an object a distance d_o from this mirror is given by $m = |f|/(d_o + |f|)$. Based

on this relation, explain why your nose looks bigger than the rest of your face when looking into a convex mirror?

8. (II) A spherical mirror of focal length f produces an image of an object with magnification m.
   (a) Prove that the object is distance $d_o = f\left(1 - \dfrac{1}{m}\right)$ from the reflecting side of the mirror.
   (b) Use the relation in part (a) to prove that, no matter where an object is placed in front of a convex mirror, its image will have a magnification in the range $0 \le m \le +1$.

9. (III) An object is placed a distance r in front of a wall, where r exactly equals the radius of curvature of a certain concave mirror. At what distance from the wall should this mirror be placed so that a real image of the object is formed on the wall? What is the magnification of the image?

Section 33–5

10. (II) The block of glass ($n = 1.5$) shown in Fig. 33–3 is surrounded by air. A ray of light enters into the block at its left-hand face with incident angle θ_1 and reemerges into the air from the right-hand face directed parallel to the block's base. Determine θ_1.

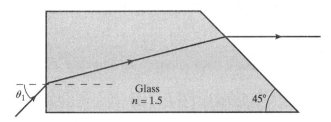

FIGURE 33–3

11. (II) A circular laser beam of diameter $d_1 = 3.0$ mm in air has an incident angle $\theta_1 = 25°$ at a flat air-glass surface. If the index of refraction of the glass is $n = 1.5$, determine the diameter d_2 of the beam after it enters the glass.

12. (II) A triangular prism made of crown glass ($n = 1.52$) with base angles of $30°$ is surrounded by air. If parallel rays are incident normally on its base as shown in Fig. 33–4, what is the angle ϕ between the two emergent rays?

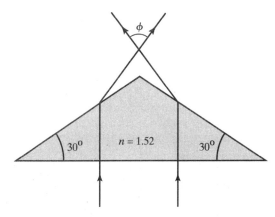

FIGURE 33–4

13. (III) A beaker full of an unknown liquid (index of refraction n) is carelessly left in an instructional laboratory and the lab instructor is concerned that the liquid might be benzene, a carcinogen. In order to identify it, the instructor uses the liquid to refract the beam from a laser pointer and obtains the data given in Table 33–1. Here, θ_1 and θ_2 are the incident and refraction angles of the laser beam as it travels from air into the liquid, respectively.

TABLE 33–1

θ_1 (deg)	15°	20°	25°	30°	35°
θ_2 (deg)	10°	13°	16°	20°	22°

(a) Explain why graphing $\sin\theta_2$ vs. $\sin\theta_1$ should yield a straight-line plot. What are the theoretical expectations for the slope and the y-intercept of this line?
(b) Using the data in Table 33–1, graph $\sin\theta_2$ vs. $\sin\theta_1$ and show that a straight line does indeed result. Use this plot to determine the liquid's index of refraction. Does this result support or refute identifying the liquid as benzene for which $n = 1.50$?

14. (III) By measurement of the incident angle θ_1 and refraction angle θ_2 of a light ray at the interface between a vacuum and a material, the material's index of refraction n can be determined via Snell's law. This commonly used method, however, is difficult in practice for materials (such as gases) whose n nearly equals one because the small difference between θ_1 and θ_2 cannot be determined accurately. Consider a light ray incident from a vacuum at angle θ_1 refracting into a near-unity index material at angle θ_2.
   (a) Noting that $\theta_2 = \theta_1 - \Delta\theta$, where the change in angle $\Delta\theta$ is small, prove that $\Delta\theta \approx (n-1)\tan\theta_1$, where $\Delta\theta$ is in radians.
   (b) To maximize $\Delta\theta$ for accurate measurement, should an experimenter use a small or a large incident angle?
   (c) For air, $n = 1.0003$. If $\theta_1 = 85°$ at a vacuum-air interface, what is the approximate value for $\Delta\theta$ in degrees?

Section 33–6

15. (III) A ray of light with wavelength λ is incident from air at $\theta_1 = 60°$ on a spherical water drop of radius R and index of refraction n (which depends on λ). When the ray reemerges into the air from the far side of the drop, it has been deflected an angle ϕ from its original direction as shown in Fig. 33–5. By how much does the value of ϕ for violet light ($n = 1.341$) differ from the value for red light ($n = 1.330$)?

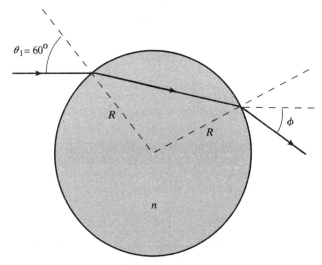

FIGURE 33–5

16. (III) For visible light, the index of refraction n of glass is roughly 1.5, although this value varies by about 1% across the visible range. Consider a ray of white light incident from air at angle θ_1 onto a flat piece of glass.

(a) Show that, upon entering the glass, the visible colors contained in this incident ray will be dispersed over a range $\Delta\theta_2$ of refracted angles given approximately by

$$\Delta\theta_2 \approx \frac{\sin\theta_1}{\sqrt{n^2 - \sin^2\theta_1}} \frac{\Delta n}{n} \cdot \left[\text{Hint: For } x \text{ in radians,} \frac{d}{dx}(\sin^{-1}x) = \frac{1}{\sqrt{1-x^2}}\cdot\right]$$

(b) If $\theta_1 = 0°$, what is $\Delta\theta_2$ in degrees?
(c) If $\theta_1 = 90°$, what is $\Delta\theta_2$ in degrees?

Section 33–7

17. (II) Consider the following explanation for a mirage: Blue light from the sky, attempting to pass from "normal" air ($n_1 = 1.00030$) into a thin layer of "heated" air ($n_2 = 1.00027$) just above a hot roadway, undergoes total internal reflection and passes into a person's eye (Fig. 33–6). Using this explanation, estimate how far down the road a 2.0-m tall person would perceive the mirage to be.

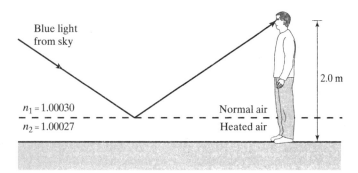

FIGURE 33–6

18. (II) A material with an index of refraction n is used to make a prism in the shape of an equilateral triangle. When a light ray is incident on side AB of this prism at angle $\alpha < 90°$ (measured relative to AB) as shown in Fig. 33–7, some of the light emerges through side BC; however, if $\alpha \geq 90°$, no light emerges from BC due to total internal reflection. Determine n.

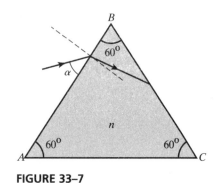

FIGURE 33–7

19. (II) A ray-like laser beam is incident at 40° on the midpoint of one end of a 1.0-km long glass fiber (Fig. 33–8). The 50-μm

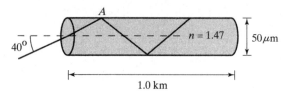

FIGURE 33–8

diameter fiber is surrounded on all sides by air and its index of refraction is 1.47.

(a) Show that the laser beam will undergo total internal reflection when it arrives at point A.
(b) Determine the approximate number of reflections the laser beam will undergo before it emerges from the opposite end of the fiber.
(c) Taking into account its zigzag path, how long (μs) does it take the laser light to journey through the fiber?

20. (II) The operation of many pieces of equipment (e.g., home appliance, spa bath, vending machine) requires the presence or absence of a liquid within a reservoir to be detected. A device to accomplish this feat consists of a light source and a light sensor housed within an enclosure with the hypotenuse of a triangular prism supported at the head of this enclosure as shown in Fig. 33–9. When situated at the bottom of a reservoir, if no liquid covers the prism's hypotenuse, total internal reflection of the beam from the light source produces a large signal in the light sensor. If liquid covers the hypotenuse, however, then some light escapes from the prism into the liquid and the light sensor's signal decreases. A large (small) signal from the light sensor thus indicates the absence (presence) of liquid in the reservoir. If this device is designed to detect the presence of water, determine the allowable range for the prism's index of refraction n. Will the device work properly if the prism is constructed from (inexpensive) lucite? For lucite, $n = 1.5$.

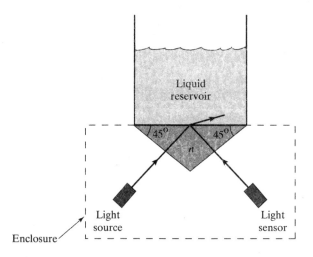

FIGURE 33–9

21. (II) An optical fiber consists of a cylindrical glass "core" surrounded by a cylindrical glass "cladding" with indices of refraction $n_1 = 1.465$ and $n_2 = 1.460$, respectively. Consider a light ray, incident from air at angle θ_a at the end of the fiber, which propagates down an optical fiber by repeated reflections

at the critical angle each time it impinges on the core-cladding interface (Fig. 33–10).

(a) Explain why only light rays from air with incident angles $\theta \leq \theta_a$ at the end of the fiber will be able to propagate via total internal reflection down the entire length of the fiber.

(b) Determine θ_a (in degrees) for this fiber. [The "acceptance angle" θ_a defines a cone within which an incident light ray must fall in order to travel down the fiber.]

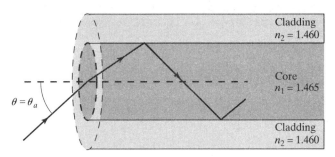

FIGURE 33–10

22. (II) Two rays A and B travel down a cylindrical optical fiber of diameter $d = 50\,\mu m$, length $l = 1.0\,km$, and index of refraction $n_1 = 1.465$. Ray A travels a straight path down the fiber's axis, while B propagates down the fiber by repeated reflections at the critical angle each time it impinges on the core's boundary. Determine the extra time Δt it takes for B to travel down the entire fiber in comparison with A,

(a) assuming the fiber is surrounded by air (Fig. 33–11a).

(b) assuming the fiber is surrounded by a cylindrical glass "cladding" with index of refraction $n_2 = 1.460$ Fig. 33–11b). [Because of this phenomenon (called "modal dispersion") optical pulses moving down a fiber must be separated by times greater than Δt. Thus, smaller pulse separations are possible by encasing a fiber "core" within a cladding.]

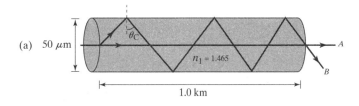

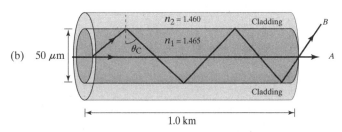

FIGURE 33–11

Section 33–8

23. (II) Within a beaker, a 2.00-cm deep layer of ethyl alcohol floats on a 4.00-cm deep layer of water. When viewed from above, how deep does the liquid in the beaker appear to be?

General Problems

24. (II) Consider the following scenario: A slab whose two faces are perfectly parallel is composed of a material with index of refraction n. A ray of light enters one face of the slab from air at incident angle θ_1, but once inside the slab, this light is unable to escape into the surrounding air due to total internal reflection as shown in Fig. 33–12.

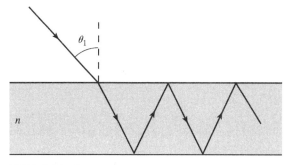

FIGURE 33–12

(a) Using Snell's law, prove that this scenario cannot take place.

(b) Light rays obey the following principle of reversibility: If a ray is a possible path in one direction, it is also a possible path in the reverse direction. Use this principle to provide a qualitative explanation for why the described scenario is prohibited.

25. (II) A slab of thickness t, whose two faces are perfectly parallel, is composed of a material with index of refraction n. A ray of light incident from air onto one face of the slab at incident angle θ_1 splits into two rays A and B. Ray A reflects directly back into the air, while B travels a total distance l within the slab before reemerging from the slab's face a distance d from its point of entry (Fig. 33–13).

(a) Derive expressions for l and d in terms of t, n, and θ_1.

(b) For normal incidence (i.e., $\theta_1 = 0°$) show that your expressions yield the expected values for l and d.

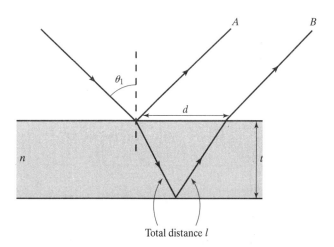

FIGURE 33–13

26. (II) A glass block ($n = 1.50$) is surrounded by water (Fig. 33–14). Find the range of incident angles θ_1 such that, when a light ray is incident on the block's left-hand face, a portion of that light will be transmitted out of the upper face of the block.

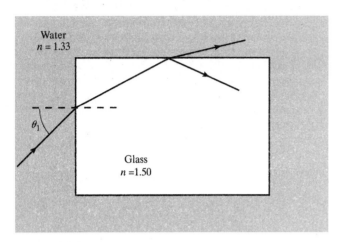

FIGURE 33–14

27. (III) Consider the refraction of a light ray at the interface between air and water.
On the same graph, plot refraction angle θ_2 vs. incident angle θ_1 for the following two situations: (1) the ray is incident from air and enters into water over the range $0 \leq \theta_1 \leq 90°$, and (2) the ray is incident from water and enters into air over the range $0 \leq \theta_1 \leq \theta_C$, where θ_C is the critical angle.
Additionally, on this graph include the "comparison" line $\theta_1 = \theta_2$. In which situation (1 or 2) is the refracted ray "bent toward the normal" and in which situation is it "bent away from the normal"?

28. (III) When light encounters an air-glass interface, a small fraction of its intensity (about 4%) is reflected. Hence, by passing a single "strong-intensity" laser beam through a glass plate with its two parallel flat surfaces angled at 45° as shown in Fig. 33–15, two "weak-intensity" laser beams can be created. These weak beams can be used, e.g., in tuning the laser light to an exact frequency. For glass, assume $n = 1.5$.
(a) Show that the two weak beams are parallel and directed at a right angle to the strong beam.
(b) If a researcher wants the distance between the two weak beams to be d, what should the thickness t of the glass plate be?

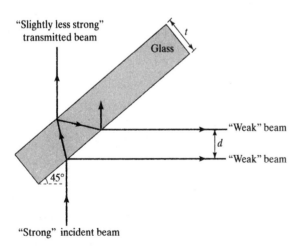

FIGURE 33–15

29. (III) The paint used on highway signs often contains small transparent spheres which provide nighttime illumination of the sign's lettering by retro-reflecting vehicle headlight beams. Consider a light ray from air incident on one such sphere of radius R and index of refraction n. Defining θ to be its incident angle, the ray will then follow the path shown in Fig. 33–16, where we have assumed that n is the value required for the ray to exit the sphere in the direction exactly antiparallel to its incoming direction. Restricting attention to rays for which $\sin \theta$ can be approximated as θ, determine n.

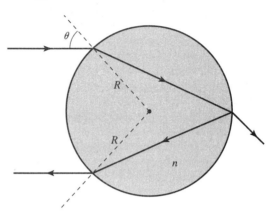

FIGURE 33–16

30. (III) Consider a ray of sunlight incident from air on a spherical raindrop of radius R and index of refraction n. Defining θ to be its incident angle, the ray will then follow the path shown in Fig. 33–17, exiting the drop in the direction that is at "scattering angle" ϕ compared with its original incoming direction.
(a) Show that $\phi = 180° + 2\theta - 4\sin^{-1}\left(\dfrac{\sin\theta}{n}\right)$.
(b) The parallel rays of sunlight illuminate a raindrop with rays of all possible incident angles from 0° to 90°. Plot ϕ vs. θ in the range $0° \leq \theta \leq 90°$, assuming $n = 1.33$ as is appropriate for water at visible-light wavelengths.
(c) From your plot, you should find that a fairly large fraction of the incident angles have nearly the same scattering angle. Approximately what fraction of the possible incident angles is within roughly 1° of $\phi = 139°$? [The collective light in this subset of incident rays creates the rainbow. Small wavelength-dependent variation in n causes the rainbow to form at slightly different ϕ for the various visible colors.]

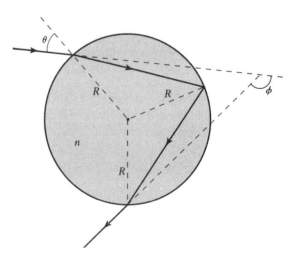

FIGURE 33–17

CHAPTER 34

Lenses and Optical Instruments

Sections 34–1 and 34–2

1. (II) The optical system of some instruments includes an "inverting" lens of focal length f, whose job is to take a focused "upside-down" real image (created by some other optical components) and create an exact "right-side-up" replica of it. For this process to work, where should the original upside-down image be relative to the inverting lens and where will the right-side-up replica be formed relative to the lens?

2. (II) To a good approximation, the "collimated" beam from a laser can be thought of as a collection of parallel light rays within a cylindrical region of radius r_o. To change its radius, an experimenter can simply pass this beam through two converging lenses of focal lengths f_1 and f_2, respectively, separated by a distance l.
   (a) In order for the output beam to be collimated (i.e., consist of parallel rays), what should l equal?
   (b) Determine the radius r of the output beam in terms of r_o, f_1, and f_2.
   (c) If the output beam is to have an expanded radius in comparison with the original beam from the laser, should f_2 be larger or smaller than f_1?

3. (II) A converging lens has focal length f. When an object is placed a distance $d_o > f$ from this lens, a real image with magnification m is formed.
   (a) Prove that $m = f/(f - d_o)$.
   (b) Sketch m vs. d_o over the range $f < d_o \leq +\infty$.
   (c) For what value of d_o will the real image have the same (lateral) size as the object?
   (d) To obtain a real image that is much larger than the object, in what general region should the object be placed relative to the lens?

4. (II) While a conventional camera employs film, a digital camera records images using a charge coupled device (CCD). A CCD is a rectangular array of light-sensing pixels, where each pixel has a lateral extent typically of about 4.0 μm. An object's image will be "highly resolved" if it is substantial enough to illuminate a very large number of pixels. At the other extreme, one can define an object's image to be "barely resolved" if it covers exactly one pixel. According to this definition, for a digital camera with a 13-mm focal length lens, how far away will a 2.0-m tall person have to stand in order for his or her image to be barely resolvable by the camera's CCD?

5. (II) An object is moving toward a converging lens of focal length f with constant speed v_o such that its distance d_o from the lens is always greater than f. Determine the velocity v_i of the image as a function of d_o. Which direction (toward or away from the lens) does the image move? For what d_o does the image's speed equal the object's speed?

6. (III) Imagine an arrow of infinitesimal length dx lying on the axis of a converging lens of focal length f so that its tip and tail are at distances x and $x + dx$, respectively, from the lens with $x > f$. A real image of the arrow will be created on the other side of the lens. This image will lie on the axis with its two ends at distances y and $y + dy$ from the lens. The longitudinal magnification of the lens is defined as $m_l = \dfrac{dy}{dx}$. Prove that $m_l = -m^2$, where m is the lateral magnification of the lens. Explain why this expression implies that the image's tail will be closest to the lens.

7. (III) In the "magnification" method, the focal length f of a converging lens is found by placing an object of known size at various locations in front of the lens and measuring the resulting real-image distances d_i and their associated magnifications m (minus sign indicates that image is inverted). The data taken in such an experiment is given in Table 34–1.

TABLE 34–1					
d_i (cm)	20	25	30	35	40
m	−0.43	−0.79	−1.14	−1.50	−1.89

   (a) Prove that, by graphing m vs. d_i, a straight line should result. What are the theoretically expected values for the slope and the y-intercept of this line? [Hint: d_o is not constant.]
   (b) Using the data in Table 34–1, graph m vs. d_i and show that a straight line does indeed result. Use the slope of this line to determine the focal length of the lens. Does the y-intercept of your plot have the expected value?
   (c) In performing such an experiment, one has the practical problem of locating the exact center of the lens since d_i must be measured from this point. Imagine, instead, that one measures the image distance d'_i from some other more convenient point on the lens axis (e.g., the surface of the lens), which is a distance l from the lens's center. Then, $d_i = d'_i + l$. Prove that, when implementing the magnification method in this fashion, a plot of m vs. d'_i will still result in a straight line. How can f be determined from this straight line?

8. (III) The focal length f of a lens can be determined through the following procedure (Fig. 34–1): With a light-emitting object separated by a distance D from a screen, the lens under investigation is placed next to the object. The lens is first moved toward the screen until the lens-position (call it point A) is found, which produces a focused image on the screen. The experimenter next determines the appropriate extra distance l the lens must be moved closer to the screen (to point B) so that another focused image appears on the screen.
(a) Find the expression for f in terms of the experimentally determined quantities D and l.
(b) In order for this method to work, what restriction is required on D?

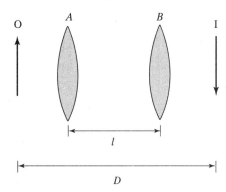

FIGURE 34–1

9. (III) Given a converging lens of focal length f, what is the smallest possible distance l between an object and its real image created by the lens? What is the object's distance from the lens when this minimum object-image separation occurs?

Section 34–3

10. (II) The focal length f of a diverging lens can be measured using the following technique: First, place a light-emitting object in front of a converging lens so that a focused real image is formed on a screen placed on the other side of this lens. Next, place the diverging lens of interest between the converging lens and the screen and measure the distance l_1 from the diverging lens to the screen. Finally, move the screen away from the two lenses until a focused image due to this two-lens system is found and measure the distance l_2 from the diverging lens to the screen.
(a) Derive the expression that allows one to find f using the experimentally determined quantities l_1 and l_2.
(b) Under what condition will this technique fail to obtain a value for f?

Section 34–4

11. (III) A converging lens made of glass with index of refraction $n = 1.5$ has two spherical surfaces with the same radius of curvature $R_1 = R_2 = +R$.
(a) If, when the lens is surrounded by air, its focal length $f = 30$ cm, determine R.
(b) Imagine that surface 1 of this lens is surrounded by water, while surface 2 is surrounded by air. By appropriately modifying the proof given for the lensmaker's equation in Section 34–4 of the text, and assuming that the incident parallel rays are in water, determine the focal length of the lens on its "air-side."
(c) Repeat the calculation in part (b), but this time assume that the incident parallel rays are in air and determine the focal length of the lens on its "water-side." Does your answer differ from that found in part (b)?

Section 34–5

12. (II) (a) For the image formed of an object at distance d_o from a converging lens of focal length f, prove that the magnification $m \approx f/d_o$, as long as $d_o \gg f$.
(b) Sports and paparazzi photographers commonly use a 1000-mm focal lens on their 35-mm cameras. About how much greater magnification is achieved by using such a lens rather than the "normal" 50-mm lens for this type of camera?

13. (II) Human vision normally covers an angle of about 40° horizontally. A "normal" camera lens then is defined as follows: When focused on a distant horizontal object which subtends an angle of 40°, the lens produces an image that extends across the full horizontal extent of the camera's light-recording medium (e.g., film or CCD). Determine the focal length f of the "normal" lens for the following types of cameras:
(a) a 35-mm camera that records images on film 36 mm in horizontal extent.
(b) a digital camera that records images on a charge coupled device (CCD) 1.0 cm in horizontal extent.

14. (II) A camera has a lens of focal length f.
(a) When the camera is focused on an object a distance d_o away, show that the distance d_i between the camera's lens and the image-recording medium (e.g., film) is $d_i = d_o f/(d_o - f)$.
(b) Sketch d_i vs. d_o over the range $f < d_o \le +\infty$.
(c) Assume that $f = 50$ mm and that the camera is designed to focus on objects at distances within the range 0.45 m $\le d_o \le +\infty$. Over what range can d_i be changed in this camera?

15. (III) (a) A point light source A is located on the axis of a camera lens of diameter D and focal length f. If this source is located a distance d_o from the lens and the camera is focused for this source (Fig. 34–2a), show that the lens-film distance is $d_i = d_o f/(d_o - f)$.

(a)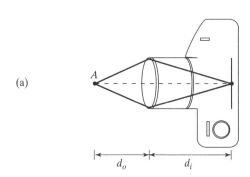

(b)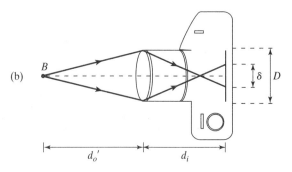

FIGURE 34–2

(b) With the camera focused on A, assume that another point source B is located a distance $d'_o > d_o$ from the lens, creating a circle of confusion of diameter δ on the film. By considering the most extreme-angle rays passing through the lens from B (Fig. 34–2b), prove that $\delta = \dfrac{fD}{d'_o} \dfrac{d'_o - d_o}{d_o - f}$.

[Note that δ increases as d'_o increases.]

(c) Let $d'_o = d_F$ be the object distance at which $\delta = 0.03$ mm. Then, given the resolution of film, both A and B will be in acceptable focus and d_F is the "far" distance limit of the recorded image's "depth of field." Defining $H = \dfrac{f^2}{\delta\, f\text{-stop}} + f$, shows that $d_F = \dfrac{d_o(H - f)}{H - d_o}$. For a lens with $f = 50$ mm and f-stop $= 11$, which is focused on an object 5.0 m away, determine d_F. [The "near" distance limit of the image's depth of field, i.e., the object distance $d'_o = d_N$ at which $\delta = 0.03$ mm, where $d_N < d_o$, can be found by a similar calculation.]

Section 34–6

16. (II) The closely packed cones in the fovea have a diameter of about 2 μm. In order for the eye to discern two images on the fovea as distinct, assume (as a best case scenario) that the images must be separated by at least one cone that is not excited. If these images are of two point-like objects at the eye's 25-cm near point, how far apart are these objects? Assume the diameter of the eye (cornea-to-fovea distance) is 2 cm.

17. (II) Sue has normal vision with a range of accommodation from 25 cm to infinity. Over what range (as measured from her eyes) would she be able to see objects distinctly when wearing her friend's glasses which have +4.0 diopter lenses? Assume a lens-eye distance of 2.0 cm.

18. (II) A drugstore owner reads in a magazine that the near point for a typical person increases from 25 cm to 40 cm to 200 cm as the person ages from 40 to 50 to 60 years old, respectively. Based on these data, the owner calculates the two powers of reading glasses expected to be the best sellers in the drugstore. What are these two powers? Assume a lens-eye distance of 2.0 cm.

Section 34–7

19. (II) A converging lens of focal length $f = 10$ cm is being used by a writer as a magnifying glass to read some fine print on his book contract. Initially, the writer holds the lens above the fine print so that its image is at infinity. To get a better look, he then moves the lens so that the image is at his 25-cm near point. How far, and in what direction (toward or away from the fine print) did the writer move the lens? Assume the writer's eye is adjusted to remain always very near the magnifying glass.

Section 34–8

20. (II) An astronomical telescope longer than about 50 cm is not easy to hold by hand. Based on this fact, estimate the maximum angular magnification achievable for a telescope designed to be handheld. Assume its eyepiece lens, if used as a magnifying glass, provides a magnification of 5× for a relaxed eye with near point $N = 25$ cm.

Section 34–9

21. (II) An inexpensive instructional lab microscope allows the user to select its objective lens to have a focal length of 32 mm, 15 mm, or 3.9 mm. It also has two possible eyepieces with magnifications 5× and 10×. Each objective forms a real image 160 mm beyond its focal point. What are the largest and smallest overall magnifications obtainable with this instrument?

22. (II) Given two 12-cm focal length lenses, you attempt to make a crude microscope using them. While holding these lenses a distance 50 cm apart, you position your microscope so that its objective lens is distance d_o from a small object. Assume your eye's near point $N = 25$ cm.
(a) For your microscope to function properly, what should d_o be?
(b) Assuming your eye is relaxed when using it, what magnification M does your microscope achieve?
(c) Since the length of your microscope is not much greater than the focal lengths of its lenses, the approximation $M \approx Nl/f_e f_o$ is not valid. If this approximation is applied to your microscope in error, how many times larger is its predicted value than your microscope's true magnification?

Section 34–10

23. (III) In an achromatic doublet (Fig. 34–3), a converging lens and a diverging lens of focal lengths f_C and f_D and (nominal) indices of refraction n_C and n_D, respectively, are placed in contact. Define $\Delta n = n_b - n_r$ to be a material's variation in index of refraction between blue and red light. The doublet's converging and diverging lenses are made of different materials so that Δn_C and Δn_D are not the same.
(a) By determining where incident parallel rays are imaged, prove that the focal length f of an achromatic doublet is given by $\dfrac{1}{f} = \dfrac{1}{f_C} + \dfrac{1}{f_D}$.
(b) For simplicity, assume that the converging lens is a symmetric double convex lens so that the radius of curvature for both its surfaces is $+R_C$. Likewise, assume that the diverging lens is a symmetric double concave lens so that the radius of curvature for both its surfaces is $-R_D$. Prove that f will be the same for blue and for red light if $\dfrac{R_C}{R_D} = \dfrac{\Delta n_C}{\Delta n_D}$.
(c) Prove that $\dfrac{1}{f} = \dfrac{2}{R_C}\left[(n_C - 1) - (n_D - 1)\dfrac{\Delta n_C}{\Delta n_D}\right]$.
(d) Assume that an achromatic doublet's converging and diverging lenses are made of silicate crown glass and silicate flint glass, respectively. Using data from Fig. 33–26 in the text, determine the values of R_C and R_D that would yield $f = 10$ cm.

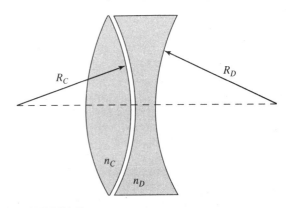

FIGURE 34–3

24. (III) (a) Consider a lens with two spherical surfaces made of glass. Show that if the index of refraction n of the glass changes by an amount Δn, the fractional change in the focal length f of the lens is given by $\dfrac{\Delta f}{f} \approx -\dfrac{\Delta n}{n-1}$.
(b) For visible light, the index of refraction n of glass is roughly 1.5, although this value increases by about 1% over the range from red to violet. What is the approximate fractional change in the focal length of a spherical glass lens for violet light in comparison with red light?

General Problems

25. (II) A sports photographer, 50 m away from a competing athlete, uses her 35-mm camera to take a photo in which the athlete's 2.0-m tall body occupies the entire 36-mm long frame of film. What is the focal length f of the lens on her camera?

26. (II) The proper functioning of certain light-handling devices (e.g., optical fibers and spectrometers) requires that the input light be a collection of diverging rays within a cone of half-angle θ (Fig. 34–4). If the light initially exists as a collimated beam (i.e., parallel rays), show that a single lens of focal length f and diameter D can be used to create the required input light if $\dfrac{D}{f} = 2 \tan \theta$. If $\theta = 3.5°$ for a certain spectrometer, what focal length lens should be used if the lens diameter is 5.0 cm?

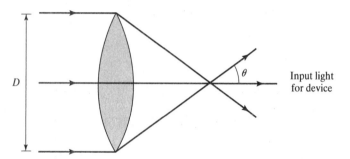

FIGURE 34–4

27. (II) Suppose some ocean-dwelling creature's eye functions underwater with a near point of 25 cm and that this creature would like to create an underwater magnifying glass out of a flexible plastic bag filled with air in order to read an inscription on a marble column in the sunken city of Atlantis. What shape should the bag have (i.e., determine radii of curvature of its surfaces) in order for it to be used by the creature as a 3× magnifying glass? Assume the eye is focused at its near point.

28. (II) A screen is initially positioned at the proper distance behind a lens (focal length f, diameter D) in order to observe the focused image of a distant point-like object in front of the lens. From this initial position, the screen is moved a distance l further away from the lens and the diameter δ of the resulting circle of confusion on the screen is measured. Find an expression for the f-stop of the lens in terms of l and δ.

29. (III) In the derivation of the lensmaker's equation, a lens with focal length f is approximated as "thin" if its thickness $t \ll f$. Consider a planoconvex lens made of glass ($n = 1.5$) whose convex surface has radius of curvature R (Fig. 34–5).
(a) Taking $t = 0.1f$ as the maximum thickness allowed by the thin lens approximation, show the maximum half-angle θ_m that can be subtended by the convex surface of a thin lens constructed of glass is about 40°.

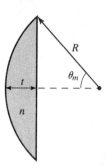

FIGURE 34–5

(b) If a glass planoconvex lens is to be constructed with a diameter D of 8.0 cm, what is the smallest radius of curvature allowed for its convex surface if it is to obey the thin lens approximation?
(c) The lens in part (b) has the smallest-allowed focal length for an 8.0-cm diameter planoconvex thin lens. Thus, this lens provides an estimate for M_m, the maximum magnification possible for an 8.0-cm diameter magnifying glass. Estimate M_m assuming an eye whose near point $N = 25$ cm.

30. (III) A telephoto lens system obtains a large magnification in a compact package. A simple such system can be constructed out of two lenses, one converging and one diverging of focal lengths f_1 and $f_2 = -f_1/2$, respectively, separated by distance $l = 3f_1/4$ as shown in Fig. 34–6.

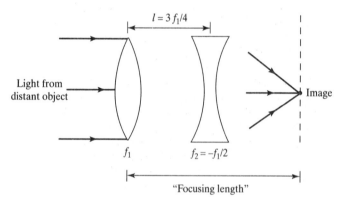

FIGURE 34–6

(a) For a distant object located at distance d_o from the first lens, show that the first lens forms an image with magnification $m_1 \approx -f_1/d_o$ located very close to its focal point. Go on to show that the total magnification for the two-lens system is $m \approx -2f_1/d_o$.
(b) For an object located at infinity, prove that the two-lens system forms an image that is a distance $5f_1/4$ behind the first lens.
(c) A single 250-mm focal length lens would have to be mounted about 250 mm from a camera's film in order to produce an image of a distant object at d_o with magnification $-(250 \text{ mm})/d_o$. To produce an image of this object with the same magnification using the two-lens system, what value of f_1 should be used and how far in front of the film should the first lens be placed? How much smaller is the "focusing length" (i.e., first lens-to-final image distance) of this two-lens system in comparison with the 250-mm "focusing length" of the equivalent single lens?

Wave Nature of Light; Interference

Section 35-3

1. (II) Light of wavelength λ passes through a pair of slits separated by 0.10 mm, forming a double-slit interference pattern on a screen located a distance 30 cm away. Suppose that the image in Fig. 35–9a of the text is an actual-size reproduction of this interference pattern. Use a ruler to measure a pertinent distance on this image; then utilize this measured value to determine λ (nm).

2. (II) Interference patterns are often used in research to measure the wavelengths of light emitted by a sample under investigation. For example, assume that the four visible wavelengths of light emitted by hot hydrogen gas pass through a double-slit arrangement, where the slits are separated by 20.0 μm, and that the four resulting interference patterns appear on a viewing screen 397 mm away. If it is found that a second-order fringe for the longest of these wavelengths is located a distance of 26.1 mm from the screen's central point, what is this hydrogen-emitted wavelength of light (nm)? To what color does it correspond (see Fig. 33–24 in the text)?

3. (II) A physics professor wants to perform a lecture demonstration of Young's double-slit experiment for her class using the 633-nm light from a helium-neon laser. Because the lecture hall is very large, the interference pattern will be projected on a wall that is 5.0 m from the slits. For easy viewing by all students in the class, the professor wants the distance between the $m = 0$ and $m = 1$ maxima to be 25 cm. What slit separation is required in order to produce the professor's desired interference pattern?

4. (II) In journeying to a certain point on the screen in Young's experiment, the ray from one slit travels an extra distance in comparison with the ray from the other slit.
(a) What is the angular location of the point(s) on the screen for which the extra distance (i.e., the path difference) is maximized? If the slit separation is d, what is the maximum value attainable for the extra distance?
(b) If $d = 500$ nm in a certain experimental set-up, for what range of visible wavelengths will the creation of interference fringes of any order be impossible? Assume the visible range extends from 400 nm to 700 nm.

5. (II) In Young's experiment, to approximate the path difference traveled by the ray from one slit in comparison with the distance traveled by the ray from the other slit, it is assumed that the rays are "essentially parallel." This assumption is made despite the fact that these rays must intersect at the screen in order to interfere. Explore the validity of the "parallel ray" approximation as follows. Let d be the distance between slits S_1 and S_2 with a screen placed a distance L from the slits. Consider point P on the screen, which is directly opposite S_1 so that the distance traveled by the ray from S_1 to P is $d_1 = L$ and angle $\alpha = 90°$. Then the ray from S_2 to P travels a distance $d_2 > L$ and is inclined at angle $\beta < 90°$ relative to the plane defined by the slits shown as in Fig. 35–1. One might plausibly define these two rays as "essentially parallel" if $\beta \geq 89.9°$. In order for this criterion to be fulfilled, what is the required restriction on the ratio L/d? If $d = 0.1$ mm, how far away must the screen be placed from the slits in order for the parallel-ray approximation to be valid according to this definition?

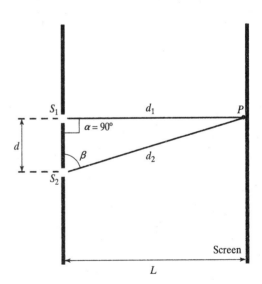

FIGURE 35–1

6. (II) In a two-slit interference experiment, slits S_1 and S_2 are above and below, respectively, the centerline between the slits that defines $\theta = 0°$. The slits are separated by distance $d = 3.0$ μm and emit light of wavelength $\lambda = 650$ nm. Assume that the light emitted from S_2 reaches its maximum a time $T/8$ later than the light emitted from S_1, where T is the period of the light wave being used. Determine the angular locations of the resulting interference pattern's first-order maximum above and below the centerline.

7. (II) Slits S_1 and S_2 are separated by a distance d of exactly 1 mm and a screen is located a distance L of exactly 2 m from the center of these slits. Consider the point P on the screen a distance x of exactly 1 m from the screen's center (Fig. 35–2). Define d_1 and d_2 to be the distances from S_1 to P and S_2 to P, respectively.
(a) Show that
$$d_2 - d_1 = \sqrt{L^2 + \left(x + \frac{d}{2}\right)^2} - \sqrt{L^2 + \left(x - \frac{d}{2}\right)^2},$$
then evaluate this exact expression for the given values of L, x, and d as accurately as your calculator allows.
(b) As long as $L \gg d$, the approximate relation $d_2 - d_1 \approx d \sin\theta$ is used to analyze the double-slit experiment. Show that this relation implies $d_2 - d_1 \approx \frac{xd}{\sqrt{L^2 + x^2}}$; then evaluate this approximate expression for the given values of L, x, and d as accurately as your calculator allows.
(c) Within the precision of your calculator, what percentage difference do you find between the exact and approximate values you found for $d_2 - d_1$?

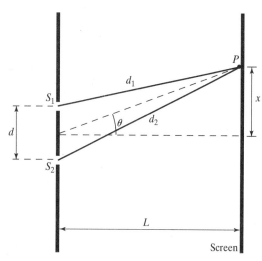

FIGURE 35–2

8. (III) The first-order maximum in a two-slit interference pattern occurs at angle θ for light of wavelength λ. The distance d between the slits then is increased by a small amount Δd.
(a) Show that the angular position of this maximum changes by an amount $\Delta\theta \approx -\tan\theta \frac{\Delta d}{d}$.
[Hint: For x in radians, $\frac{d}{dx}(\sin^{-1} x) = 1/\sqrt{1-x^2}$.]
(b) If $d = 1.5\lambda$ and the slit separation is increased by 1.0%, what is the fractional change in θ? Does the first maximum move toward or away from the center of the interference pattern as the slit separation is increased?

9. (III) As long as the observation distance x is much greater than the slit separation d, the position y of the two-slit interference pattern's m^{th} order maximum lies on a line inclined at angle θ, where θ is given by the condition $d \sin\theta = m\lambda$ (see Fig. 35–3).
(a) Prove that $d \sin\theta = m\lambda$ does indeed imply that y and x are related by a "straight-line" relation of the form $y = ax + b$, where a and b are constants. What are the slope and the y-intercept of this line in terms of d, λ, and m?

(b) In a certain two-slit interference experiment, $d = 3.0\,\mu\text{m}$ and $\lambda = 700$ nm. If, after the interference pattern has been established on a distant screen, the slit-screen distance is increased by 0.10 m, by what distance (along the screen) will the second-order maximum increase its distance from the zero-order maximum?

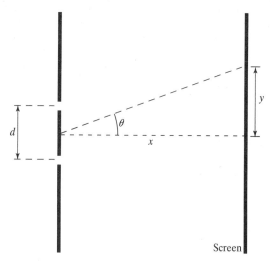

FIGURE 35–3

10. (III) Young's double-slit experiment consists of slits S_1 and S_2 separated by distance d with a screen whose center is a distance L from the center of the slits. Consider the point P on the screen a distance x from the screen's center (Fig. 35–3). Define d_1 and d_2 to be the distances from S_1 to P and S_2 to P, respectively.
(a) Show that $d_2 - d_1 =$
$$\sqrt{L^2 + x^2}\left[\sqrt{1 + \left(\frac{d^2/4 + xd}{L^2 + x^2}\right)} - \sqrt{1 + \left(\frac{d^2/4 - xd}{L^2 + x^2}\right)}\right].$$
(b) By considering the two cases $x \leq d$ and $x > d$, explain why the inequality $L^2 + x^2 \gg d^2/4 + xd$ holds as long as $L \gg d$.
(c) Show that for $L \gg d$, $d_2 - d_1 \approx \frac{xd}{\sqrt{L^2 + x^2}} = d\sin\theta$.

Section 35–5

11. (II) At a certain point in space, $E_1 = (5000\,\text{N/m})\sin\omega t$ and $E_2 = (3000\,\text{N/m})\sin(\omega t + 50°)$ are the time-dependent electric fields of the two light waves that interfere there. The resultant electric field at that location is given by $E = E_o \sin(\omega t + \phi)$. Using phasors, determine E_o and ϕ.

12. (II) In a two-slit interference experiment, the path length to a certain point P on the screen differs for one slit in comparison with the other by 1.25λ.
(a) What is the phase difference between the two waves arriving at P?
(b) Determine the intensity at P, expressed as a fraction of the maximum intensity I_o on the screen.

13. (II) Each of three identical slits is separated from its adjacent neighbor by distance d (Fig. 35–4a). After light passes through these slits, it creates a three-slit interference pattern on a distant screen. The phasor diagram for a minimum in this interference pattern is shown in Fig. 35–4b. Use this diagram to predict the angles θ at which minima occur on the screen.

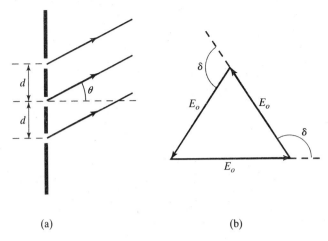

FIGURE 35–4

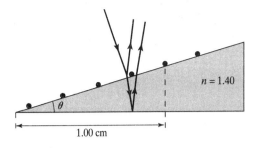

FIGURE 35–6

Section 35–6

14. (II) A thin film of oil $(n_o = 1.50)$ with varying thickness floats on water $(n_w = 1.33)$. When it is illuminated from above by white light, the reflected colors are as shown in Fig. 35–5. In air, the wavelength of yellow light is 580 nm.
    (a) Why are there no reflected colors at point A?
    (b) What is the oil's thickness t at point B?

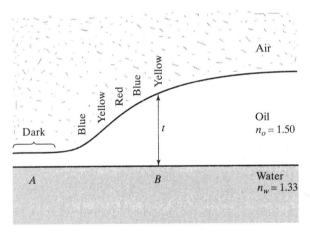

FIGURE 35–5

15. (II) A thin oil slick $(n_o = 1.50)$ floats on water $(n_w = 1.33)$. When a beam of white light strikes this film at normal incidence from air, the only enhanced reflected colors are red (650 nm) and violet (390 nm). From this information, deduce the (minimum) thickness t of the oil slick.

16. (II) A piece of transparent material with an index of refraction $n = 1.40$ is fabricated in the shape of a wedge as shown in Fig. 35–6. The person who fabricated the wedge knows that its angle θ is very small, but wishes to determine θ precisely. When 650-nm red light is incident normally on the wedge from the air above, a sequence of bright red fringes appears along the length of the wedge when viewed from above and it is found that the fifth bright fringe is 1.00 cm from the wedge's tip. Use this information to find θ.

17. (II) Let's explore why only "thin" layers exhibit thin-film interference. Assume a layer of water, sitting atop a flat glass surface, is illuminated from the air above by white light (all wavelengths from 400 nm to 700 nm). Further, assume that the water-layer's thickness t is much greater than a micron $(= 1000 \text{ nm})$; in particular, let $t = 200{,}000$ nm. Take the index of refraction for water to be $n = 1.33$ for all visible wavelengths.
    (a) Prove that a visible color will be reflected from the water layer if its wavelength $\lambda = 2nt/m$, where m is an integer.
    (b) Show that the two extremes in wavelengths (400 nm and 700 nm) of the incident light are both reflected from the water layer and determine the m-value associated with each.
    (c) How many other visible wavelengths, besides $\lambda = 400$ nm and 700 nm, are reflected from the "thick" layer of water? [Since the thick film reflects not one, but a multitude of colors across the visible range, the reflected light appears white.]

18. (III) A layer of water with thickness t sits atop a flat glass surface and is illuminated from the air above by white light. If t is "thin," only a certain color will be reflected from the water; if t is "thick," many wavelengths will be reflected and their combination will appear white. As a plausible dividing line between these two regimes, define the upper limit of a "thin" layer to be that threshold water-layer thickness for which more than one wavelength in the visible range (400 nm and 700 nm) begins to appear in the reflected light. Using this definition, determine the maximum thickness for the "thin film" regime. Take the index of refraction for water and glass to be $n = 1.33$ and $n = 1.50$, respectively, for all visible wavelengths.

Section 35–7

19. (III) A Michelson interferometer is commonly used to measure the index of refraction n for materials with $n \approx 1$ (e.g., gases). In this scheme, a hollow glass container of (interior) length l is placed in one beam of the interferometer as shown in Fig. 35–7. Starting with the cell containing a good vacuum and the interferometer outputting a bright fringe, gas is slowly leaked into the cell while counting the number N of bright-to-dark-to-bright fringe transitions that take place at the output until the cell contains gas at a certain pressure.
    (a) Prove that $n - 1 = \dfrac{N\lambda}{2l}$, where λ is the wavelength in vacuum of the light used.
    (b) Assume λ and l are very accurately known so that there is negligible uncertainty in these quantities. Show then that
    $$\dfrac{\Delta(n-1)}{n-1} \approx \dfrac{\Delta N}{N}.$$
    (c) Take $\lambda = 632.8$ nm and $l = 5.000$ cm. If air is leaked in until atmospheric pressure is attained within the cell and it is found that $N = 54$ with $\Delta N = 1$, what values are found for the deviation of n from unity and its fractional uncertainty for air at atmospheric pressure?

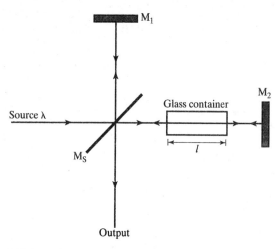

FIGURE 35–7

General Problems

20. (II) In a double-slit set-up, 650-nm light from a diode laser passes through two slits that are separated by 0.25 mm, and forms an interference pattern on a very large distant screen. (a) Chapter 35 of the text presents an explanation for what will happen in this experiment based on the wave theory of light. Show that straightforward application of this theory leads one to expect that the interference pattern appearing on the (assumed infinitely large) screen will consist of 769 bright fringes. (b) When one actually carries out this experiment, although many bright fringes do appear on the screen, they are nowhere near 769 in number. In fact, the bright fringes that do appear are confined within a restricted range of angles. Thus, one of the assumptions made in the theory of Chapter 35 is violated in the actual experimental situation. Which assumption is the culprit for the observed departure of experimental observation from theory?

21. (II) In a compact disc (CD), digital information is stored as a sequence of raised surfaces called "pits" and recessed surfaces called "lands." Both pits and lands are highly reflective and are embedded in a thick plastic material with index of refraction $n = 1.55$ (Fig. 35–8). As a 780-nm wavelength (in air) readout laser scans across the pit-land sequence, the transition between a neighboring pit and land is sensed by monitoring the intensity of reflected laser light from the CD. If, at the moment that half of the finite width of the laser beam is reflected from the pit and the other half from the land, one wishes the two reflected halves of the beam to be 180° out of phase with each other, what should the (minimum) height difference t between a pit and land be? [When this light is subsequently focused on a light detector, cancellation of the two out-of-phase halves of the beam produces a minimum detector output.]

22. (II) A two-slit interference set-up with slit separation $d_L = 0.10$ mm produces interference fringes at a particular set of angles θ_m (where $m = 0, 1, 2, \ldots$) for red light of wavelength $f_L = 4.6 \times 10^{14}$ Hz. If one wishes to construct an analogous two-slit interference set-up that produces interference fringes at the same set of angles θ_m for room-temperature sound of middle-C frequency $f_S = 262$ Hz, what should the slit separation d_S be for this analogous set-up?

23. (II) In Example 35–8 of the text, a nonreflective coating with index of refraction n_c is placed on top of glass with index of refraction n_g. Advanced analysis shows that when light of electric-field amplitude E_o from a material of refractive index n reflects at normal incidence from material of refractive index n', the ratio of the reflected light's amplitude in comparison with E_o is given by $(n - n')/(n + n')$. In Fig. 35–23 of the text, the amplitudes E_1 and E_2 of rays 1 and 2, respectively, must be equal in order for them to interfere destructively.

(a) Estimate the amplitudes E_1 and E_2 as follows, taking $n = 1$ for air: Ray 1 is created directly by the reflection of the incident ray of amplitude E_o at the air-coating interface. Determine E_1. Assume $E_1 \ll E_o$, so that the amplitude of light refracted into the coating can be approximated as E_o. Then, taking E_o as the incident light amplitude at the coating-glass interface, determine the amplitude of the reflected light created and use this amplitude as an estimate for E_2. This estimation method is justified in that most of the reflected light from the coating-glass interface will subsequently emerge into the air as ray 2 (i.e., only a small amount will be lost due to reflection at the coating-air interface).

(b) Equate your estimates for E_1 and E_2 and show that complete destructive interference of rays 1 and 2 requires that $n_c = \sqrt{n_g}$. For glass with $n_g = 1.5$, show that magnesium fluoride with a refractive index of 1.38 is a reasonably good choice for use as a coating.

24. (II) A radio telescope, whose two antennas are separated by 50 m, is designed to receive 3.0-MHz radio waves produced by astronomical objects. The received radio waves create 3.0-MHz electronic signals in the telescope's left and right antennas. These signals then travel by equal-length cables to a centrally located amplifier, where they are added together. The telescope can be "pointed" to a certain region of the sky by adding the present signal from the right antenna to a "time-delayed" signal received by the left antenna a time Δt ago (this time delay of the left signal can be easily accomplished with the proper electronic circuit). If a radio astronomer wishes to "view" radio signals arriving from an object oriented at a 12° angle to the vertical as in Fig. 35–9, what time delay Δt is necessary?

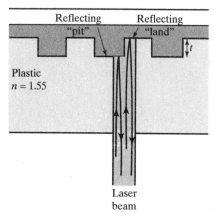

FIGURE 35–8

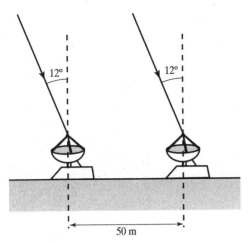

FIGURE 35–9

Over the range of possible incident angles $0 \le \theta_1 \le 90°$, find the range of visible wavelengths that will produce constructive interference at the point of focus. Assume visible wavelengths are 400 nm to 700 nm.

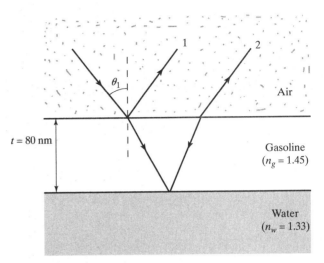

FIGURE 35–11

25. (II) The frequency of light emitted by a diode laser can be easily scanned over a small range of frequencies. Consider a pulse of light emitted from such a laser whose frequency f varies linearly with time t according to $f = f_o + \alpha t$, where f_o is the frequency at the start of the pulse and α is a constant called the "chirp rate." Imagine pointing this laser at a reflecting object located a distance x away. If the light currently emitted by the laser is then allowed to interfere with the reflected light received from the object, the intensity of the resultant light will oscillate at a "beat frequency" f_{beat}. Show that by measuring this beat frequency (using an appropriate light detector) the distance to the reflecting object can be found by $x = c f_{\text{beat}} / 2\alpha$, where c is the speed of light. [This "chirped lidar" method is used by some laser rangefinders.]

26. (III) Two identical sources S_1 and S_2, separated by distance d, coherently emit light of wavelength λ uniformly in all directions. Defining the x-axis with its origin at S_1 as shown in Fig. 35–10, find the locations (expressed as multiples of λ) of the interference minima along this axis for $x > 0$, if $d = 3\lambda$.

FIGURE 35–10

27. (III) A gasoline layer $(n_g = 1.45)$ of thickness $t = 80$ nm floats on water $(n_w = 1.33)$. When a light beam of wavelength λ strikes the gasoline layer at an incident angle θ_1 from air, reflections for the air-gasoline and gasoline-water surfaces create rays 1 and 2, respectively, as shown in Fig. 35–11. Assume these rays subsequently pass through a lens (e.g., in a person's eye) and are focused to a point where they interfere.

28. (III) Two closely spaced wavelengths of light λ and $\lambda + \Delta\lambda$ create interference patterns on a distant screen after passing through two slits separated by distance d.
    (a) Let θ_m and $\theta_m + \Delta\theta$ be the angles at which the mth order fringes for the wavelengths λ and $\lambda + \Delta\lambda$ occur, respectively, where $\Delta\theta$ and $\Delta\lambda$ are small. Show that $\Delta\lambda \approx \Delta\theta \sqrt{\left(\dfrac{d}{m}\right)^2 - \lambda^2}$, where $\Delta\theta$ is in radians.
    (b) Imagine that you want to use two slits separated by 0.02 mm as a spectrometer to make measurements of visible wavelengths near 550 nm. If you operate this spectrometer by observing the first-order fringes and you are able to measure angles to an accuracy of 0.2°, to what accuracy will you be able to determine wavelengths?
    (c) To what accuracy will you be able to determine wavelengths if you, instead, observe tenth-order fringes?

29. (III) In Example 35–7 of the text, the interference of two rays created by reflections from the outer and inner surfaces of a soap bubble of thickness t and refractive index n is analyzed. At the location where the two rays overlap and interfere, assume that the electric-field amplitude for both rays is E_o. Then, measuring phase relative to the incident light, the interfering light created by reflection from the outer and inner surfaces can be represented as $E_{\text{out}} = E_o \sin(\omega t + \pi)$ and $E_{\text{in}} = E_o \sin(\omega t + \delta)$, respectively. Here, ω is the frequency of the incident light with wavelength λ (in air).
    (a) Explain the presence of "π" in the expression for E_{out}.
    (b) The extra distance (i.e., the path difference) traversed by the "inner" ray through the bubble is responsible for the presence of "δ" in the expression for E_{in}. Determine δ in terms of t and λ.
    (c) The "reflecting" thicknesses t for incident light of wavelength λ maximizes E, where E is the amplitude of the sum of E_{out} and E_{in}. Use a phasor diagram to predict these "reflecting" thicknesses t.

30. (III) In Example 35–7 of the text, the interference of two rays created by reflections from the outer and inner surfaces of a soap bubble of thickness $t = 100\,\text{nm}$ and refractive index $n = 1.35$ is analyzed. At the location where the two rays overlap and interfere, assume that the electric-field amplitude for both rays is E_o. Then, measuring phase relative to the incident light, the interfering light created by reflection from the outer and inner surfaces can be represented as $E_{out} = E_o \sin(\omega t + \pi)$ and $E_{in} = E_o \sin(\omega t + \delta)$, respectively. Here, ω is the frequency of the incident light with wavelength λ (in air).

(a) Explain the presence of "π" in the expression for E_{out}.

(b) The extra distance (i.e., the path difference) traversed by the "inner" ray through the bubble is responsible for the presence of "δ" in the expression for E_{in}. Determine δ in terms of t and λ.

(c) Determine E through the use of a phasor diagram, where E is the amplitude of the sum of E_{out} and E_{in}. Then, defining I_o as the maximum reflected light intensity, show that the intensity of the total reflected light as a function of incident wavelength is given by $I_R = I_o \sin^2\left(\dfrac{2\pi nt}{\lambda}\right)$.

(d) Show that I_R takes on its maximum value for $\lambda = 540\,\text{nm}$ (as found in Example 35–7). Find I_R (in terms of I_o) at the extremes of the visible spectrum (i.e., for $\lambda = 400\,\text{nm}$ and $\lambda = 700\,\text{nm}$).

CHAPTER 36

Diffraction and Polarization

Section 36–1

1. (II) Red light passes through a slit whose width is ten times larger than its 650-nm wavelength (i.e., $a = 10\lambda = 6500$ nm) and a diffraction pattern is created on a screen 30 cm away. Most of the diffracted light energy will be contained within the central peak of this pattern. How much larger is the width of this central peak on the screen (measured from first minimum to first minimum) than the width of the slit?

2. (II) Coherent light created in the thin, semiconducting layers of a laser diode is emitted out of a rectangular area of 3.0 μm × 1.5 μm (horizontal-by-vertical) on the laser-facet's face. If the laser light has a wavelength of 780 nm, determine the (total) angle between the first diffraction minima in both the horizontal and vertical directions for the emitted light. [The highly divergent light from a laser diode must be collimated by a lens.]

Section 36–2

3. (II) Light of wavelength 750 nm passes through a slit 1.0 μm wide and a single-slit diffraction pattern is formed on a screen 20 cm away. Let the x-axis be aligned along the diffraction pattern on the screen with its origin at the central maximum. Determine the light intensity I at $x = 15$ cm, expressed as a fraction of the central maximum's intensity I_o.

4. (II) The "full-width at half-maximum" (FWHM) of the single-slit diffraction central peak is defined as the angle $\Delta\theta$ between the two points $\pm \theta_{1/2}$ where the intensity is one-half that at $\theta = 0$. That is, $\Delta\theta = \theta_{1/2} - (-\theta_{1/2})$.

   (a) Prove that $\Delta\theta = 2\sin^{-1}\left(0.442\dfrac{\lambda}{a}\right)$. [Hint: The solution to the transcendental equation $\sin\alpha = \dfrac{\alpha}{\sqrt{2}}$ is $\alpha = 1.39$ rad.]

   (b) Determine $\Delta\theta$ (in degrees) for a "small slit" with $a = 1\lambda$ and for a "large slit" with $a = 100\lambda$.

5. (III) (a) When a wave passes through a single slit, the diffracted wave's intensity I vs. angle θ is given by $I = I_o\left[\dfrac{\sin(\beta/2)}{\beta/2}\right]^2$, where $\beta = 2\pi a \sin\theta/\lambda$.

   (a) By taking the derivative $dI/d\beta$, show that extremal values for this diffraction pattern occur when either $\sin(\beta/2) = 0$ or $\tan(\beta/2) = \beta/2$.

   (b) Show that $a\sin\theta = m\lambda$, where $m = 1, 2, 3, \ldots$, satisfies the condition $\sin(\beta/2) = 0$. What are the names of these extremal values?

   (c) Show that $a\sin\theta = m\lambda$, where $m = 0, 1.4303, 2.4590, 3.4707$, satisfies the condition $\tan(\beta/2) = \beta/2$. What are the names of these extremal values?

6. (III) Defining $x = a\sin\theta/\lambda$, the intensity of single-slit diffraction pattern can be written as $\dfrac{I}{I_o} = \left[\dfrac{\sin(\pi x)}{\pi x}\right]^2$.
As investigated in Example 36–3 in the text, the approximate locations of the first and second secondary maxima are predicted by $x = a\sin\theta/\lambda = 1.50$ and $x = a\sin\theta/\lambda = 2.50$, respectively. Determine the location of these maxima more precisely through the following numerical method:

   (a) Using a computer or a graphing calculator, evaluate $\left[\dfrac{\sin(\pi x)}{\pi x}\right]^2$ for the 101 x-values in the range from $x = 1.00$ to $x = 2.00$ in steps of $\Delta x = 0.01$. Find the maximum value for I/I_o in this range and its corresponding x-value. How do your findings compare with the approximate values found in Example 36–3 (i.e., $I/I_o = 0.045$ at $x = 1.50$)?

   (b) Repeat this method to find I/I_o and its corresponding x-value for the second secondary maximum. Here, bracket the approximate peak location of $x = 2.50$ with 101 x-values in the range from $x = 2.00$ to $x = 3.00$ in steps of $\Delta x = 0.01$. Compare your finding with the approximate value of $I/I_o = 0.016$ found in Example 36–3.

7. (III) Most of a diffracted wave's energy is contained within its central peak. Prove the veracity of this statement by investigating the diffraction pattern created by a wave of wavelength λ passing through a slit whose width $a = 5\lambda$. Assume the resulting diffraction pattern lies along an x-axis on a screen 1/2 m away with $x = 0$ at the central maximum (Fig. 36–1). Let I_o be the intensity at $x = 0$. Since the pattern is symmetric about the origin, we will concentrate our attention only on the positive x-direction.

   (a) Show that, as a function of x, the diffraction pattern's intensity I relative to I_o is given by $\dfrac{I}{I_o} = \left[\dfrac{\sin(5\pi\sin\theta)}{5\pi\sin\theta}\right]^2$ with $\theta = \tan^{-1}(2x)$, where x is in meters.

   (b) Use a computer or a graphing calculator to evaluate the quantity (I/I_o) for 101 equally spaced positive values of x in the range $0 \leq x \leq 1.00$ m (i.e., at $x = 0.00, 0.01, 0.02, 0.03, \ldots$). Note: For $x = 0.00, I/I_o = 1.00$. Display your results by plotting (I/I_o) vs. x for $0 \leq x \leq 1.00$ m. On your plot, identify the first, second, and third secondary maxima as well as the first, second, third, and fourth minima.

   (c) If the height (perpendicular to the page in Fig. 36–1) of the diffraction pattern is H, the light energy contained in a slice along the x-axis of length dx is $dE = I(H dx) = I_o H\left(\dfrac{I}{I_o}dx\right)$.
Thus, the energy in the diffraction pattern is proportional to the integral $\int (I/I_o)\, dx \approx \Sigma(I/I_o)\,\Delta x$. You evaluated (I/I_o)

for 101 different x at a constant spacing $\Delta x = 0.01$ m. Thus, $\Sigma(I/I_o)\Delta x = \Delta x \Sigma(I/I_o)$, so that a value proportional to the total light energy E_T from $x = 0$ to $x = 1.00$ m can be found by simply summing the 101 values of (I/I_o). Additionally, a value proportional to the light energy E_c contained within the central peak can be found by summing the (I/I_o) from $x = 0$ to the x-value associated with the first minimum. Determine these two summations; then use them to estimate the ratio E_c/E_T, the fraction of the diffraction pattern's total light energy contained within the central peak.

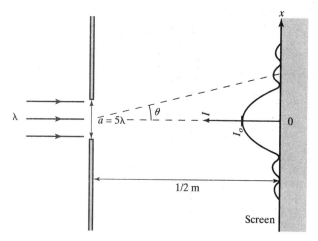

FIGURE 36–1

Section 36–3

8. (II) In a double-slit experiment, if the central diffraction peak contains 13 interference fringes, how many fringes are contained within both of the first secondary diffraction peaks?

9. (II) 600-nm light passes through a pair of slits and creates an interference pattern on a screen 2.0 m behind the slits. The slits are separated by 0.120 mm and each slit is 0.040 mm wide. How many constructive interference fringes are formed on the screen? [Many of these fringes will be of very low intensity.]

10. (II) In a double-slit experiment, the slit separation is ten times greater than each slit width. Compare (as a ratio) the intensity of the seventh-order interference maximum with that of the zero-order maximum.

Sections 36–4 and 36–5

11. (II) If you shine a flashlight beam toward the Moon, estimate the diameter of the beam when it reaches the Moon. Assume that the beam leaves the flashlight through a 5.0 cm aperture, that its white light has an average wavelength of 550 nm, and that the beam spreads due to diffraction only.

12. (II) The nearest neighboring star to the Sun is about 4 light-years away. If a planet happened to be orbiting this star at an orbital radius equal to that of the Earth-Sun distance, what minimum diameter would an Earth-based telescope's aperture have to have in order to obtain an image that resolved this star-planet system? Assume the light emitted by the star and planet has a wavelength of 550 nm.

13. (II) Mizar, the second star from the end of the Big Dipper's handle (Ursa Major), is one of the most famous stars in the sky.

(a) Part of Mizar's fame is derived from the fact that pairing it with a star nearby on the nighttime sky, called Alcor, provides a good test of minimal vision. From Earth, Mizar and Alcor have an angular separation of 12 minutes of arc (60 min = 1 deg). Show that this angular separation is near the threshold of diffraction-limited vision for the human eye, assuming the minimum pupil diameter of 1 mm and light of wavelength 550 nm.

(b) Even without Alcor, Mizar would be celebrated as the first "binary star" system to be discovered. In 1650, it was found that Mizar is actually a pair of stars that orbit each other. From Earth, these two stars are separated by 14 seconds of arc (60 s = 1 min). The Mizar binary system is a prime viewing target for an amateur astronomer with a new telescope. Estimate the minimum telescope aperture required to resolve this pair, if limited only by diffraction. [Even more remarkably, each of these two stars has been found to be a binary star, so Mizar is actually four stars.]

Section 36–7

14. (II) The known wavelength of red light from a helium-neon laser ($\lambda = 633$ nm) is used to calibrate a diffraction grating. If this light creates a second-order fringe at 53.2° after passing through the grating, and light of an unknown wavelength λ creates a first-order fringe at 20.6°, find λ.

15. (II) Laser light passes through a diffraction grating with 6000 lines/cm and a pattern of bright red fringes is observed on a wall, as shown in Fig. 36–2. What wavelength of light does this laser emit?

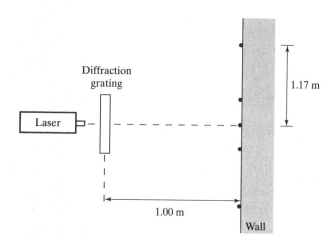

FIGURE 36–2

16. (II) A tungsten-halogen bulb emits a continuous spectrum of ultraviolet, visible, and infrared light in the wavelength range 360 nm to 2000 nm. Assume that the light from a tungsten-halogen bulb is incident on a diffraction grating with slit spacing d and that the first-order brightness maximum for the wavelength of 1200 nm occurs at angle θ. What other wavelengths within the spectrum of incident light will produce a brightness maximum at this same angle θ? [Optical filters are used to deal with this bothersome effect when a continuous spectrum of light is measured by a spectrometer.]

17. (II) White light passes through a 600-line/mm diffraction grating. First-order and second-order visible spectra ("rainbows") appear on the wall at distance 30 cm away as shown in Fig. 36–3 (the visible spectrum ranges from 400 nm to 700 nm). In which order is the "rainbow" dispersed over a larger distance? That is, calculate l_1 and l_2; then state which is the larger.

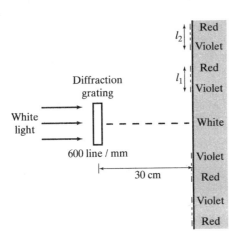

FIGURE 36-3

Section 36-9

18. (II) Hot hydrogen emits four visible colors; however, if a diffraction grating with sufficient resolution is used, it is found that each of these colors consists of a pair of closely spaced wavelengths (due to magnetic effects within the atom). For example, the "H-alpha" (H_α) color is two wavelengths of approximately $\lambda = 656.28$ nm, separated by an amount $\Delta \lambda = 0.013$ nm.
    (a) In order for a grating to resolve these two wavelengths in the first order, what minimum number of lines must it have?
    (b) If a square grating with side length 12 cm has this minimum number of lines, at what nominal angle will the first-order fringes for the H_α pair appear?

19. (III) Two closely spaced wavelengths of light λ and $\lambda + \Delta\lambda$ pass through a diffraction grating with grating line separation d. A well-known pair of such wavelengths is the "sodium doublet" for which $\lambda = 589.00$ nm and $\Delta\lambda = 0.59$ nm. In this problem, assume a 1200-line/mm grating that is 3.0 cm wide.
    (a) Let θ_m and $\theta_m + \Delta\theta$ be the angles at which the mth order fringes for the wavelengths λ and $\lambda + \Delta\lambda$ occur, respectively, where $\Delta\theta$ and $\Delta\lambda$ are small. Using the trigonometric identity for $\sin(A + B)$, show that $\Delta\theta \approx \dfrac{m\,\Delta\lambda}{d\cos\theta_m}$, where $\Delta\theta$ is in radians. For the sodium doublet passing through the given grating, determine $\Delta\theta$ for the first-order fringes.
    (b) Each of these fringes itself has an angular half-width $\Delta\theta_m$ (see Eq. 36–17 in the text). For the given grating, find $\Delta\theta_m$ (in radians) for each of the first-order sodium-doublet fringes.
    (c) Using the numerical values you obtained for $\Delta\theta$ and $\Delta\theta_m$, calculate the ratio $\Delta\theta/\Delta\theta_m$. Explain why your result indicates that the given grating clearly resolves the sodium doublet in the first order.

20. (III) Two closely spaced wavelengths of light λ and $\lambda + \Delta\lambda$ pass through a diffraction grating with grating line separation d and N total lines.
    (a) Let θ_m and $\theta_m + \Delta\theta$ be the angles at which the maxima for the mth order fringes of the wavelengths λ and $\lambda + \Delta\lambda$ occur, respectively, where $\Delta\theta$ and $\Delta\lambda$ are small. Using the trigonometric identity for $\sin(A + B)$, show that
    $$\Delta\theta \approx \frac{m\,\Delta\lambda}{d\cos\theta_m},$$ where $\Delta\theta$ is in radians.
    (b) Each of these fringes itself is a peak with an angular half-width $\Delta\theta_m$ from its maximum to minimum (see Eq. 36–17 in the text). As a plausible criterion, define the two fringes as "barely resolved" if the maximum of one is located at the angular position of the minimum for the other. Use this definition to prove the grating's resolving power
    $$R = \frac{\lambda}{\Delta\lambda} = Nm.$$

Section 36-10

21. (II) (a) If one wishes to diffract an EM wave from the atomic planes in crystalline solids, show that the maximum EM wavelength λ_m that can be used is $2d$, where d is the spacing between the planes.
    (b) For crystalline solids, d is on the order of 0.1 nm. Use this fact to estimate λ_m. To what region of the EM spectrum does this wavelength correspond?

Section 36-11

22. (II) Two polarizers A and B are aligned so that their transmission axes are vertical and horizontal, respectively. A third polarizer is placed between these two with its axis aligned at angle θ with respect to the vertical. Assuming vertically polarized light of intensity I_o is incident upon polarizer A, find an expression for the light intensity I transmitted out of the other side of this three-polarizer sequence. Calculate the derivative $dI/d\theta$; then use it to find the angle θ that maximizes I.

23. (II) (a) Light passes through a sequence of ten polarizers, where the transmission axis of each polarizer is oriented at the same angle θ relative to its preceding neighbor's axis. Before entering the polarizer sequence, the light is vertically polarized and has intensity I_o. The transmission axis of the first polarizer is vertical and θ is chosen so that when the light exits the sequence it will be horizontally polarized and have intensity I. Determine I, expressed as a fraction of I_o.
    (b) Generalize the situation in part (a) to a sequence of $(N + 1)$ polarizers and determine an expression for I. In the limit $N \to \infty$, what is I? [Analogously, a sequence of liquid crystal molecules, each slightly rotated from its neighbor, is used to rotate the polarization plane of light in a liquid crystal display (LCD).]

24. (II) In certain types of lasers, a gas-filled laser tube is sealed at both ends by a glass window ($n = 1.5$) and this tube is placed between two reflecting mirrors. When the laser is in operation, vertically polarized laser light passes through the tube as it bounces back and forth between the mirrors. So that none of the light is lost to reflections at the window faces as it enters and exits the tube, these windows are inclined at angle θ as shown in Fig. 36–4. Determine θ. Assume the index of refraction of the gas within the tube is approximately unity, so that the light travels in the same direction both inside and outside the tube.

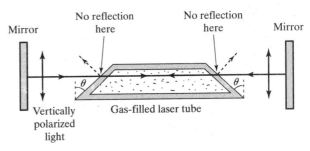

FIGURE 36-4

General Problems

25. (II) Sound travels at 340 m/s and has a typical frequency of 600 Hz. Light travels at 3×10^8 m/s and has a typical frequency of 6×10^{14} Hz. Explain why sound bends around a corner when passing through a 1-m wide door, while light appears to travel in a straight line when passing through this same opening.

26. (II) A certain spectrometer allows the spectrum produced by a diffraction grating to be observed up to an angle of 45°. If a researcher intends to input light with wavelengths in the visible to infrared range of 400 nm to 1000 nm and the entire first-order spectrum is to be observable, what maximum number of lines per mm should the spectrometer's grating have? This grating will also be able to observe the second-order spectrum of a subset of the 400 nm to 1000 nm wavelengths. Determine this subset.

27. (II) The entrance to a boy's bedroom consists of two doorways, each 1.0 m wide, which are separated by a distance of 3.0 m. The boy's mother yells at him through the two doors as shown in Fig. 36–5, telling him to clean up his room. Her voice has a frequency of 400 Hz. Later, when the mother discovers the room is still a mess, the boy says he never heard her telling him to clean his room. The velocity of sound is 340 m/s.
    (a) Find all of the angles θ at which no sound will be heard within the bedroom when the mother yells. Assume sound is completely absorbed when it strikes a bedroom wall (i.e., there are no reflections of sound from the walls).
    (b) If the boy was at the position shown when his mother yelled, does he have a good explanation for not having heard her?

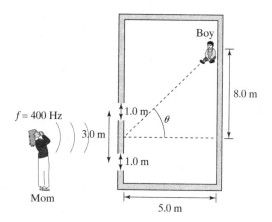

FIGURE 36–5

28. (III) A slit of width $a = 20\,\mu$m is cut through a thin aluminum plate. Light with wavelength $\lambda = 650$ nm passes through this slit and forms a single-slit diffraction pattern on a screen a distance $L = 2.0$ m away. Defining x to be the distance between the first minima in this diffraction pattern, find the change Δx in this distance when the temperature T of the metal plate is changed by amount $\Delta T = 50\,\text{C}°$. [Hint: Since $\lambda \ll a$, the first minima occur at a small angle.]

29. (III) Most modern spectrometers use reflection (rather than transmission) diffraction gratings. In one common form, a "blazed" reflection grating consists of a glass block with a highly reflective sawtooth-shaped surface, where adjacent "teeth" are a small distance d apart and each "tooth" is inclined at the "blaze angle" γ relative to the grating's base (Fig. 36–6). Consider two parallel rays of light with wavelength λ incident on adjacent teeth at angle θ_1 relative to the normal to the grating's base. Define the direction of the parallel rays after reflection from the grating to be angle θ_2 relative to the normal to the grating's base. Assume that, subsequently, these reflected parallel rays will be focused to a point by a spherical mirror and allowed to interfere.
    (a) Prove that constructive interference will occur if $d(\sin\theta_2 - \sin\theta_1) = m\lambda$, where $m = 0, 1, 2, \ldots$
    (b) Prove that $\theta_2 = \theta_1 + 2\gamma$.
    (c) The "blaze wavelength" λ_b is defined as the wavelength for which zero incident angle $(\theta_1 = 0°)$ yields a first-order constructive interference fringe. If a designer wishes a grating with 1200 lines/mm to have a blaze wavelength of 650 nm, what should the grating's blaze angle be? [Blazed gratings work most efficiently for wavelengths near the blaze wavelength.]

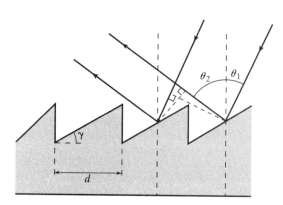

FIGURE 36–6

30. (III) The faces of a transparent Nd:YAG laser rod $(n = 1.818)$ are cut so that they are at an angle $90° - \theta_f$ to the rod's axis. This rod is placed between two vertically aligned mirrors with its axis at angle θ_a to the horizontal (Fig. 36–7). When used as the light-producing medium in a laser, vertically polarized light passes through this rod, directed along its axis, as it bounces back and forth between the mirrors. Find the proper choice for θ_f and θ_a so that there will be no reflections each time the light enters the face of the rod as it travels back and forth between the mirrors. Assume the rod is surrounded by air. [This geometry is used in solid-state laser design to eliminate losses due to reflections.]

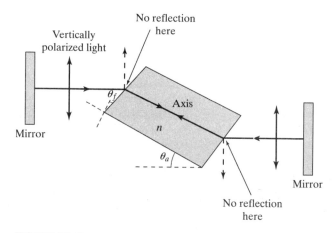

FIGURE 36–7

CHAPTER 37

Special Theory of Relativity

Sections 37–4 to 37–6

1. (II) An unstable particle produced in an accelerator experiment travels at constant velocity, covering 1.00 m in 3.40 ns in the lab frame before changing ("decaying") into other particles. In the rest frame of the particle,
   (a) how long did it live before decaying?
   (b) how far did it move before decaying?

2. (II) An Earth-dweller observes a rocket ship fly overhead with speed $0.60\,c$ at exactly noon on his wristwatch. In the overhead rocket ship, the pilot sets her clock so that it coincides with the setting of the Earth-dweller's wristwatch. Later, at 12:40 P.M. as determined by the pilot, the rocket ship passes a space station that is stationary with respect to Earth. What time is it at the station when the rocket passes, assuming the space-station clocks are set to the same time zone as the Earth-dweller's wristwatch?

3. (II) When it is stationary, the half-life of a certain subatomic particle is T_o. That is, if N of these particles are present at a certain time, then a time T_o later, only $N/2$ particles will be present, assuming the particles are at rest. A beam carrying N such particles per second is created at position $x = 0$ in a high-energy physics laboratory. This beam travels along the x-axis at speed v in the laboratory reference frame and it is found that only $N/2$ particles per second travel in the beam at $x = 2cT_o$, where c is the speed of light. Find the speed v of the particles within the beam.

4. (II) In its own reference frame, a box has the shape of a cube 2.0 m on a side. This box is loaded onto the flat floor of a spaceship and the spaceship then flies past us with a horizontal speed of $0.80\,c$. What is the volume of the box as we observe it?

5. (II) An astronaut in a spaceship travels 1.3×10^{12} m from Earth to Saturn at a constant velocity v. Clocks on Earth measure the journey to take 5.0×10^3 s.
   (a) According to Earth-based observers, what is the spaceship's speed during its journey to Saturn?
   (b) At the end of the journey, how much time has passed on the astronaut's clock?
   (c) Suppose that the spaceship has an odometer to measure distance traveled. What does the odometer read at the end of the journey, assuming it was set to zero at Earth?
   (d) Show that the readings from the spaceship's odometer and the astronaut's clock yield the same value for the spaceship's speed as that determined by Earth-based observers.

6. (II) The factor $\gamma = 1/\sqrt{1 - v^2/c^2}$, which appears in many relativistic relations, becomes difficult (or impossible) to evaluate on a calculator when v is very close to the speed of light (e.g., $v/c = 0.999\,999\,999\,999$). Show that in this extreme speed limit of $v \approx c$, $\gamma \approx 1/\sqrt{2(1 - v/c)}$. Evaluate γ when $v/c = 0.999\,999\,999\,999$.

7. (II) An object of proper length L_o, when moving past an observer at speed v, is measured by that observer to have length L. Define the "shortening" of the object to be $\Delta L = L_o - L$.
   (a) Show that an object's fractional shortening
   $$\frac{\Delta L}{L_o} \approx \frac{1}{2}\left(\frac{v}{c}\right)^2, \text{ if } v \ll c.$$
   (b) According to a person standing by the side of a highway, what is the fractional shortening of a car traveling down the road at 100 km/h?

8. (II) When at rest, a spaceship has the form of an isosceles triangle whose two equal sides have length $2l$ and whose base has length l. If this ship flies past an observer with a relative velocity of $v = 0.95\,c$ directed along its base, what are the lengths of the ship's three sides according to the observer?

9. (III) A spaceship, of proper length 100 m, passes by a certain point on Earth. The ship's speed is such that only $1/3\,\mu\text{s}$ is required for it to pass fully by the point, as measured by clocks on Earth.
   (a) How fast is the ship moving relative to Earth?
   (b) How long is the ship according to an observer on Earth?

10. (III) When they are both at rest, a Limousine Town Car is twice as long as a Volkswagen Beetle. As both of these vehicles pass a police officer who is standing at the side of the roadway, the officer observes that they both have the same length. If the Beetle is traveling at one-half the speed of light relative to the officer, how fast is the Limousine traveling?

11. (III) Two space ships A and B journey from Earth to a star that is exactly 10.00 light-years away from Earth. According to Earth-based observers, ship A makes this journey in 10.01 y, while ship B takes 10.10 y. How many times longer is the trip-time according to passengers on ship B compared with passengers on ship A?

Section 37–8

12. (II) Reference frame S' moves at speed $v = 12/13\,c$ in the $+x$ direction with respect to reference frame S. An object is stationary in S' at position $x' = 100$ m. What is the position of the object in S when the clock in S reads $1.00\,\mu\text{s}$ according to the (a) Galilean and (b) Lorentz transformation equations?

13. (II) In a space rodeo, a cowboy tries to throw a rope lasso around the neck of a fleeing bull. The bull is released from a pen and flees on a space scooter with a speed of 4/5 c; the cowboy follows close behind with a speed of 3/5 c. Both speeds are with respect to the rodeo spectators. If the cowboy launches the rope at a speed of 1/3 c (with respect to himself), is it possible for the rope (if aimed correctly) to overtake the bull's neck according to (a) Galileo's relativity? (b) Einstein's relativity?

14. (II) An extraterrestrial flies its spaceship at low altitude over Earth's surface at speed v relative to Earth. When directly overhead an Earth-based observer, the extraterrestrial turns on the ship's headlamps, emitting light at a speed c relative to the ship. Using the relativistic velocity transformation equation, show that the Earth-based observer also observes this light traveling at speed c.

15. (III) An observer in reference frame S notes that two events are separated in space by 200 m and in time by 0.80 μs. How fast must reference frame S' be moving relative to S in order for an observer in S' to detect the two events as occurring at the same location in space?

16. (III) Two light bulbs A and B are placed at rest with respect to each other on the x-axis at positions $x_A = 0$ and $x_B = +L$, respectively, where L is a positive constant. In this reference frame, the bulbs are turned on simultaneously. Use the Lorentz transformations to find an expression for the time interval between when the bulbs are turned on as measured by an observer moving at velocity v in the $+x$ direction. According to this observer, which bulb is turned on first?

17. (III) An observer is located at the origin in each of two inertial reference frames S and S', where S' moves at speed v along the $+x$ direction relative to S. Assume the two observers pass each other at time $t = t' = 0$ and that a light flash is produced at their common location at this instant. In reference frame S, the position of this light flash travels away from the origin at the speed c. Use this fact, along with the Lorentz transformation equations, to show that the position of the light flash also moves away from the origin in S' at the speed c.

Section 37–9

18. (II) An unstable particle is at rest and suddenly decays into two fragments. No external forces act on the particle or its fragments. One of the fragments has a velocity of 0.60 c and a mass of 6.68×10^{-27} kg, while the other has a mass of 1.67×10^{-27} kg. What is the velocity of the less massive fragment?

Section 37–11

19. (II) At present, the rate of worldwide energy consumption is 15×10^{12} W. If energy were added to a 50-kg person at this rate, how many days would it take to accelerate that person from rest to a speed of 0.99 c?

20. (II) A free particle has a rest mass m. In a process that requires work W_1, this particle is accelerated from rest to speed 0.90 c. Next, the particle is accelerated from speed 0.90 c to 0.99 c, a process requiring work W_2. Determine how much greater W_2 is than W_1 (i.e., find the ratio W_2/W_1).

21. (II) When a free proton and a free electron with rest masses m_p and m_e, respectively, are brought together to form a hydrogen atom of rest mass m_H, 13.6 eV of energy is released as a result of the electrical attraction between the oppositely charged particles. Determine the fractional difference between the sum of the proton and electron rest masses and the rest mass of a hydrogen atom. That is, defining $m = m_p + m_e$, evaluate $(m - m_H)/m$.

22. (III) A particle whose rest energy is mc^2, travels with a velocity v very near the speed of light. Write $v = c(1 - \delta)$, where $\delta \ll 1$; then show that $\delta \approx \frac{1}{2}\left(\frac{mc^2}{E}\right)^2$, where E is the particle's energy. In the latest generation of particle accelerators called the Large Hadron Collider, a fast-moving proton will have $E = 5$ TeV (1 TeV $= 10^{12}$ eV). For these protons, what is δ?

Section 37–12

23. (III) A radar "speed gun" emits microwaves of frequency $f_o = 36.0$ GHz. When pointed at an object moving toward it at speed v, the object "observes" the microwave at the Doppler-shifted frequency f and thus, due to reflection, becomes a moving source of microwaves at this frequency. The stationary radar gun then detects these reflected waves at the Doppler-shifted frequency f'. The gun combines its emitted wave at f_o and its detected wave at f'. These waves interfere, creating a beat pattern whose beat frequency $f_{beat} = f' - f_o$ the gun measures.

(a) Show that $v \approx \dfrac{c f_{beat}}{2 f_o}$, if $f_{beat} \ll f_o$. If $f_{beat} = 6670$ Hz, what is v (km/h)?

(b) If the object's speed changes by Δv, show that the change in beat frequency Δf_{beat} is given by $\Delta f_{beat} = \dfrac{2 f_o \Delta v}{c}$. If the accuracy of the speed gun is to be 1 km/h, to what accuracy must the beat frequency be measured?

24. (III) Imagine that a certain atom emits light of frequency f_o when at rest. Imagine further that a monatomic gas composed of this particular atom is at temperature T, so that some of the gas atoms move toward and others away from an observer due to their random thermal motion. Using the rms speed of the thermal motion as an estimate for the atomic speeds, show that, in comparison with f_o, the difference Δf between the Doppler-shifted frequencies for atoms moving directly toward the observer and atoms moving directly away from the observer is $\dfrac{\Delta f}{f_o} \approx 2\sqrt{\dfrac{3kT}{mc^2}}$, assuming $mc^2 \gg 3kT$. Evaluate $\Delta f/f_o$ for a gas of hydrogen atoms at 500 K. [This "Doppler-broadening" effect is commonly used to measure gas temperature in, e.g., astronomy.]

General Problems

25. (II) Equal electric charges q are glued all along the x-axis at a uniform distances d apart. Given that charge is invariant (i.e., the same in all reference frames), what is the charge per unit length λ', according to an observer traveling in the $+x$ direction at 0.80 c relative to the x-axis?

26. (II) As observed on Earth, a certain type of bacteria is known to double in number every 24 h. Two cultures of these bacteria are prepared, each consisting initially of one bacterium. One culture is left on Earth and the other is placed on a rocket that travels at a speed of 0.866 c relative to Earth. At a time when the earthbound culture has grown to 256 bacteria, how many bacteria are in the culture on the rocket?

27. (II) If E is the total energy of a particle with zero potential energy, prove that $\dfrac{dE}{dp} = v$, where p and v are the momentum and velocity of the particle, respectively.

28. (II) Two electrons approach each other, each traveling at speed $0.9500\,c$ with respect to a laboratory reference frame. In the reference frame of one electron,
(a) what is the speed of the other electron?
(b) what is the kinetic energy of the other electron in terms of the electron rest energy mc^2?

29. (III) A stationary particle of rest mass m decays into two "daughter" particles, each of rest mass $0.40\,m$. What is the magnitude of the relativistic momentum of each daughter particle?

30. (III) An incident proton of rest mass m strikes a stationary proton. The two protons disappear and are replaced by a particle of rest mass $4m$. Find the kinetic energy of the incident proton.
[*Hint*: $E^2 = (pc)^2 + (mc^2)^2.$]

CHAPTER 38

Early Quantum Theory and Models of the Atom

Section 38–1

1. (II) In a diatomic hydrogen (H_2) molecule, the chemical bond between the two atoms behaves as a spring, so that the atoms oscillate about their equilibrium positions with a frequency of 1.3×10^{14} Hz.
   (a) Find the minimum energy required to commence the oscillation of an H_2 molecule according to Planck's quantum hypothesis.
   (b) According to the kinetic theory of gases, the average oscillation energy per molecule in a gas at temperature T equals kT. Use this fact to estimate the temperature at which a diatomic hydrogen gas must be maintained, if one wishes most of the molecules to be oscillating with the oscillation energy found in part (a). [In a real hydrogen gas, the H_2 molecule dissociates well below this temperature.]

2. (III) What fraction of sunlight is in the visible range? Planck found that the light intensity per unit wavelength interval emitted by a blackbody at temperature T is given by
$$I(\lambda, T) = \alpha \frac{\lambda^{-5}}{e^{\beta/\lambda} - 1},$$
where the constants $\alpha = 2\pi hc^2$ and $\beta = hc/kT$.
   (a) The temperature of the Sun's surface is 5800 K. Assuming the Sun's surface emits radiation as a blackbody, show that $\beta = 2500$ nm.
   (b) Define $y = I(\lambda, T)/\alpha$. Using a computer or a graphing calculator, evaluate $y = \dfrac{\lambda^{-5}}{e^{\beta/\lambda} - 1}$ with $\beta = 2500$ nm for the 100 λ-values in the range from $\lambda = 20$ nm to $\lambda = 2000$ nm in steps of $\Delta\lambda = 20$ nm. Display your results by plotting y vs. λ for $0 \leq \lambda \leq 2000$ nm.
   (c) The intensity of sunlight, which contains a range of wavelengths from λ_1 to λ_2, is found from
$$\int_{\lambda_1}^{\lambda_2} I(\lambda, T)\, d\lambda \approx \Sigma I(\lambda, T)\, \Delta\lambda,$$
which is proportional to $\Sigma y\, \Delta\lambda$. Since $\Sigma y\, \Delta\lambda = \Delta\lambda \Sigma y$ for constant spacing $\Delta\lambda$, a value proportional to the total sunlight intensity I_T from $\lambda = 20$ nm to $\lambda = 2000$ nm can be found by simply summing your 100 values of y. Additionally, a value proportional to the sunlight intensity I_v contained within the visible range can be found by summing the y-values from $\lambda = 400$ nm to $\lambda = 700$ nm. Evaluate these two summations; then use them to estimate the ratio I_v/I_T, the fraction of sunlight's total light energy contained within the visible range.

Section 38–2

3. (II) The range of visible light wavelengths extends from about 400 nm to 700 nm.
   (a) Estimate the minimum energy (eV) necessary to initiate the chemical process on the retina that is responsible for vision.
   (b) Speculate as to why, at the other end of the visible range, there is a threshold photon energy beyond which the eye registers no sensation of sight. Determine this threshold photon energy (eV).

4. (II) In insulating and semiconducting solids, all of the energy levels associated with chemically bound electrons are less energetic than the energy states associated with mobile conduction electrons. The energy difference between the lowest "conduction electron" energy level and the highest "bound electron" energy level is called the band gap of the solid. When a photon with energy equal to or greater than the band gap is incident on the solid, the photon will be absorbed by the solid through the excitation of a bound electron to a conducting state. If the photon has less energy than the band gap, the photon will pass through the solid without being absorbed.
   (a) The band gap of silicon is 1.1 eV. For what range of wavelengths will silicon be transparent? In what region of the electromagnetic spectrum does this transparent range begin?
   (b) Window glass is transparent for all visible wavelengths (400 nm to 700 nm). Based on this observation, what is the minimum possible band gap value for glass?

5. (II) Many practical devices function based on the photoelectric effect. In a photomultiplier tube (a very sensitive light sensor), incident photons strike a metal surface and the resulting ejected electrons are collected. By counting the number of collected electrons, the number of incident photons (i.e., the incident light intensity) can be determined.
   (a) If a photomultiplier tube is to respond properly for incident wavelengths throughout the visible range (400 nm to 700 nm), what is the maximum value for the work function W_o (eV) of its metal surface?
   (b) If W_o for its metal surface is above a certain threshold value, the photomultiplier will only function for incident ultraviolet wavelengths and be unresponsive to visible light. Determine this threshold value (eV).

6. (III) Potassium has one of the lowest work functions of all metals and so is useful in photoelectric devices that respond to visible wavelengths. Light from a variable-wavelength source is incident on a potassium surface in a photoelectric effect experimental set-up. The acquired data of stopping voltage V_o as a function of wavelength λ are shown in Table 38–1 (1 μm = 1000 nm).

TABLE 38–1					
$\lambda\,(\mu m)$	0.400	0.430	0.460	0.490	0.520
V_o (V)	0.803	0.578	0.402	0.229	0.083

(a) Explain why a graph of V_o vs. $1/\lambda$ is expected to yield a straight line. What are the theoretical expectations for the slope and the y-intercept of this line?
(b) Using the data in Table 38-1, graph V_o (V) vs. $1/\lambda$ (μm^{-1}) and show that a straight-line plot does indeed result. Determine the slope a and y-intercept b of this line.
(c) Assume, as was true for Millikan when he performed his photoelectric effect experiments, that the values for the charge of the electron e and the speed of light c are known. Using your values for a and b, determine potassium's work function (eV) and Planck's constant h (J·s).

7. (III) A group of atoms is confined to a very small (point-like) volume in a laser-based atom trap. Because of its interaction with the laser light, each atom emits 1.0×10^6 photons of wavelength 780 nm every second (through a process called fluorescence). A light sensor of area $1.0\,cm^2$ measures the light intensity emanating from the trap to be 1.6 nW when placed a distance 25 cm away from the trapped atoms. Assuming each atom emits photons with equal probability in all directions, determine the number of trapped atoms.

8. (III) Imagine a hypothetical "photoelectric-like" process in which an incident photon of wavelength λ is absorbed by a free electron, which is initially at rest. After the photon has been absorbed, the electron moves off with velocity v. Prove that this hypothetical process is inconsistent with energy and momentum conservation.

9. (III) Could a photoelectric effect experimental set-up be used to measure a visible-light wavelength accurately? Assume the wavelength λ is incident on a metal surface, whose work function is known precisely (i.e., its uncertainty is less than 0.1% and so can be ignored). Show that if the stopping voltage can be determined to an accuracy of ΔV_o, the fractional uncertainty (magnitude) in wavelength is $\dfrac{\Delta \lambda}{\lambda} = \dfrac{\lambda e}{hc} \Delta V_o$. Determine this fractional uncertainty if $\Delta V_o = 0.01$ V and $\lambda = 550$ nm.

Section 38–3

10. (II) In the Compton effect, compare (as a ratio) the maximum change $\Delta \lambda$ in a photon's wavelength with the photon's initial wavelength λ, if the photon is
(a) a visible-light photon with $\lambda = 550$ nm.
(b) an X-ray photon with $\lambda = 0.10$ nm.

11. (II) A photon, which is initially a 1.0-MeV gamma ray photon, undergoes a sequence of Compton-scattering events. If the photon is scattered at an angle of 0.50° in each event, estimate the number of such events required to convert the photon into a visible-light photon with wavelength 555 nm. [Gamma rays created near the center of the Sun are transformed to visible wavelengths as they travel to the Sun's surface through a sequence of small-angle Compton scattering events.]

Section 38–6

12. (II) In addition to X-rays and electrons, neutrons can be used in diffraction experiments to probe the lattice structure of crystalline solids. This experimental technique requires that the neutron's wavelength be on the order of the spacing between atoms in the lattice. Given that this spacing is about 0.3 nm, what should the speed of the neutrons be?

13. (II) Conduction electrons in a semiconductor behave as an ideal gas (this is not true for conduction electrons in a metal). Taking mass $m = 9 \times 10^{-31}$ kg and temperature $T = 300$ K, determine the de Broglie wavelength of a semiconductor's conduction electron. Given that the spacing between atoms in a semiconductor's atomic lattice is on the order of 0.3 nm, would you expect room-temperature conduction electrons to travel in straight lines or diffract when traveling through this lattice?

14. (II) After passing through two slits separated by a distance of 3.0 μm, a beam of electrons creates an interference pattern with its second-order maximum at angle 50°. Find the speed of the electrons within this beam.

Sections 38–9 and 38–10

15. (II) A hydrogen atom has an angular momentum of $5.273 \times 10^{-34}\,kg \cdot m^2/s$. According to the Bohr model, what is the energy (eV) associated with this state?

16. (II) Compare (as a ratio) the wavelength of the lowest-energy photon in the Balmer series with that of the lowest-energy photon in the Lyman series.

17. (II) The "lifetime" of the first excited energy state in hydrogen is 1.6 ns. Assume that the electron in a Bohr hydrogen atom has been excited to the $n = 2$ energy state and that it stays in that state for 1.6 ns before jumping to the ground state. While in the excited state, how many times does the electron orbit the nucleus?

18. (II) Assume hydrogen atoms in a gas are initially in their ground state. If free electrons with kinetic energy 12.75 eV collide with these atoms, what photon wavelengths will be emitted by the gas?

19. (III) When an atom with mass m emits a photon of wavelength λ, it recoils with a (nonrelativistic) velocity v.
(a) Show that the kinetic energy E_R associated with this atomic recoil in comparison with the energy E_P of the emitted photon is given by $E_R/E_P = h/2\lambda mc$.
(b) When an electron jumps from a higher to a lower energy state, the Bohr model assumes that all of the energy lost by the electron is transformed into the emitted photon (i.e., the atom's recoil energy is assumed to be zero). For the smallest-wavelength photon that a hydrogen atom can emit, evaluate E_R/E_P. Is it safe to neglect E_R in comparison with E_P as is done in the Bohr model?

General Problems

20. (II) For atomic-scale phenomena, the electron volt and nanometer are well-suited units for energy and distance, respectively.
(a) Show that the energy E in eV of a photon, whose wavelength λ is known in nm, is found by $E = \dfrac{1240}{\lambda}$. How much energy (eV) is possessed by a 650-nm photon?
(b) Show the wavelength λ in nm of an electron, whose (nonrelativistic) kinetic energy K is known in eV, is given by $\lambda = \dfrac{1.227}{\sqrt{K}}$. What is the wavelength (nm) of an electron that has been accelerated across a potential difference of 1000 V?

21. (II) Three fundamental constants of nature—the Gravitational constant G, Planck's constant h, and the speed of light c—have the dimensions of $[L^3/MT^2]$, $[ML^2/T]$, and $[L/T]$, respectively.
    (a) Find the mathematical combination of these fundamental constants that has the dimension of time. This combination is called the "Planck time" t_P and is thought to be the earliest time, after the creation of the universe, at which the currently known laws of physics can be applied. Determine the numerical value of t_P.
    (b) Find the mathematical combination of these fundamental constants that has the dimension of length. This combination is called the "Planck length" λ_P and is thought to be the smallest length over which the currently known laws of physics can be applied. Determine the numerical value of λ_P.

22. (II) Imagine a free particle of mass m bouncing back and forth between two perfectly reflecting walls, separated by distance L. Imagine further that the two oppositely directed matter waves associated with this particle interfere to create a standing wave with a node at each of the walls. Show that the ground (first harmonic) state and first excited (second harmonic) state have (non-relativistic) kinetic energies $\dfrac{h^2}{8mL^2}$ and $\dfrac{h^2}{2mL^2}$, respectively.

23. (II) (a) A rubidium atom ($m = 85$ u) is at rest with one of its electrons in an excited state. When the electron jumps to the ground state, the atom emits a photon of wavelength $\lambda = 780$ nm. Determine the resulting (nonrelativistic) recoil speed v of the atom.
    (b) The recoil velocity sets the lower limit on the temperature to which an ideal gas of rubidium atoms can be cooled in a laser-based atom trap. Using the kinetic theory of gases (Chapter 18 in the text), estimate this "lowest-achievable" temperature.

24. (II) While space walking, an untethered astronaut becomes separated from the Space Shuttle by 10 m, holding nothing other than a flashlight. The mass of the suited astronaut is 120 kg.
    (a) If the astronaut points the lit flashlight away from the Shuttle, how many days will it take for the astronaut to be pushed (by the force due to the emitted light) back to the Shuttle? Assume the flashlight emits 30 mW of light at an average wavelength of 555 nm.
    (b) If the astronaut throws the flashlight in the direction away from the Shuttle, how many minutes will it take for the astronaut to travel back to the Shuttle? Assume the flashlight has a mass of 0.7 kg and that the astronaut can throw it at 10 m/s.

25. (III) A rubidium atom (atomic mass 85) is initially at room temperature and so has a velocity of $v = 300$ m/s due to its thermal motion. Consider the absorption of photons by this atom from a laser beam of wavelength $\lambda = 780$ nm. Assume that the rubidium atom's initial velocity v is directed into the laser beam (e.g., the photons are moving right and the atom is moving left) and that the atom can absorb a new photon every 25 ns. How long will it take for this process to completely stop ("cool") the rubidium atom? [Note: a more detailed analysis predicts that the atom can be slowed to about 1 cm/s by this light absorption process, but it cannot be completely stopped.]

26. (III) In the Raman effect, an incident photon of wavelength λ scatters from an object that possesses an internal oscillator (e.g., molecule with spring-like chemical bond). The oscillator gains one quantum of energy during this scattering process, hence the wavelength of the scattered photon is $\lambda' = \lambda + \Delta\lambda$, where $\Delta\lambda$ is called the Stokes shift. Show that by measuring λ and λ', the frequency f of the oscillator can be found from $f \approx c\,\Delta\lambda/\lambda^2$, if $\Delta\lambda \ll \lambda$.

27. (III) A mass m is constrained to move on a circular ring of radius R where its potential energy is zero.
    (a) Assuming the allowed states for this mass are circular standing waves, use the de Broglie relation to deduce its allowed values of angular momentum L and (nonrelativistic) kinetic energy E.
    (b) The particle, initially in the $n = 6$ energy state, falls to the next lower energy state by emitting a photon of wavelength λ. Determine λ.

28. (III) In a nitrogen molecule, two nitrogen atoms, each of mass $m = 14$ u, are chemically bound to each other with an equilibrium separation $d = 0.11$ nm (called the "bond length").
    (a) Assume Bohr's quantum condition holds for the angular momentum of this molecule (about its central axis) and use it to estimate the minimum energy required to excite the molecule from rest to its first excited rotational kinetic energy state.
    (b) By shining light of wavelength λ on a gas of nitrogen molecules, one wishes to excite the molecules from rest into their first excited rotational energy state. What value of λ should be used? To what region of the electromagnetic spectrum does this wavelength belong? [Further developments in quantum theory show that this "back-of-the-envelope" calculation misses the correct angular momentum value by a factor of two.]

29. (III) A Space Shuttle ($m = 103{,}000$ kg) is in a circular orbit of radius $r = 6550$ km from the center of Earth (mass M).
    (a) Assume that Bohr's quantum condition for angular momentum applies to this situation and use it to determine (to one significant figure) the quantum number n for the Shuttle's orbit.
    (b) Show that, since $n \gg 1$, the radial distance between possible orbits (i.e., the change in radius when the quantum number is increased by one) is given by $\Delta r \approx h^2 n / 2\pi^2 Gm^2 M$. Show that Δr is immeasurable by evaluating this expression.

30. (III) In a collection of atoms, when the de Broglie wavelength of an individual atom is as large as the spacing between neighboring atoms, the atomic matter waves can interfere and produce interesting quantum effects. Consider an ideal gas of rubidium atoms with a dilute (so it does not condense into a liquid) atomic density of 5×10^{12} atoms/cm^3. Estimate the spacing between atoms. Then estimate the maximum temperature at which the de Broglie wavelength of a typical atom in this gas equals the atomic spacing. [The interesting quantum effect that occurs in this situation is called Bose-Einstein condensation.]

Answers to Odd-Numbered Problems

CHAPTER 1
1. (a) 1840; (b) 1836.1528
3. 12°
5. 4.00×10^7 m, 6.37×10^6 m, 0.2%
7. (a) 1.10 acre; (b) 10%
9. 4×10^{51} kg
11. 4.2 L
13. about 4000 for a 40-kg child
15. 0.1 mm
17. 6.8×10^6 m
19. $F_D = k\rho A v^2$
21. (a) $k\sqrt{g\lambda}$; (b) $k\sqrt{gh}$
23. (a) 81.2; (b) 390
25. 1.18×10^9 atoms/m²
27. 3420 y
29. (a) 90 m³, 90,000 kg; (b) 20

CHAPTER 2
1. 24 km/h
3. 6.0 m/s
5. (b) 0.17 m; (c) 8.6 m
7. (b) 6.8 m
9. 0.78 m/s²
11. 4.0 m/s, 12 m/s²
13. (a) +17%; (b) −17%; (c) 2.0 m/s
17. 3.5 d
19. (250, 346), (250, 353), (250, 364), (250, 379), (250, 398), (250, 400)
21. (c) 0.4 s, 0.8 m
23. (a) 4.32×10^6 bits; (b) 67%
25. 27 car lengths
27. about 100 m
29. (a) 5 m; (b) 5 cm (taking $a = 9.75$ m/s²)

CHAPTER 3
1. 68, 17°
3. $(-16\mathbf{i} - 8.2\mathbf{j})$ m
5. (a) 8.9 m/s; (b) makes it across
9. 5.1 s
13. (a) 5.0×10^{-8} s; (b) 5.0 cm; (c) 18°
15. (a) $\dfrac{v_o^2 \sin 2\theta_o}{2g} \times \left[1 + \sqrt{1 + \dfrac{2gh}{v_o^2 \sin^2 \theta_o}}\right]$; (b) 42°
17. (a) 5.4 g; (b) 33%
21. 25 s, 25 m
25. 2.2 m/s
29. (a) 31 m/s; (b) southwest

CHAPTER 4
1. 2×10^6 y
3. (b) 180 N; (c) 900 N
5. $(-4.0\mathbf{i} + 34\mathbf{j})$ m/s
7. 420 N, 1400 times the weight of a paintball
9. 15°
11. (a) 4.9 m/s²; (b) 12 m/s
15. (a) 0.71; (b) 60°
17. 400 m
19. (b) 54°
21. (a) $\dfrac{F_{N_1}}{mg} = 1 - \dfrac{a \tan \theta}{g}$, $\dfrac{F_{N_2}}{mg} = \dfrac{a}{g \cos \theta}$; (b) 0.58
23. yes, lose 6.5 kg
25. 2.7×10^{-23} m
27. no, needs 1.6 m to stop
29. $0.68 < a < 0.93$

CHAPTER 5
1. 0.21 m/s²
3. 600 N
5. 0.42mg
7. (a) 160 N; (b) 150 N
9. (b) 0.15
11. (b) 6°, 22°, 45°
13. (c) 6°, 22°, 45°
15. 0.17 rev/s
17. 1300 N, continue circling
19. (b) 0.21mg, toward center of circle; (c) 0.19mg, away from center of circle; (d) 0.81R
23. (b) 1.4 m/s²; (c) 100 km/h
25. (a) 27°; (b) 4.0 m/s²; (c) $F_P \geq 10$ N
27. 40 m
29. 84 min

CHAPTER 6
1. 1×10^{-10} N·m²/kg²
5. $-GMm\left(\dfrac{R}{2}\right)(4d^2 + R^2)^{-3/2}$, $GMm[-d^{-2} + d(4d^2 + R^2)^{-3/2}]$
7. (b) $0.3\, R_E$; (c) 200 m
11. 2.98×10^4 m/s, 428 s
13. (b) 29.6 d
17. 0.0945, 0.0434, 9.40×10^{-4}, 5.15×10^{-6}
19. (b) 35 AU, no
23. 1×10^{28} N·m²/kg², required 2×10^{38} times larger than actual
25. 3.28 AU, 2.96 AU, 2.82 AU, 2.50 AU
27. $n = 2$
29. can't discern if hollow

CHAPTER 7
1. 150 J
3. $(-0.80\mathbf{i} + 0.60\mathbf{j})$ or $(0.80\mathbf{i} - 0.60\mathbf{j})$
7. (a) 4.0 J; (b) ∞
9. (a) along straight line $y = -0.40x$; (b) along straight line $y = +2.5x$, 54 J
11. 1230 J
13. (a) 39 kJ; (b) 36 kJ; (c) yes
15. (b) 0.59, 8.8
17. (a) $\dfrac{1}{2}mv^2 + mvV, \dfrac{1}{2}mv^2$; (b) $\dfrac{1}{2}mv^2 + mvV, \dfrac{1}{2}mv^2$
19. 100 m
21. (b) 0.1
23. 4.74×10^{21} J, 10
25. (a) 180 N; (b) 180 N
27. (a) mgh, $\sqrt{v_o^2 + 2gh}$
29. (a) −7170 J, −23,300 J, +30,500 J; (b) 24%, 76%; (c) 175 Food Calories

CHAPTER 8
1. $\dfrac{k}{8}\left[1 - \dfrac{4}{x^2}\right]$
3. (a) $\sqrt{2gh}$; (b) $\sqrt{2gh}$
5. (a) $5g/9 = 5.4$ m/s²; (b) 2.9 m/s; (c) 2.9 m/s
7. 9.2 m, makes it to the other side
9. 100 m
11. (a) $\sqrt{2L/g \sin \theta}$; (b) $a = g \sin \theta, t = \sqrt{2L/g \sin \theta}$
13. (a) 210 m; (b) 210 m
15. $2g(m_2 - \mu m_1)/k$
17. 0.72 m/s
19. (d) 8.4 min
21. (a) 3300 hp
23. (a) $\sqrt{2Pt/m}$; (b) $\sqrt{P/2mt}$, $t = 0$; (c) $\sqrt{8P/9m}\, t^{3/2}$; (d) 2600 hp; (e) calculated is about half of real
25. (c) $\dfrac{A}{b}\left[\dfrac{\sin(bx)}{x^2} - \dfrac{b\cos(bx)}{x}\right]$

27. (a) 60 kJ; (b) 40%, 17%, 43%; (c) you provide 0.14 hp
29. 1.9 cm

CHAPTER 9

1. (a) 3.1 s; (b) 600 kg·m/s, 190 N; (c) 900 N
3. $\left(4\mathbf{i} + \dfrac{10}{3}\mathbf{j} - \dfrac{10}{3}\mathbf{k}\right)$ m/s
5. 0.3 m/s, in direction 80-kg skater moved originally
7. 1.2 ms
9. 1/3
11. (a) 20 m/s, 110°; (b) 40%
13. $2\left(1 + \dfrac{M}{m}\right)\sqrt{gL}$
15. (b) 23.3 m/s; (c) 201 kJ; (d) 29%, 71%
17. (b) 27 m/s, 13 m/s; (c) m_1 was speeding
23. 0.50 m/s
25. 1.1×10^{-4} m/s, 3.5×10^{-5} m/s
27. (c) 2.5×10^{-24} m/s
29. (a) 7.0 J; (c) 11 m/s

CHAPTER 10

1. 460 rpm, 200 rpm
3. (a) slope is α in rev/y/My, 365 rev/y; (b) 6.2×10^{-22} rad/s², 363 rev/y
5. (a) 27.4 d; (b) 29.6 d
7. (b) 1.9×10^{-12} rad/s/century, 21 h; (c) close agreement
9. (a) 6.3 rad; (b) nighttime
11. 3.1 s
13. $0.027M$ kg·m², $0.0079M$ kg·m², 3.4
21. (b) 2.5 rev/s, 6.5 rev/s
23. (a) $\sqrt{4gh/3}$; (b) $\sqrt{4gh/3}$
25. (a) xg/y; (b) small, large, stand up and back; (c) 2.9 m/s²
27. (a) 1×10^{32} rad/s; (b) 1×10^{14} m/s; (c) 3×10^5 times c
31. (a) 0.40 kg·m²; (b) 0.16 kg·m²; (c) 2.5

CHAPTER 11

3. 45°
5. $mv_oR\Big\{[\sin\alpha - \alpha\cos\alpha]\mathbf{i}$
   $- [\cos\alpha + \alpha\sin\alpha]\mathbf{j} + \dfrac{2\pi R}{p}\mathbf{k}\Big\}$
   where $\alpha = \dfrac{2\pi z}{p}$
7. (a) no; (b) zero; (c) v_o
9. (b) 46 m vs. 61 m

11. (a) $mRz_o\omega[-\cos\omega t\,\mathbf{i} + \sin\omega t\,\mathbf{j}] + mR^2\omega\,\mathbf{k}$;
    (b) $\cos^{-1}[R/\sqrt{z_o^2 + R^2}]$; (c) 0
13. (a) 0.074 rad/s; (b) 1.2%
17. 94 d
19. (a) y-axis; (b) $\pi MR\omega/5F$
21. 17 N, 3.5 times the weight of the hand
23. $-R[\alpha\sin\theta + (\alpha t)^2\cos\theta]\mathbf{i}$
    $+ R[\alpha\cos\theta - (\alpha t)^2\sin\theta]\mathbf{j}, mR^2\alpha\mathbf{k}$
27. (a) -4.67×10^6 m, 3.79×10^8 m, 3.47×10^{32} kg·m²/s, 2.81×10^{34} kg·m²/s;
    (b) 7.07×10^{33} kg·m²/s, 2.37×10^{29} 10^{29} kg·m²/s; (c) 0.26; (d) 0.049, 1.2×10^5
29. (a) 8.8 kg·m²/s; (b) -0.39 kg·m²/s; (c) -4.4%, fair approximation

CHAPTER 12

1. 23 N–46 N
3. $0.64w, 0.36w$, bottom-side friend
5. 260 N
7. (a) $\sqrt{3}w/2\cos(60° - \theta)$; (b) 60°; (c) 0.87 times smaller
9. $0.42 < \alpha \le 1.0$
11. 3460 N, 3390 N, 4.9, 4.8
13. (a) A; (b) B; (c) 1/7, 7, 1.75 m/s
15. $y = 2.5$ m
17. 2600 N
19. $\sqrt{17}w/2 = 2.1w$, 76° above horizontal
21. (a) $w\left[1 - \dfrac{R\cot\theta}{h}\right]$;
    (b) $\mu = \dfrac{h_S}{R} - \cot\theta_S$
23. (a) $5 \times 10^{-6}\%$; (b) 1×10^{-17} m
25. k/d, 20 N/m
27. 5 km, no
29. (c) 0.11 mm

CHAPTER 13

3. 2×10^{78} protons and neutrons
7. (a) 1.0×10^5 N/m², toward outside; (b) 1.5
9. 88 Pa/s, 3.8 min
11. (b) $x_2 = x_1\left(\dfrac{R_2}{R_1}\right)^2$
13. yes
15. $\rho < 130$ kg/m³, gases
17. 4800 helium-filled balloons
21. (b) 46%, 47%, no
23. (c) 9.1 m
25. (d) 14 cm, 13 cm, 1.5 cm
27. 6.6 atm
29. (b) 200 atm; (c) 0.01 atm

CHAPTER 14

1. 13 cm/s, 12 cm/s², 1.2%
3. 2.6×10^{13} Hz
5. 0.39 m
7. (a) $4A/T$; (b) $4A/T$
9. (a) $0.5D - L$; (b) $2\pi\sqrt{m/2k}$
11. (a) $d^2x_R/dt^2 = -(k/m)(x_R - x_L)$, $d^2x_L/dt^2 = +(k/m)(x_R - x_L)$; (c) 2200 N/m, 220
13. (c) 80, 1
15. (a) $4\pi^2/\text{slope}$, 0; (b) 0.128 s²/kg, 0.139 s²; (c) 308 N/m, 1.09 kg
19. (b) 5 mm
21. (b) 180 N/m
25. (a) 4.0 Hz; (b) 0.36 m; (c) 5°
29. (a) 5.2 cm; (b) 5.2 cm
31. (b) $k = GMm/R^3$; (c) $\sqrt{GM/R^3}/2\pi$

CHAPTER 15

1. 140 m, yes
3. 7.0×10^{10} N/m², aluminum
5. (a) $k\sqrt{g\lambda}$; (b) $k\sqrt{gh}$
7. (a) 0.042 m; (b) 0.63 rad
9. $v'_{\max} = 2\pi vD_M/\lambda$, 0.063
11. (a) $T = \sqrt{x^2 + 4D^2}/v$; (b) $D = 0.5\sqrt{b/a}$, where $a = $ slope, $b = $ y-intercept
13. $R = \left(\dfrac{v_1 - v_2}{v_1 + v_2}\right)^2$; (b) 3.9%; (c) $v_1 \gg v_2$ or $v_2 \gg v_1$; (d) $v_1 = v_2$
15. (a) 720 m/s; (b) 1200 Hz; (c) 2.25
17. 390 m/s
19. $\pm\pi/4$ m, $\pm 3\pi/4$ m, $\pm 5\pi/4$ m, etc.
21. $N = -a$, $A = 2Lf10^{-b}$, where $a = $ slope, $b = $ y-intercept
23. 351 Hz, 8 nodes
25. 1.07

CHAPTER 16

1. (a) 0.17 m; (b) 8.6 m; (c) 0.5%
3. $-0.824\,\Delta P_M$, $+0.284\,\Delta P_M$
5. 6.02 dB
7. 32
9. row 37
11. (a) 1.31 m; (b) 3, 4, 5, 6
13. 2100 N
15. 3.65 cm, 7.09 cm, 10.3 cm, 13.4 cm, 16.3 cm, 19.0 cm
17. 21.44 Hz, 42.88 Hz
19. 83.5 μm, 5
23. 444 Hz and 8 Hz, 436 Hz and 8 Hz
25. $D^\#$

27. 10 m/s
29. 1100 N

CHAPTER 17
1. 2×10^4 molecules
3. 5 mm
5. 400 C°, no
7. 22%
9. 33%
11. (a) 240 min; (b) 240 m
13. $PV^2 = n^2RT^{4/3}$ with $R = 0.283$ L·atm/mol²·K
15. after every four days of use
17. 8.0 L
19. (a) 110 N; (b) 6.7×10^{-4} mol; (c) 4.5 cm³
21. 3.1 atm gauge pressure, cork won't pop
23. 0.03
25. 1.5 cm
29. (d) 80 m

CHAPTER 18
1. 2×10^{-5} s
3. $730 \gg 1$
5. (c) 3×10^{28} collisions/s
7. (a) 642 m/s; (b) 199 K; (c) 595 m/s, 201 K, yes
11. 70 g
13. 3000 m
15. 30 months
19. 0.01 Pa
21. 3×10^{-7} m
23. (a) 9×10^9 K; (b) no
25. (d) 2.8×10^{-8} s, 7.9×10^{-7} N/m², 7.8×10^{-12}
27. 9.9
29. (d) $x_o = 30$, $N_{esc} = 3 \times 10^{-346}$

CHAPTER 19
1. 0.04 candy bar
3. 0.0067 C°
5. 12.7°C, no
7. 3 ice cubes
9. (a) 106 s, 92%; (b) 23 s, 61%; (c) 4.6
13. 5.9 atm
15. $P = \left(1 + \dfrac{3.8 \text{ cm}}{H}\right)^{1.4}$, for P in atm, H in cm
17. (c) 7.3 d_E
19. (b) 390 J/s; (c) 280 J/s; (d) 370 J/s
21. (a) 1.25×10^{28} molecules; (b) 34 C°
23. (a) 638 W; (b) 4.2 g

25. 7.4 J
27. (d) 34°C

CHAPTER 20
1. decrease T_L
3. (a) no change; (b) +0.046 cm³, −0.073 cm³
5. 8.5
7. 71 K, 227 K
9. (b) 0.55; (c) −7.5°C
11. $0.51 (per hour)
13. $\Delta S = -0.45$ J/K
15. +14 cal/K
17. T/nC_V
19. (a) +2.2 cal/K, yes; (b) −2.2 cal/K, no; (c) +0.26 cal/K, yes
25. $nR\ln(2)$
27. (a) $+7.6 \times 10^{-23}$ J/K, yes; (b) -7.6×10^{-23} J/K, no
29. (a) 0.19; (b) 0.23

CHAPTER 21
1. 1 mC
3. (b) $\sqrt{4\pi/c^2\mu_o} = 1.055 \times 10^{-5}$ C
5. (b) 6.4×10^8 N·m²/C², yes
7. (a) $\dfrac{Q}{2\pi\epsilon_o Ly}\left[\dfrac{a+L}{\sqrt{(a+L)^2+y^2}} - \dfrac{a}{\sqrt{a^2+y^2}}\right]\mathbf{j}$; (b) $m\dfrac{d^2y}{dt^2} = \dfrac{qQ}{2\pi\epsilon_o Ly}\left[\dfrac{a+L}{\sqrt{(a+L)^2+y^2}} - \dfrac{a}{\sqrt{a^2+y^2}}\right]$
9. $\dfrac{\lambda}{2\pi\epsilon_o}\sqrt{\dfrac{1}{x^2}+\dfrac{1}{y^2}}$, $\theta = \tan^{-1}\dfrac{x}{y}$ relative to x-axis
11. $\dfrac{1}{4\pi\epsilon_o}\dfrac{Q}{L}\left[\dfrac{1}{\sqrt{(x_o - L/2)^2 + a^2}} - \dfrac{1}{\sqrt{(x_o + L/2)^2 + a^2}}\right]\mathbf{i}$
15. (a) 9×10^{-7} m; (b) 2×10^{18} m/s²; (c) 1×10^7 N/C, factor of 3 too large
17. $\alpha = Q^2/k$
19. (a) $\dfrac{2\alpha EL^3}{3}\mathbf{k}$; (b) $\dfrac{2\alpha L^3}{3}\mathbf{i}$; (c) $\dfrac{2\alpha EL^3}{3}\mathbf{k}$

21. $\dfrac{2\lambda p}{\pi\epsilon_o}\dfrac{1}{4y^2 - l^2}$, toward line charge
23. 1.00002 C
25. (c) 0.2 rad
27. (b) $\dfrac{k\lambda}{d}\left[\dfrac{x}{\sqrt{x^2+d^2}} - \dfrac{x-L}{\sqrt{(x-L)^2+d^2}}\right]$; (c) $\dfrac{2kQ^2}{L^2}\left[\sqrt{1+\dfrac{L^2}{d^2}} - 1\right]$
29. (c) 2.0×10^{-4} N

CHAPTER 22
3. $Q/2\epsilon_o$
7. (a) $2\pi r^2 E$; (b) $+Q/2$; (c) $Q/4\pi\epsilon_o r^2$
9. $\rho_E R_1^2/2\epsilon_o r$
11. $+4\pi\epsilon_o Cr^5$
15. (a) $\rho_E d/2\epsilon_o$, away from slab
17. (a) 0; (b) $-\dfrac{\rho_o}{2\epsilon_o d}(d^2 - x^2)\mathbf{i}$
19. 0.053 mm
21. (a) $\rho_E a^3/\epsilon_o$; (b) no
23. $\dfrac{1}{2\pi}\sqrt{\dfrac{Q\sigma}{\epsilon_o mL}}$
25. (a) $ed/4\pi\epsilon_o R^3$, toward sphere's center; (b) $4\pi\epsilon_o R^3 E_{ext}$; (d) 1×10^{-41} C²·m/N
29. (a) $\sigma/2\epsilon_o$; (b) $\sigma^2/2\epsilon_o$; (c) 0.092 N/m²

CHAPTER 23
1. (a) 30 m
3. (a) +1.9 V; (b) +1.9 V; (c) +3.8 V; (d) yes
5. $\dfrac{-C}{2x^2} + \dfrac{C}{32 \text{ m}^2}$
7. (a) $\dfrac{kQ}{r_o}\left[\dfrac{3}{2} - \dfrac{1}{2}\dfrac{r^2}{r_o^2}\right]$; (b) $\dfrac{kQ}{2r_o}$; (c) $-\dfrac{kQ}{2r_o^3}r^2$ (d) $\dfrac{kQ}{2r_o}$
9. $x = \pm 2a$
11. (a) 18 V; (b) 5.9×10^4 m/s
13. (a) $\dfrac{2kQ}{L} \times \ln\left[\dfrac{a+L+\sqrt{(a+L)^2+b^2}}{a+\sqrt{a^2+b^2}}\right]$; (b) $\left\{\dfrac{4kQ}{mL} \times \ln\left[\dfrac{a+L+\sqrt{(a+L)^2+b^2}}{a+\sqrt{a^2+b^2}}\right]\right\}^{1/2}$

17. 5.8 min
19. 5×10^6 m/s
21. 4.47 m/s, 2.24 m/s
23. kqQ/R
25. $v = \sqrt{ke^2/mr}$
27. $\sqrt{\dfrac{md}{kQ^2}}\left[\sqrt{2R(d-2R)} + d\sin^{-1}\sqrt{1-\dfrac{2R}{d}}\right]$
29. 5×10^{10} V/m, 4×10^7 V/m, Model B

CHAPTER 24

1. 69 d
3. 6000 V
5. 9300 V
11. (a) $Q^2/8\pi\epsilon_o r_b$; (b) 1000
13. (a) $\dfrac{\lambda^2 \ln(R_a/R_b)}{4\pi\epsilon_o}$; (b) $\dfrac{\lambda^2 \ln(R_a/R_b)}{4\pi\epsilon_o}$
15. (a) 2, 2; (b) $Q/3$; (c) 1/2
17. (b) 1.5 nF, 1.1 nF
19. (b) 100 pF; (c) 2 μC
21. $V = \dfrac{A\epsilon_o}{C}E$; (b) 0.04 nF
23. (a) 4×10^{-7} C; (b) 1×10^{-10} F; (c) 4000 V
25. 3C
27. (a) $Q^2/8\pi\epsilon_o R$; (e) 6×10^{-11} C
29. $b = \dfrac{\ln\left(1+\dfrac{t}{R}\right)}{2\pi\epsilon_o K}$, $a = -x_o$

CHAPTER 25

1. (a) slope = $1/R$, y-intercept = 0; (b) 1.39 Ω; (c) 1.02 $\mu\Omega \cdot$m, nichrome
3. (b) $1 \times 10^8\ \Omega$
5. 966 turns
7. (a) $\alpha T/[1 + \alpha(T - T_o)]$; (b) $1.4 > S > 1.0$ over entire range
11. (b) 700 V
15. 156 V
17. $j^2\rho$
19. (a) $\rho = m/ne^2\tau$; (b) 2.5×10^{-14} s
23. (b) 0.6 μC, 6000 V
25. $F = 1 - \dfrac{V}{V_o}$
27. (b) $A = \rho I/t$
29. (a) 4.6δ; (b) 0.3 mm

CHAPTER 26

1. $R_o/\sqrt{3}$
3. $R = r$

7. (b) $y_{max} = 1$, $y_{min} = 0.5$, $x_{max} = 0$, $x_{min} = \infty$;
  (c) $P_{max} = \mathcal{E}^2/R_o$, $P_{min} = \mathcal{E}^2/2R_o$, $R_{max} = 0$, $R_{min} = \infty$
9. $R/3$
11. 40 s
13. (a) $2\mathcal{E}/3R$, $\mathcal{E}/3R$, $\mathcal{E}/3R$
  (b) $\mathcal{E}/2R$, $\mathcal{E}/2R$, 0; (c) $\mathcal{E}/2$
15. (a) $(R_1 + R_2)C \ln(2)$; (b) $R_2 C \ln(2)$; (d) 0.05 μF
17. (b) 6 L/min; (c) 2 s
19. (b) $\tau = 3RC/2$, $Q_{max} = \mathcal{E}C/2$
21. 200 MΩ
23. $5 \times 10^9\ \Omega$
25. 1 mV, 2 mV, 4 mV, 8 mV
27. (b) 15 mA, 150 mA
29. (b) $Q_o = \rho u RC$, $V_o = \rho u R$; (c) 6 L/min

CHAPTER 27

1. $-2rIB \sin\theta_o \mathbf{j}$
3. $m = 2.0 \times 10^{-2}$ N/A, $b = 4.0 \times 10^{-3}$ N
7. $qBy_o/2m \sin\theta$
9. μ_1/N
11. (b) 0.02 A$\cdot$m^2; (c) 40 A
13. (d) 1.4×10^{10} Hz
15. (a) 0.53 T; (b) 0.45 T
17. (a) slope = I/ent, y-intercept = 0; (b) 5.5×10^{22} m^{-3}
19. (b) 2000 V
21. w/IB
23. (a) downward; (b) 0.026 T; (c) 0.15 T
25. (a) 260 mV/T; (b) 1.4 cm, 2.1 mV
27. 642 m
29. (c) proton moves in circle with $r = 0.43\ \mu$m, which moves at $v = 2.0$ mm/s in the $\mathbf{i}$ direction (horizontal direction perpendicular to $\mathbf{B}$)

CHAPTER 28

1. (b) 39 A; (c) 0.55
3. $\dfrac{I'I\mu_o}{2\pi}\left[\dfrac{L}{d} - \dfrac{2}{\sqrt{3}}\ln\left(1 + \dfrac{\sqrt{3}L}{2d}\right)\right]$, toward wire
5. (a) $\pm\dfrac{j\mu_o t}{2}\mathbf{k}$; (b) $+j\mu_o t \mathbf{k}$ (between), 0 (outside); (c) $+j\mu_o t \mathbf{k}$
7. (a) 2.00×10^{-5} T, 1.00×10^{-5} T; (b) 0.628 m, 1.26 m; (c) both are 1.26×10^{-5} T$\cdot$m; (d) $\mu_o I = 1.26 \times 10^{-5}$ T; (e) 1.26×10^{-5} T$\cdot$m, yes
9. (a) $3I/2\pi R^3$; (b) $\mu_o I/8\pi R$, $4R$

11. $12R$
13. $\dfrac{\mu_o I}{2R}\left(\dfrac{1}{\pi} + \dfrac{1}{2}\right)$
15. $\pi/3$, $1/3$
17. $\dfrac{\mu_o I}{4\pi R}(\sin\theta_1 + \sin\theta_2)$, $\sqrt{5/2} \approx 1.6$
19. (a) 2.7 A CW, 0.89 A CW, 1.8 A CCW; (b) 3.3 A CW, $(8.0\mathbf{i} + 2.7\mathbf{j} - 5.3\mathbf{k})$ cm
21. (b) $12R$; (c) $<8.3 \times 10^{-3}$, $<6.5 \times 10^{-5}$
23. (a) $+\dfrac{j\mu_o t}{2}\mathbf{k}$; (b) $-\dfrac{j^2\mu_o t^2}{2}\mathbf{j}$, attract
25. $2R/\pi d$, 16
27. 250 turns/cm
29. (c) 1.5 A

CHAPTER 29

1. -0.021 V
3. (a) 0.16 J; (b) 0.083° C
5. (a) 0; (b) 0.99 A clockwise
7. (a) $\dfrac{\mu_o Il}{2\pi}\ln\left[\dfrac{(r+l)}{r}\right]$, $\dfrac{\mu_o Ivl^2}{2\pi R}\dfrac{1}{r(r+l)}$; (b) clockwise; (c) $\dfrac{\mu_o^2 I^2 v l^4}{4\pi^2 R}\dfrac{1}{r^2(r+l)^2}$
9. (b) $\pm R/2$, $\dfrac{24\mu_o \mu Nv}{5^{5/2}R^2}$
11. $(M+m)\dfrac{dv}{dt} + \dfrac{B^2 l^2}{R}v = Mg$, $v_T = \dfrac{MgR}{B^2 l^2}$, $\tau = \dfrac{(M+m)R}{B^2 l^2}$
13. (a) R; (b) $\left(\dfrac{N_P}{N_S}\right)^2 R$
15. (b) $1/\sqrt{2}$, $\sqrt{2}$
17. $\dfrac{\mu_o Ivl}{2\pi(r_o - vt)}$, $\dfrac{\mu_o Iv}{2\pi(r_o - vt)}$
19. $V_o/\pi r^2$, smaller
21. (a) $\dfrac{\pi d^2 B^2 lv}{16\rho}$; (c) 3.7 cm/s
23. (b) 18, 36

CHAPTER 30

1. 67 m, 26 km, 850 Ω
3. $\mu_o I^2/16\pi$
7. $(10\ \text{V})e^{-(1000\ \text{s}^{-1})t}$, $t = 0$, 10 V
9. (b) 5000 V
11. $L \ln(2)/R$
13. (a) 0.13 μF; (b) 21 kHz
15. (b) 2×10^{-4}%

17. (a) slope = $4\pi^2 L$,
    y-intercept = $4\pi^2 LC$;
    (b) 24.6 mH, 2.13 μF
19. (a) 36 mH, 0.70 pF
21. 2.0 mA, 0, 2.0 mA
23. (a) 2×10^{-16} Ω;
    (b) 8×10^{-6} s
27. $\dfrac{\mu_o I^2 \Delta l}{4\pi}\left[\ln\left(\dfrac{l}{2R}\right) + \dfrac{1}{2}\right]$
29. 5 Hz

CHAPTER 31

1. (a) $R' + R$; (b) R'
3. 180 Hz
5. (a) 0.62 A; (b) $-21°$;
   (c) 93 V, $+21°$; (d) 47 V, $+111°$;
   (e) 82 V, $-69°$
7. (a) 17 W; (b) 0
9. 4800
11. (a) slope = $4\pi^2 R^2 C^2$,
    y-intercept = 1;
    (b) 3.00×10^{-9} F
13. 12 Ω, 111 V rms
15. (c) $V_o/\sqrt{2}$, 8 Ω
17. (b) 160 Ω
19. (a) $f < 30$ kHz; (b) $f > 300$ kHz
23. (a) ω_o; (b) both increase by factor of α
25. (a) 2.2 pF; (b) 7.2 MΩ; (c) 0.42 mA, no; (d) 42 mA, yes
27. $C = I_c/V_o\omega$, $R_L = V_o/I_s$
29. (a) 7.1 kHz, V_{rms};
    (b) $0.91 V_{rms}$

CHAPTER 32

1. 1.9×10^{-12} T, 4×10^{-8}
3. $\dfrac{\mu_o \epsilon_o \pi f V_o}{d}r$, $\dfrac{\mu_o \epsilon_o \pi R^2 f V_o}{d}\dfrac{1}{r}$
5. (a) $\dfrac{10^{-9}}{4\pi(2.99792458)^2}$;
   (b) 8.85419×10^{-12}
7. $k_x^2 + k_y^2 + k_z^2 = \dfrac{\omega^2}{v^2}$
9. (a) $\sqrt{\mu_o \sigma \omega/2}$; (b) 5.3 μm
11. 532 nm, visible
13. (a) 5 cm^2, yes; (b) 15 m^2, yes; (c) 150 m^2, no
15. 3×10^5 W/m^2
17. (a) 2×10^{-8} N; (b) 2×10^{-6} N; (c) gravitation is 100 $\times$ larger
19. (b) 377 Ω
21. 5 nodes, 6.1 cm
23. 2.25×10^8 m/s, 75% of c
25. (a) $\pi R^2 B$; (b) $\pi R^2 B$
27. (a) $E_y = 2E_o \sin(kx)\cos(\omega t)$,
    $B_z = 2B_o \cos(kx)\sin(\omega t)$;
    (b) $\dfrac{E_o B_o}{\mu_o}\sin(2kx)\sin(2\omega t)$, $n\pi/2k$,
    where $n = 0, \pm 1, \pm 2$, etc.
29. (b) 1.5×10^{-13} A;
    (c) 3.0×10^{-17} T, 1.5×10^{-13}

CHAPTER 33

1. 1.7 cm
3. 150 m
5. θ
9. $\left(1 + \dfrac{1}{\sqrt{2}}\right)r = 1.7r$,
   $-(1 + \sqrt{2}) = -2.4$
11. 3.2 mm
13. (a) slope = $1/n$, y-intercept = 0;
    (b) 1.5, yes
15. 0.81°
17. 260 m
19. (a) $64° > 43°$; (b) 9.7×10^6;
    (c) 3.7 μs
21. (b) 6.95°
23. 4.48 cm
25. (a) $l = 2t/\cos\theta_2$, $d = 2t \sin\theta_2$,
    where $\theta_2 = \sin^{-1}[\sin\theta_1/n]$;
    (b) $2t$, 0
29. 2

CHAPTER 34

1. $d_o = 2f$, $d_i = 2f$
3. (c) $2f$
5. $f^2 v/(d_o - f)^2$, away, $2f$
7. (a) slope = $-1/f$, y-intercept = $+1$;
   (b) 14 cm; (c) $f = -1/$slope
9. $l = 4f$, $d_o = 2f$
11. (a) 30 cm; (b) 45 cm; (c) 45 cm, no
13. (a) 49 mm; (b) 14 mm
15. (c) 15 m
17. 14–27 cm
19. 2.9 cm, toward
21. 400, 20
23. (d) 5 cm, 13 cm
25. 880 mm
27. $R_1 = R_2 = -8.3$ cm
29. (b) 6.2 cm; (c) 5.0

CHAPTER 35

1. 550 nm (assuming $m = 6$ is 1.0 cm from $m = 0$)
3. 0.013 mm
5. $L/d \geq 573$, 6 cm
7. (a) $4.472135866 \times 10^{-4}$ m;
   (b) $4.472135955 \times 10^{-4}$ m;
   (c) 2×10^{-6}%
9. (a) slope = $1/\sqrt{(d/m\lambda)^2 - 1}$,
   y-intercept = 0; (b) 5.3 cm
11. 7300 N/m, 18°
13. $d\sin\theta = \lambda/3, 4\lambda/3, 7\lambda/3, 10\lambda/3$, etc.
15. 325 nm
17. (b) $m_{400} = 1330$, $m_{700} = 760$;
    (c) 569
19. (c) 3.4×10^{-4}, 0.019
21. 126 nm
23. (a) $E_1 = \dfrac{(1 - n_c)}{(1 + n_c)}E_o$,
    $E_2 = \dfrac{(n_c - n_g)}{(n_c + n_g)}E_o$
27. 400 nm $\leq \lambda \leq$ 640 nm
29. (b) $4\pi nt/\lambda$; (c) $\lambda/4n, 3\lambda/4n, 5\lambda/4n$, etc.

CHAPTER 36

1. 9300 $\times$ larger
3. $0.057 I_o$
7. (c) 5.045, 5.658, 0.89
9. 269
11. 10 km
13. (a) $5 \times$ eye's resolution limit;
    (b) 1.0 cm
15. 635 nm
17. 6.4 cm, 30 cm, l_2 is larger
19. (a) 1.0×10^{-3} rad;
    (b) 2.8×10^{-5} rad; (c) 36
21. (b) 0.2 nm, X-ray
23. (a) $0.76 I_o$;
    (b) $I = I_o \cos^{2N}(90°/N)$, 1.0
27. (a) 8°, 25°, 45°, 58°, 83°; (b) yes
29. (c) 25.6°

CHAPTER 37

1. (a) 0.68 ns; (b) 0.20 m
3. $0.894c$
5. (a) 2.6×10^8 m/s; (b) 2500 s;
   (c) 6.5×10^{11} m; (d) 2.6×10^8 m/s
7. (b) 4.3×10^{-15}
9. (a) $0.707 c$; (b) 70.7 m
11. 3.17
13. (a) $\dfrac{14}{15}c > \dfrac{12}{15}c$, yes;
    (b) $\dfrac{35}{45}c < \dfrac{36}{45}c$, no
15. 2.5×10^8 m/s
19. 25 d
21. 1.4×10^{-8}
23. (a) 100 km/h; (b) 66.7 Hz
25. $5q/3d$
29. $0.30 mc$

Answers to Odd-Numbered Problems

CHAPTER 38

1. (a) 8.6×10^{-20} J; (b) 6200 K
3. (a) 1.78 eV; (b) 3.11 eV
5. (a) 1.78 eV; (b) 3.11 eV
7. 5×10^7 atoms
9. 4×10^{-3}
11. 6×10^9
13. 6 nm, diffract
15. -0.544 eV
17. 1.3×10^6
19. (b) 7.3×10^{-9}
21. (a) $\sqrt{Gh/c^5}$, 1.34×10^{-43} s; (b) $\sqrt{Gh/c^3}$, 4.05×10^{-35} m
23. (a) 6.0×10^{-3} m/s; (b) 1.2×10^{-7} K
25. 1.3 ms
27. (a) $nh/2\pi$, $h^2n^2/8\pi^2mR^2$ where $n = 1, 2, 3$, etc.; (b) $8\pi^2mR^2c/11h$
29. (a) 5×10^{49}; (b) 3×10^{-43} m